Integrated Regional Risk Assessment, Vol. I

Environmental Science and Technology Library

VOLUME 4/1

Integrated Regional Risk Assessment, Vol. I

Continuous and Non-Point Source Emissions: Air, Water, Soil

by

Adrian V. Gheorghe

ETHZ, Swiss Federal Insitute of Technology, Zürich, Switzerland

and

Michel Nicolet-Monnier

PSI, Paul Scherrer Institute, Villigen, Switzerland

SPRINGER SCIENCE+BUSINESS MEDIA, B.V.

A C.I.P. Catalogue record for this book is available from the Library of Congress

ISBN 978-94-010-4192-8 ISBN 978-94-011-0431-9 (eBook)
DOI 10.1007/978-94-011-0431-9

Printed on acid-free paper

CONTENTS

PREFACE

Industrial development is essential to improvement of the standard of living in all countries. In a given region, old and new plants, processes, and technologies have to coexist. Technological penetration and substitution processes are generally taking place; they are entirely dynamic and this trend is going to stay like this.

People's health and the environment can be affected, directly or indirectly by routine waste discharges or by accidents. A series of recent major industrial accidents and the effect of pollution highlighted, once again, the need for better management of routine and accidental risks. Moreover, the existence of natural hazards complicate even more the situation in any given region. In the past effort to cope with these risks, if made at all, have been largely on a plant by plant basis; some plants are well equipped to manage environmental hazards while others are not.

Managing the hazards of modern technological systems has become a key activity in highly industrialized countries. Decision makers are often confronted with complex issues concerning economic and social development, industrialization and associated infrastructure needs, population and land use planning. Such issues have to be addressed in such a way that ensures that public health will not be disrupted or substantially degraded.

While hazard managers and risk assessors have been successful at identifying hazards and reducing overall risk exposure, economic growth and technological development have led to a new risk situation characterized by:

- an increasing number and variety of hazards
- hazards giving rise to a broad range of partial and temporal risks
- public dissatisfaction with hazards managers and hazards' owners.

Due to the increasing complexity of technological systems and the higher geographical density of punctual hazard sources, new methodologies and a novel approach to these problems are challenging risk managers and regional planers. Risks from these new complex technological systems are inherenttly different from those addressed by the risk managers of the 1960's and 70's. Recent awareness of environmental problems by a large public has led worldwide to dissatisfaction and to the formation of all kind of pressure groups that do exert a strong political influence and are quite often not ready to accept any compromise. As a result of such approach, interesting and vital research project were stopped (Swiss nuclear moratorium), plans are buried and decisions delayed for many years (i.e., repository for nuclear wastes). Nowadays it becomes increasingly difficult to site new plant facilities which are perceived as risky or undesirable by the local population. Project of national interest cannot be launched or even achieved, once started. On the other hand distrust has become so problematic, that at times, risk managers are no more considered as impartial person, or as reliable source of information regarding risk protection aspects.

It was felt that existing hazard management techniques need to be supplemented with concepts and methods that are integrative at a regional level. Integrated regional risk assessment and safety/hazard management (IRRASM) represents a coordinated strategy for risk reduction and safety/hazard management in a spatially-defined region across a broad range of hazard sources (during normal operation and accidental situations) that includes synergistic effects.

Integrated Regional Risk Assessment and Safety/Hazard Management (IRRASM)
IRRASM is a multi-disciplinary process: engineers, computer scientists and modelers play a central role in the risk assessment stage. Social scientists can contribute with practical advice to the embedding process concerning hazard sources and help communal organizations to deal with such problems, taking into account local economic conditions and political reality. They can make important contributions towards a better understanding of how the practice of regional risk management can meet the needs and expectations of society/community. In this way other kinds of regional planing tasks (for emergency preparedness or remedial actions) or the implementing of utilities as part of the infrastructure (such as canalization network, sewage plants, waste disposal units, power plants and drinking water distribution, etc.) can be integrated more easily after reaching the consensus of agreement of the population.

In view of the above mentioned optic a joint project, *PPR&S (Polyproject on "Risk and Safety of Technical Systems)*, was launched together with the participation of the following institutions:

- Swiss Federal Institute of Technology, ETHZ (Zürich, Switzerland)
- Paul Scherrer Institute, PSI (Villigen / AG, Switzerland)
- EAWAG (Dübendorf / ZH, Switzerland)

There are a number of national and international efforts to deal with risk assessment at the regional level. The ETHZ (Swiss Federal Institute of Technology - Zürich) - "Poly-project on Risk and Safety of Technical Systems" took the initiative to research on various aspects related to regional risk assessment and safety management. A series of basic questions had been asked. Answers had to be given; they had to comprehend the following main issues:

- what is integrated area risk assessment and safety management
- how to define a region/area for study
- type of activities and targets at risk
- objectives and scope
- the need for risk impact indicators
- the need for a comprehensive methodology.

Regional risk assessment and safety management seems to be a medium which helps to integrate people, issues and decisions in area risk assessment.

PPR&S is the discursive description of knowledge in addition to the development of interdisciplinary and practical methods for the application of risk management for a large variety of technological systems (e.g., rail and road transportation of dangerous goods, chemical plants, nuclear power plants, biotechnology, landfill, etc.). The scope of these applications are to be presented as a book series and is intended to be an integrated regional risk assessment and safety management guideline manual complemented by specialized software, databases, literature reviews and a novel methodological framework with due regard to the existing conditions prevailing in Switzerland. Further goals aimed at in this project are namely:

i) Encouraging and promoting multi-disciplinary work among the different departments and institutes at the Swiss Federal Institute of Technology (ETHZ), Zürich, and among other research institutions;

ii) Establish and confirm the technical competence of the ETHZ in the field of risk and safety of technological systems including their relationship with natural disasters;

iii) Encouraging scientific and academic contacts to other polytechnic institutes universities, industry, governmental agencies, and political institutions within Switzerland and internationally;

iv) Supporting the teaching aims in the interdisciplinary field of risk analysis at ETHZ.

Plans are underway to develop together with the Swiss Federal Institute of Technology of Lausanne, EPFL, a postgraduate study program in the field of "Risk and Safety". For the most part, the PPR&S has now developed from a local to a national interest in disasters' prevention and emergency planing activities.

In the framework of the PPR&S it was decided to collect and review basic technical information and topics concerning the *Integrated Regional Risk Assessment Process* and to publish the results in book form, as being part of a books' series presented under the auspices of PPR&S. This series forms a whole and covers different aspects of risk assessment, management, risk acceptance, as well as legal and societal aspects thereof. The present book, entitled " INTEGRATED REGIONAL RISK ASSESSMENT " is subdivided into two volumes:

- Vol. I : "Continuous and Non-point Source Emissions: Air, Water, Soil", and
- Vol. II : "Consequence Assessment of Accidental releases"

Assessing the risks of a region implies the use of a complex methodology dealing with risks to health and to the environment, normal operation and accidental situations, a large variety of industries, impacts, regulations and actors involved in the decision making process.

As opposed to other existing approaches (e.g., UN- interagency project on risk assessment of large industrial complexes), the PPR&S project took the initiative to design procedural guidelines for IARASM implementation by means of identifying tasks and integrating them into a comprehensive and systematic approach. By contrast, other existing guidelines take a problem solving oriented approach which is too global and does not always assist systematically the analyst or the project manager.

This approach (i.e., task oriented approach) allows a systematic analysis of the problem of regional risk assessment, offers flexibility and efficiency in the implementation process, allows initiatives and ad hoc modeling and simulation.

Integration of risk can not be done through a single risk indicator. Integrated regional risk assessment should be considered as a process whereby decision aiding techniques (ranging from simple brainstorming, the Delphi method to Multicriteria Decision Analysis, and Decision and Knowledge Support Systems) should play an important role. Various techniques have to complement expert judgment, public participation and risk communication. It is a need to balance hard approaches (models, calculations) versus soft approaches (acceptability) in regional risk analysis. In the process of risk integration an important role should be played by the task of comparative risk assessment. PPR&S made successful experiments in working with such tools and approaches.

The advent of new information technology, e.g., artificial intelligence (expert systems, fuzzy logic, neural networks), multimedia, virtual reality, GIS, specialized relational databases, computer graphics, or ISDN technology, would play a significant role in the future of regional risk assessment and safety management practice. The experiments made within the PPR&S with some of these techniques are showing promising results.

The PPR&S project has had important inputs from existing Swiss practice and legislation. The need for a comprehensive regional risk assessment methodology has been highlighted in different occasions during the time of the Poly-project.

Some of the Poly-project lessons we have learned, when dealing with above issues are:

- When getting involved in a regional risk assessment do not take a simplistic approach
- On regional risk assessment, try not to exclude political or human interactions at all stages. Develop a risk triplex, namely: *"safety culture, environmental awareness, and emergency culture"*.
- When running a research or a case study on regional risk assessment do not entirely rely on a self-organizing effect within the project / case study. A strong interactive project management framework is needed from the beginning.
- Due to technical accidents or natural disasters, or their synergism, there is an emerging need for national and international organized research and activities in the above field. Risks from normal operation or from accidents may have some transboundary effects.
- Legal issues at the local level / national level have to be solved and harmonized before any implementation of a comprehensive regional risk assessment methodology.
- There is a need for specialized databases; their use might diminish the uncertainty in results.
- Recent advancements in information and telecommunication technologies (GIS, ISDN), multimedia, virtual reality, neural networks) could play an important contribution to the modeling of various risks.

- In regional risk assessment, all risks (local, regional, global) should be taken into consideration.
- Safety culture, public participation and risk communication are relevant issues in the overall landscape of the regional risk assessment process. Emergency culture, preparedness, and planing is an integral part of regional safety management.

Within the Poly-project we experimented various aspects as previously highlighted. As compared with similar projects in the world (e.g., the UN Inter-Agency on Risk Management), the present work brought new answers to this interdisciplinary subject. Work done within the PPR&S is complementary to the numerous activities developed in Switzerland.

Further information on the Poly-project and its publication series can be obtained from:

Poly-project "Risk and Safety of Technical Systems"
ETH-Center
CH-8092 Zürich
Switzerland
Phone: +41 1 632 2356
Fax: +41 1 632 1094

FOREWORD

In recent years, the community has become increasingly aware of the risks of locating hazardous industries near heavily populated environmentally sensitive areas. This new awareness means a novel approach to safety planing for hazardous industries, looking at the problem from the point of view of integrated regional risk assessment, which should include beside the risks arising from natural events (like earthquake, flood, forest fires, etc.) also the risks arising from processing plants, storage and transportation of dangerous goods.

The purpose of *Volume I* is to highlight the main procedures for risk assessment of health and environmental impacts from continuous emissions of pollutants into air, water, and soil referring to normal operation conditions.

Volume II is concerned with the assessment of the consequences of accidental releases. The matter treated should help to find an answer to questions, such as: - what can go wrong?; - what are the effects and consequences?; - how often will it happen?. The main procedural steps are supported by relevant methods of risk assessment recognized on an international level; this document gives also an overview of criteria and guidelines for implementation of risk assessment and management at different stages.

Information contained in Vol. I and II is based on a wide range of scientific publications and references, and particularly on contributions provided by the *Biomedical and Environmental Assessment Division of Brookhaven National Laboratory, USA; UN Inter-Agency Program (UNEP / WHO / IAEA / UNIDO) on the Assessment and Management of Health and Environmental Risks from Energy and other Complex Industrial Systems.*

Both volumes shall be valuable to students, engineers, and scientists in charge of developing new methodologies for hazard analysis and risk assessment; practitioners active in the field of environmental protection; local or governmental Authorities in charge of implementing environmental risk impact assessment procedures and guidelines.

It should be noted that, although consideration of the continuous emissions from nuclear power stations and other nuclear facilities form an important part of the assessment of the integrated risks from large industrial areas which contain nuclear facilities, they are not considered here. This is because nuclear risk assessments are currently carried out at a higher level than that used for other facilities (e.g., in Switzerland by the HSK, *Hauptabteilung für die Sicherheit der Kernanlagen*, Villigen, CH) and would be available for use in integrated risk assessment at community level.

Complementary readings which are strongly suggested are: "Management and Control of the Environment, (WHO, 1989)" and "Rapid Risk Assessment of Sources of Air, Water and Land Pollution, (WHO, 1982, 1993)", Guidelines for "Integrated Risk

Assessment and Safety Management for Large Industrial Complexes and Energy Generating Systems" (IAEA, 1995).

ACKNOWLEDGEMENTS

The authors wish to express their sincere gratitude to Professor Wolfgang Kröger, Chairman of the Executive Committee of the "Polyproject, Risk and Safety of Technical Systems (PPR&S)", ETHZ - Swiss Federal Institute of Technology, Zürich, for his guidance and critical reviews during the different phases of this work.

We are also greatly indebted to all people, who through their support and many valuable suggestions for corrections and improvements of the manuscripts, helped us to finalize this work. We wish also to acknowledge more specifically the following individuals:

Prof. R. Hütter, Vice-President Research, ETH Zürich.
S. Chakraborty, HSK.
Dr. Hans-Jörg Seiler, Project Manager for the PPR&S.
Prof. J. Schneider, Institut für Baustatik/Konstruktion, ETH, Hönggerberg, Zürich.
Mr. H.A. Merz, EBP, Ernst Basler & Partner Ingenieurunternehmen.
Prof. Dr. B. Böhlen, former Director of BUWAL (Bundesamt für Umwelt, Wald und Landschaft), Bern.
H.R. Wasmer, Deputy Director, EAWAG, Dübendorf (ZH).
Prof. K. Hungerbuehler, Institut für Technische Chemie, ETHZ.
Dr. H. Künzi, Konzern Sicherheit und Umweltschutz, Hoffmann-La Roche AG., Basel.
Mr. K. Cassidy, Head Major Hazard Assessment Unit, Health and Safety Executive, London.

One of us (AG) would like to express special consideration and high appreciation to Mrs. Françoise Bordier for her exquisite support and distinguished encouragement in his professional activity while in Switzerland. Finally, in the preparation of this book, the authors are greatly indebted to Mrs. I. Kusar (PSI), who skillfully prepared the drawings and pictures for illustrating the manuscript.

LIST OF ABBREVIATIONS

API	:	American Petroleum Institute
BOD	:	Biological Oxygen Demand
BUWAL	:	Bundesamt für Umwelt, Wald und Landschaft, Bern (Switzerland)
CEAM	:	Center for Exposure Assessment Modeling
CEC	:	Commission of the European Communities
COD	:	Chemical Oxygen Demand
CRWM	:	Civilian Radioactive Waste Management
DOE	:	US Department of Energy (USA)
DOT	:	Department of Transportation
EC	:	European Communities
EGES	:	Expert Group for Energy Scenarios, Bern (Switzerland)
EPA	:	Environmental Protection Agency (USA)
FEMA	:	Federal Emergency Management Administration (USA)
GIS	:	Geographical Information System
HEC	:	Hydrologic Engineering Center in California (USA)
HMIX	:	Hazardous Materials Information Exchange (USA)
HSE	:	Health and Safety Executive (UK)
IARC	:	International Agency for Research on Cancer
IGWMC	:	International Groundwater Modeling Center (USA)
NCASI	:	National Council of the Paper Industry for Air and Stream Improvement
NRC	:	National Research Council
OECD	:	Organization for Economic Co-operation and Development
PM	:	Particulate Matter
TDS	:	Total Dissolved Solids
UNEP	:	United Nations Environment Program
US EPA	:	Environmental Protection Agency (USA)
WHO	:	World Health Organization

CHAPTER 1

FRAMEWORK FOR THE ANALYSIS AND ASSESSMENT OF REGIONAL EMISSIONS

1.1. Introduction

All human activities are possible sources of risk. In general it is widely accepted that the term risk implies the consideration of the measure of some form of loss in terms of both the likelihood and magnitude of that loss. In the context of an integrated area risk assessment and management, the following constitute the most relevant types and sources of emissions injurious to the environment to be considered:

■ *Continuous emissions* to air, water, and land from:

- *Industrial and associated activities.*
- *Energy supply, distribution and production.*
- *Waste generation and disposal (incinerators for household filth and industrial wastes; municipal sewage).*

■ *Non-point source emissions* are the result of:

- *Urbanization itself and its associated infrastructure.* Large cities and industrial areas are a source of environmental risk. Such risk include surface water contaminated runoff, air pollution from transportation systems (in particular road traffic).

- *Large scale agricultural activities.* Fertilizers, insecticides and herbicides may contaminate lakes, rivers, ground water, and soils. These form a potential serious risk to the environment and to the public health. Large scale agricultural activities may also consume large amounts of water, causing water shortage in periods of dryness, and soil erosion; and

Products releases to the environment which result in a degradation of water or air quality are considered as pollutants. These include:

– *Pollutants routinely released into the atmosphere* from stacks, tailpipes, and fugitive emissions from vents, open burning, etc.;
– *Pollutants discharged to surface water* from outfall pipes or channels, routine overflow from waste ponds, and spread sources such as run-off from urban roadways products or agriculture; and

- *Emissions to ground water* from landfill leachate, percolation from surface ponds and lagoons, leakage from pipelines, and discharges from injection wells.

Continuous emissions generally lead to exposures that create chronic, long-term effects. Acute health effects may also result. Extended meteorological inversions, for example lead to acute exposures and acute effects from routine emissions if adequate provisions are not taken. *Continuous emissions to water more generally yield only chronic effects, but there can be exceptions.* For example, contaminant concentrations built-up in river sediments over long periods may be released during storms that stir up sediments, resulting in acute, high-level exposure.

Emission sources have to be characterized and compared with the relevant emission standards concerning air, water and grid pollution. As far as the *effects* are concerned, once the receptors have been identified, the transport of the pollutants from the source to the receptor has to be defined and calculated and the dose-response to the receptor assessed. Owing to the probabilistic character of the transport process, *exposures* to the receptors are obtained in terms of figures with probabilities. The estimated risk to the receptor, i.e., certain damage affected by a certain probability, is calculated and compared with the relevant risk acceptance criteria. If the criteria are not met, the pollution abatement process in the plant has to be adapted in a iterative procedure. The general assessment framework is outlined below.

Methodology for the Analysis and Assessment of Continuous Emissions

The first step in analyzing continuous emissions is to identify their sources and to characterize their quantities and their physical and chemical properties. The second step is to identify receptors and characterize the movement of pollutants from source to receptor, generally through the use of mathematical models. It requires that *receptors*, i.e. the human populations or sensitive environments, be identified and located, and *pathways* from source to receptor be determined. Appropriate models are then established and exposures estimated. Ambient monitoring of pollution levels is helpful in guiding this process and in validating results of modeling. Modeling the transport of pollutants from source to receptor provides an estimate of exposure.

The next steps are meant to identify or develop dose-response relationships between exposure and effects so that effects or risk may be determined. *Figure 1.1* gives an overview of the steps involved in the analysis and assessment of continuous emissions.

1.2. Identification of Continuous Emission Sources

Estimates of sources, types and quantities of gaseous, liquid, and solid emissions from industrial activities and energy systems are needed to evaluate their risk to health and the environment.

1.2.1. GENERAL SOURCES OF INFORMATION

The necessary information can be collected from government and private organizations, from compiled literature, from new engineering estimates, or from new measurements.

As noted by WHO (1982),[1] *"A major task of the study team is to locate all major government information sources and to extract the required data from them."*

TABLE 1.1 presents a partial list of possible sources of information. Undoubtedly, a sizable portion of the required information available from these organizations will be in unpublished form. Therefore, some efforts will be needed to extract, process and classify useful information.

The major difficulties with unpublished data are to determine which ones are needed and then to interpret them properly. Often there is a danger of omitting important information if screening is not done carefully. But, complexity and resource requirements increase considerably if relatively unimportant data are retrieved and processed. *Cross-checking* collected data with information from other sources is often possible and highly desirable, since it is one way of insuring accuracy of the results. If important data from various sources are in significant disagreement, investigation of their original derivation often provides a good basis for formulation of the most accurate estimates.

U.S. Toxic Chemical Release Inventory Data Base
The Toxic Chemical Release Inventory, mandated by the U.S. Superfund Amendments and Reauthaurization Act of 1986, will be the most comprehensive source of information on toxic releases once companies begin to report in 1988 (309 chemicals, 20 chemical categories).

In the event that the agencies listed in

TABLE 1.1 have not compiled the needed information, first-order approximations of the engineering and environmental characteristics for most energy systems and for most conventional air (e.g., PM, SO_x , NO_x) and water (e.g., TDS, BOD, pH) pollutants can be derived from several *summary documents* (TABLE 1.2).

■ *Emission data on many industrial processes* for conventional pollutants have been evaluated by *WHO*[1] *(1982)* for the "Rapid Assessment of Sources of Air, Water, and Land Pollution", and

■ *Air and water pollutant emission standards* have been compiled and published by the *EPA (US Environmental Protection Agency)*. Such data were used to prepare the tables presented in *Appendix A* (TABLES A-1 and A-2), which summarize the industries and pollutants examined by the EPA. EPA's efforts have also focused on some *toxic chemicals* (TABLE A-3).

More detailed characterization efforts may be required for any of the following reasons:

- Development of site-specific case studies
- Analysis of site specific energy systems (e.g., peat or dung) or industrial activities that are not widely used; or
- Emission coefficients for non conventional (e.g., toxic or hazardous) pollutants.

TABLE 1.1. Possible sources of information concerning emissions

Type of Data	Possible Sources of Information
Industrial activity	Ministry of industry or commerce National planning/economic development agencies
Electric energy ministry, authority or company	Internal revenue agencies, Local governments Industry associations Ministry of animal production Air, water and solid waste pollution control authorities
Fuel consumption	Ministry of energy Ministry of industry Internal revenue agencies Refineries or oil distribution companies
Rail & road traffic activity	Ministry of transportation
Air traffic activity	Airport authorities Ministry of transportation
Shipping activity	Port authorities Ministry of transportation
Water emissions	Oceanographic institute Ministry of health or environment River authorities Water pollution control authorities Ministry of fisheries Area planning agencies Local health departments Universities
Air emissions	Ministry of health or environment Air pollution control authorities Universities
Solid wastes	Local authorities Ministry of environment Private refuse disposal companies Area planning or development agencies
Occupational health	Ministry of health Local health departments Universities
Public health	Ministry of health Local health departments Universities

Note: *WHO[1] (1982).*

Step 1	*Identify the sources of continuous emissions and characterize their quantities, and their physical and chemical properties.*
Step 2	*Characterize the Emission Source Inventory*
2.1	*If monitoring is available estimate pollution from different source terms by direct measurement*
2.2	*If monitoring is not available or monitoring is not technically and environmentally feasible, then calculate emissions of different pollutant by means of conversion factors and the efficiency figures of the controlled pollution equipment.*
2.3	*If no measurement values are available, and monitoring not at hand, use comparative values from similar situations or based on existing literature, in order to estimate emission values; check if the results of using these values are applicable to the facility under investigation.*
2.4	*Expert judgment.*
Step 3	*Select a pathway for analysis organized by a given receiving media; air, water, soil.*
Step 4	*Using models calculate dispersion values in the receiving media.*
4.1	*If air is the media where dispersion of pollutants occurs, then calculate concentration of pollutants under given weather conditions Go to step 5.*
4.2	*If water is the media where dispersion of pollutants occurs, then calculate concentration of different pollutants at some time instance and distance from the source of pollution. Go to step 6.*
4.3	*If soil is the media, evaluate the critical load and the concentration of pollutants in the given environment. Go to step 6.*
Step 5	*For evaluating the concentration of pollutants as a time-distance function use atmospheric dispersion models.*
5.1	*For distances between one and about 50-80 km dispersion from a point source use simple Gaussian Plume Models.*
5.2	*For complex orographical and meteorological conditions use Complex Plume Models.*
Step 6	*Use dose-response relationships to estimate the risk to the population; evaluate the health impacts.*
Step 7	*Use analytical methods or expert judgment for environmental impact assessment.*

Figure 1.1. Methodology for the analysis and assessment of continuous emissions

For these characterizations, data gathering efforts may need to focus on technical literature published by various research (e.g., US Department of Energy) or regulatory organizations (e.g., US Environmental Protection Agency), as well as by equipment manufacturers. In Switzerland such agencies are to be found under the heading: "Swiss Federal Department of ..."; Swiss Federal Office of Statistics; Swiss Federal Office for

Environment, Forestry, and Landscapes (BUWAL, Bundesamt für Umwelt, Wald und Landschaft, Bern).

TABLE 1.2. Source documents for data on emissions

Citation	Title
Hittman (1974)	• Environmental Impacts, Efficiency and Cost of Energy Supply and End Use
Hubert et al., (1981)	• Les Impacts Sanitaires et Ecologiques de la Production D'électricité - Le Cas Français
Manthey et al. (1980)	• Energy Technology Data Handbook I, Conversion Technologies
OECD[7] (1984)	• Emission Standards for Major Air Pollutants from Energy Facilities in OECD Member Countries
The Aerospace Corporation (1981)	• Energy Technologies and the Environment
The Aerospace Corporation and Mueller Associates, Inc., (1983)	• Energy Technology Characterizations: Handbook - Environmental Pollution Control Factors
The Science and Public Policy Program (1975)	• Energy Alternatives: A Comparative Analysis
UNEP[6] (1985)	• The Environmental Impacts of Production and Use of Energy, Part IV: The Comparative Assessment of the Environmental Impacts of Energy Sources, Phase I: Data on the Emissions, Residuals and Health Hazards of Energy Sources
USEPA (1977)	• Compilation of Air Pollutant Emission Factors - Third Edition (Including Supplements 1-7)
USEPA (1980)	• Environmental, Operational, and Economic Aspects of Thirteen Selected Energy Technologies
USEPA (1986)	• Superfund Public Health Evaluation Manual
USEPA (1988)	• Superfund Exposure Assessment Manual
USEPA (1989)	• Risk Assessment Guidance for Superfund - Human Health Evaluation Manual Part A
WHO[1] (1982)	• Rapid Assessment of Sources of Air, Water, and Land Pollution
WHO[2] (1983a)	• Compendium of Environmental Guidelines and Standards for Industrial Discharges
WHO[3] (1983b)	• Selected Techniques for Environmental Management - Training Manual
WHO[4] (1988)	• Emissions, Environmental Transport, and Dose-Response Models: Guidelines for Case Studies
WHO[5] (1989)	• Management and Control of the Environment

1.2.2. INFORMATION AND LITERATURE ON EMISSION SOURCES

Estimates of sources, types, and quantities of gaseous, liquid, and solid emissions from industrial activities and energy systems are needed to evaluate their risks to health and the environment. Although there is a large compiled literature on a range of technologies and emission types (TABLE 1.2), the *World Health Organization* (WHO[2], 1983a; WHO[3], 1983b; WHO[4], 1988; WHO[5], 1989), the *United Nations Environment Program* (UNEP[6], 1985) and others (e.g., OECD[7], 1984) have found that in developing quantitative assessments of health and environmental effects, emissions data for a given technology in different countries vary.

Principal reasons for these variations include differences in operating characteristics of fuel or material consuming devices, in fuel or material quality, or in regulatory-based pollution control requirements.

Owing to these variations, country/technology specific emissions estimates are needed to increase the accuracy of any risk assessment effort. There are four principal approaches (TABLE 1.3) that can be used to develop estimates of routine or continuous emissions from a source; each has its own unique strengths and weaknesses (see also section 1.2.3).

TABLE 1.3. Estimation of routine or continuous emissions from a source

Methods For the Estimation of Continuous Emissions
• Monitoring of Data or Emission Sources
• Use of Correlation Models
• Comparison of Data
• Expert Judgment

1.2.3. METHODS FOR ESTIMATING THE SOURCE STRENGTH

First Method

The first method consists of collecting *monitoring data* from an operational source. In such a case, monitoring equipment are used either on a continuous or intermittent bases to provide data specific to the process unit in question. Monitoring, however, requires substantial time and effort. Furthermore, if data are not collected over a long enough period of time, they may not be representative of the true emission characteristics because of time and process dependent variations. Monitoring may not be technically nor economically feasible in many cases.

Second Method

A second approach is based on using *theoretical or empirical equations correlating operating parameters to pollutant emissions rates.* Stoichiometric estimates may be, however, erroneous because of inadequate specification or understanding of the process or knowledge.

Third Method

A third approach is to *use data compiled from other facilities* or based on existing literature and assume that the results are applicable to the facility in question. In this approach questions will always remain about the accuracy and precision of the extrapolation.

Fourth Method
Finally in addition to the preceding approaches or when information is completely lacking, *expert judgment* must be used.

The following sections highlight information relevant to these methods and include the following:

- Data reporting protocols
- Potential information sources
- Sample emission coefficients for some energy processes
- Collecting monitoring data from an operational source
- National emission standards in OECD member countries
- Concensus on how to estimate coefficients of emissions.

1.2.4. DATA REPORTING PROTOCOLS

Scope of Data
The scope of data to be compiled can vary according to the *process* and the *pollutant*. Processes being evaluated may need to be treated as one unified system, or as many independent subsystems. The degree of aggregation depends on the complexity of the facility in question, as well as on the degree of dependence among process operations. Facilities which tend to be more complex and are composed of many semi-independent operations require from the investigator a different approach than this would be the case with simple integrated manufacturing operations. In general, this dichotomy pinpoints the difference between *energy-related* vs. *industry-related activities*. Energy-related activities tend to focus on the processing or combustion of a fuel in a unified way. Industrial operations, however, may include many loosely aggregated activities that must be evaluated independently.

Definition of the System Boundaries
As collection efforts have started, some thought should be given to *defining the system boundaries of interest* (i.e., the back- and front-ends of the fuel and material supply cycles). In some instances, these contribute most of the emissions. Hence, the potential consequences of including or excluding them should be considered. As a general guide, complete cycles are often evaluated when systems are being compared, or when regional or national-scale analyses are being conducted. As the geographic or technologic scales of the analysis decrease, the value of including complete cycles diminishes.

Identification of the Pollutants
Similarly, in assessing risks from these processes it might be appropriate to identify *all pollutants from all alternatives*. Practical limitations, however, quickly demand that effort be focused.

Data collection could focus on any or all of the following:

- Process
- Activity
- Trajectory
- System

The level of complexity increasing from top to bottom.

Specificity of the Pollutant
Moreover, data collection should pay due consideration to the *specificity of the products* released from the emission source, their environmental impacts and to existing standards, e.g.:

1) Pollutants that *quantitatively dominate* the waste streams (e.g., carbon dioxide from oil- or gas-fired steam electric power plants)
2) *Index pollutants* (e.g., BOD or sulfur oxides)
3) Pollutants for which there are *environmental standards* (e.g., Lead in the atmosphere); and
4) Pollutants for which there are *acute* (e.g., hydrogen sulfide) or *chronic* (benzo-α -pyrene) *health's effects*
5) Pollutants that are *emitted routinely* (e.g., noble gases from nuclear steam electric power plant).

How to Establish Emission Coefficients
Emission coefficients may derive from simple point estimates to complex models. In generating simple and complex coefficients for specific activities, many underlying predictors may need to be defined. In *combustion-based systems*, for example, the following types of information must often be specified:

- Energy content of fuel
- Moisture, sulfur, ash, and trace element content of fuel
- Thermal efficiency of boiler
- Temperature of exiting gases
- Type and characteristics of emission-control equipment.

In *industrial-based systems*, all the aforementioned information must be examined. In addition, rate of feed stock input and rate of product output may also need to be identified.

Customary Units
In the technical literature, different unit formats may be used to report emissions data for different processes, as the following examples show:

a) Mass of pollutant per mass of product (kg/kg)

b) Mass of pollutant per mass of fuel (kg/kg)
c) Mass of pollutant per unit time (kg/s)
d) Mass of pollutant per unit of volume (kg/t)
e) Mass of pollutant per unit of energy input or output (kg/J).

Protocols
Reporting protocols for emission data differ, in part, for historical and regulatory reasons. In the US and elsewhere (see OECD[8], 1984), emissions are regulated as pollutant mass per unit of energy input (i.e., kg/J) or as pollutant mass per unit of volume (kg/m^3). Emission standards for non-combustion sources associated with industrial activities span the range of reporting protocols listed above.

1.3. Non-point Source Emissions

Beside continuous emission sources, the non-point source emissions should not be neglected during risk assessment, since they contribute to a great deal to surface and ground water pollution as a result of agricultural activities or urban runoff, etc. Water supplies are particularly vulnerable to non-point sources pollution.

1.3.1. PERSPECTIVES ON NON-POINT SOURCE POLLUTION

Progress toward attaining the Nation's Water quality goals have been made in Switzerland. Much of the results could be accomplished by controlling industrial and municipal pollution point sources. Given the nature of non point source pollution, cantonal and communal management, in Switzerland, are a key for improving the situation, since only at this level does enough flexibility exist to make site-specific and source specific decisions that really work. *Non-point source concerns should be incorporated into all aspects of water management.* Non-point sources of pollution are undoubtedly the most difficult to control because the variety of sources is large and their dispersion is extreme. Improving things in this matter involves changing individuals and corporate behavior patterns. Major difficulties must be overcome, particularly as non-point source control affects land use decisions and individual actions. *Non-point source pollution is best solved at the local level.*

Pollution from Agricultural Activities and from Urban Runoff
Long-distance transport of pollutants are generally regarded as the biggest threat to the environment. Forest damage, acidification of soils and water, as well as eutrophication of lakes are well known. The other major source of non-point source pollution is urban runoff.

Agricultural Activities. The use of fertilizers is of great importance for the farming activities. *Losses of nitrogen from agriculture to the atmosphere* are mainly due to ammonia evaporation from manure and from commercial fertilizers. The following re-deposition of ammonia can lead to increased acidification of soils and waters. The ammonia is absorbed by the roots - a process whereby hydrogen ion is released. *Acidification of soils* might induce a leakage of potassium, calcium, and magnesium. *Acidification of waters* can drastically change the ecosystem leading to the extinction of plants and organisms. Ammonia deposited directly to water surfaces is responsible for large amount of the nitrogen input, thereby enhancing the eutrophication process. Among the factors influencing the extent of nutrient leakage to water from arable land, the following play a role. Climate, type of soil, and crops, the way of planning and harvesting, and the quantity of fertilizer used. *Losses of nutrients from arable land are due to surface runoff and transport through sub-drainage systems.*

In terms of mass, sediments are the major water pollutant from agricultural activities. Soil lost seriously affects not just productivity of the land, but also the quality of the waters into which most of the soil and agrochemicals flow and the bottom habitat in our aquatic ecosystems. It is important to realize that a very small percentage of the land contributes to the total annual situation loss. *Controlling erosion* from a surprisingly small number of areas, and some of them rather small in size, can result in most of the pollutant reduction needed to protect water quality and habitat for fish and wildlife. In addition to soil, we have lots of toxic materials, nutrients, and other materials adsorbed onto the soil particles. Thus the non-point source pollution that washes out the land can be loaded with all sorts of bad contributing factors.

Urban Runoff. Pollution in urban runoff includes air pollutants that have settled in streets, erosion from construction sites, slat and other deicing chemicals, litter and animal refuses. Similar to agricultural non-point source pollution, most of the urban runoff pollution comes from limited areas, such as the industrialized and highly urbanized sections of a city. Appropriate management practices targeted at those areas can control the pollution. Such practice include more frequent street cleanings, use of porous pavement, and suitably designed and maintained sedimentation and catchment basins to reduce the amount of soil and adsorbed toxic and other non-point source pollutants carried in runoff. Much effort is still to be done for removing nutrients in municipal waste treatment units and for promoting changes of habits concerning the use of laundry detergents. Most of the pollutants contributing to water turbidity are entering the water system as non-point source pollutants and are not part of the municipal waste water treatment plants effluents. As an example, the water quality of the "Lake of Annecy" (F), could be drastically improved, after a ring canalization catching all domestic and agricultural water effluents of the region was built around the lake. Such remedial action is fare more efficient that trying to improve the water quality of an eutrophic lake by aeration and oxygen transfer (by means of mechanical agitators, pumping, or other aeration equipment).

Non-point Source Pollution. They includes not only conventional pollutants like sediments, oxygen-demanding wastes, coliform bacteria, and nutrients (excess nitrogen), but toxic substances like heavy metals, pesticides, herbicides, And a lot of other products. Greatest stream nutrient concentrations can be linked to snow melting periods or spring storm flow periods when cultivation is active and vegetation cover is poor. Lake drinking water supplies are particularly vulnerable to non-point sources pollution because of cause-effect relationships linking watershed characteristics, nutrient loading, lake eutrophiction, sewage treatment plant disturbances, and quality of water supplied to the consumer.

Pollution from Industrial Areas
Hazardous wastes have to be controlled at their origin. Organic halogens AOX, and heavy metals (e.g., Pb, Cd, Cr, Cr^{VI}, Cu, Ni, Zn) account for a large part of the pollutants released to water effluents. The major portion of the heavy metal load of the treatment plant inlets is discharged by industries without any or with insufficient pre-treatment. Other sources are the discharge from municipal sewage plants, precipitation, natural ore deposits, runoff from fertilized land.

■ *Phosphorus Abatement:* algae growth (phytoplankton, microphytes) is stimulated by the increase of phosphate in rivers and lakes. Phosphorus can be precipitated in activated sludge plants. There is a tendency for biological processes to operate at constant reduction rates under changing loading conditions. Substitution of phosphates in synthetic detergents is also contributing to the total phosphorus reduction.

■ *Ammonia Abatement:* from the point of view of potable water supply ammonia in raw water is very detrimental as it results in disinfecting problems. The German standards for desirable ammonia value in water are: 0.2 mg/l NH_4, respectively 0.15 mg/l NH_4-N in surface water.

1.3.2. MONITORING NON-POINT SOURCE POLLUTION

The broad application and continued use of herbicides and pesticides have resulted in a major diffuse source input of toxic material into aquatic ecosystems.[9] Unlike point source input, where levels, quantity, and consistency of loadings are known, diffuse source input must be estimated using assessment procedures. The different environment processes and fates that potentially regulate a compound's effect in the aquatic ecosystem emphasizes the need for system behavior information. Approaches to the monitoring and assessment of water quality, as well as water quality management, will vary according to the relative significance of non-point and point source pollution. Management of water quality within an aquatic ecosystem system involves measurement, evaluation, and remedial action. Measurement pertains to the collection of biological, chemical, and physical water quality data. Evaluation necessitates a set of criteria with which the water quality can be realistically compared.

Point source inputs to an aquatic ecosystem are usually a consistent load of a given set of materials or chemicals. Datasets can be generated in specific areas of a river basin, and areas of noncompliance established with remedial actions confined to specific sources. *Diffuse loadings from land use or atmospheric inputs* tend to be more event oriented without a quantifiable area of effect in the aquatic environment. To determine the state of aquatic resources requires a general monitoring program that will yield data describing the presence, level, and change over time of specific chemicals entering the aquatic system as a result of man's activities. *By virtue of its diffuse nature, understanding of non-point source pollution relies more on monitoring and assessment than does point source pollution.* Process assessment requires measuring the system variability and examining the physical-chemical and biological processes that determine environmental quality. Variability should consider statistical estimates of variance as well as include the comparison and analysis of the different sets of physical-chemical conditions.

The *monitoring programs* for lakes, streams and rivers should, at minimum, include precipitation and stream discharge recording and water sampling for suspended solids, phosphorus, and nitrogen analysis. In parallel a land-use monitoring program should be established as well. To this end, land-use and agricultural activities in the watershed should be monitored on an intensive, field-by-field basis. The data collected can be processed through a geographical information system (GIS). The *cumulative frequency distribution* is useful for basic data description and for assessing year-to-year changes in water quality.

Frequency distribution of concentration, discharge, and export data must be generated for each year. Shifts in the plots between years can indicate trends in the data; differences between the distributions are tested for significance using the Kolmogorov-Smirnov 2-Sample Test. Extremes in the data may be evaluated by comparing the probability of exceeding (Pe), a particular critical value in succeeding years, where Pe is defined as 100 percent minus the cumulative frequency percentage of the critical value. The shape of such plots may indicate a trend. The critical values selected may be water quality standards or some other specific concentration or load. *Another method is based on paired watersheds.* This approach compares the behavior of two watersheds, where one is assumed to be the control (untreated) and the other the experimental (treated). A linear regression is performed on paired data, when both watersheds are treated similarly (during the calibration period), with the control watershed serving as the independent variable. Following treatment of the experimental watershed, a second regression is performed. Significant differences between the calibration and treatment regressions is a measure of the treatment effects. (Hewlett and Pienaar,[10] 1973). Furthermore the probability of exceeding some specified level in a given year can be easily derived from the distribution.

Concentration Patterns During Runoff Events

Nutrient and sediment transport in area rivers follows a typical pattern. During a runoff event, stream flow increases very rapidly on the rising limb of the hydrograph, reaches a peak value, and then decreases more slowly on the falling limb of the hydrograph. Sediment concentrations peaks early in the runoff event and is already decreasing well before the peak discharge occurs. Since most of the *phosphorus* transported during storms is attached to sediment, the phosphorus concentration closely follows the concentration pattern for suspended sediments. *Nitrate* concentrations increases during the falling limb of the hydrograph, since most of the nitrate enter the streams as part of the tile drainage and inter flow. The concentration patterns of soluble herbicides are distinct from both the sediments and the nitrates. The large changes in sediment and nutrient concentrations with stream flow necessitate using two distinct methods for averaging concentration data:

a) *Flux-weighted average concentration:* the method is used to characterize pollutant concentrations relative to pollutant loading.

$$\text{Average time - weighted concentration } = \frac{\sum c_i t_i}{\sum t_i} \tag{1.1}$$

b) *Time-weighted average concentration:* the method is used to characterize pollutant concentrations relative to ambient water quality at the sampling station

$$\text{Average flux - weighted concentration } = \frac{\sum c_i q_i t_i}{\sum q_i t_i} \tag{1.2}$$

where

c_i = concentration of ith sample
q_i = instantaneous discharge when i^{th} sample was collected
t_i = time i^{th} sample was used to characterize a stream (with a range from 2 to 24 hr)

Storm-to-Storm Variability. Considerable efforts goes into characterizing nutrient and pesticide export during a single storm event. One way to characterize concentrations during a single storm event is to calculate *flux-weighted average concentrations* for that storm. There is always a great concentration variability observed, even for storms with similar peak discharges. Part of the variability observed in the storms is associated with seasonal effects. Similar storm to storm variability can be observed for nitrates and suspended matter.

Annual Variability. Nutrient loads derived from cropland runoff generally have large year to year variability. The large amounts of annual variability in loading from non-point sources complicate comparing point source and non-point source loads, thus it is important to estimate a mean annual load from cropland sources, rather than using loads for any single year. Seasonal variations in concentrations and loading can account for some of the variability in cropland runoff.

The *seasonal occurrences of herbicides* in stream systems also result in seasonal pesticide exposures in drinking water derived from river systems, since most soluble pesticides pass through conventional water treatment plants with very little attenuation in concentration. The highest pesticide concentrations in streams and rivers, as well as the highest loads, generally occur during the first runoff event following pesticide application.

Seasonal pattern are also present for nitrate concentrations. The bulk of the nitrate loading occurs during the winter and early spring period. The effect of the watershed size is also noticeable on the peak concentration of chemicals. One convenient way to describe concentration patterns at a measuring station is to use concentration-duration curves. Such curves depict the percent of time particular concentrations are exceeded. Sediment concentration-curves also show systematic differences with watershed size.

Critical Area Selection within Watersheds

Guidelines for selecting critical areas within watersheds are generally not available. A publication by Maas[11] et al., (1984) proposes criteria for selecting sediment, nitrogen, phosphorus, and pathogen critical areas and summarizes selection approaches presently being used by different non-point source pollution control projects in the United States.

The criteria for selecting pesticide critical areas can be grouped into broad categories:

■ *Type of Water Resource Impairment: critical areas should be selected to reduce peak or ambient pesticide concentrations.* The impairment is related to the toxic effects of pesticides on the ecosystem and are due more to concentrations than to total loading. It is important to know that remedy measures such as conservation tillage, terraces, and contouring reduce pesticide surface loading primarily by reducing surface runoff volume. Such practices often do not reduce edge-of-field pesticide concentrations. Runoff reducing measures will generally reduce pesticide loading more than concentrations. In the case of ground water, such practices will increase infiltration and may actually increase ground water pesticide concentrations. Critical areas should be selected on the basis of the water resource impairment. The severity of impairment affects the extent of areas or sources that should be designated as critical. *Determining which pesticide is impairing a water resource use is another important consideration. This determination is usually made on the basis of monitoring results.*

■ *Source Magnitude Criteria:* this will be the most important criterion for selecting critical areas for pesticide control. The selection process for pesticide critical areas should focus on the usage pattern and application rate, since pesticide losses are roughly proportional to the application rate. The erosion rate will be also another important criterion for pesticides lost primarily in the sediment-adsorbed phase of surface runoff. Organo-chlorine and organo-phosphorus insecticides are strongly adsorbed on soil particles. Carbamates, anilides, and triazines are lost predominantly in the dissolved phase of surface runoff.

■ *Transport-related Criteria:* there exists a relationship between the distance to watercourse and the percentage of applied pesticide that reaches the impaired water resource. However, the dominant transport mechanism may vary greatly with pesticide class, thus affecting the importance of this criterion. The distance between the nearest water course from the pesticide source and the site of the actual water resource impairment is also an important criterion.

■ *Contaminant Stagnation Areas:* The impact effects of contaminant stagnation areas, in case of ground water contamination, are an important yet often overlooked factor in the design of site remediation systems.

■ *Other Criteria:* the *on-site evaluation* for determining pesticide critical areas should focus on:

- pesticide usage patterns (type, rates, frequency, timing, method, equipment)
- disposal practices
- presence of gullies or sinkholes which short-circuit transport.

A *farm-level critical area rating scheme* based on the above criteria has been proposed by Maas[11] et al., (1985). Young[12] et al., (1985) have proposed a *method for prioritizing water quality problem areas* based on a computer simulation model, AGNPS I and II,[13] to analyze non-point source pollution from agricultural watersheds. The simulation is based on single storm events defined in terms of frequency and duration. The model uses geographic cells of data units. the model inputs existing (or proposed) land management conditions and simulates the transport of sediments, nutrients, and flow from the headwaters of the watershed to the outlet in a stepwise manner so that an assessment can be made for land parcels in the watershed. The model runs on a handhold programmable (HP)-calculator. The basic components of the model are *hydrology* (runoff volume and peak flow), *erosion* (modified soil loss equation), *sediment transport* (transportation and deposition) and *nutrient transport* (nitrogen, phosphorus, chemical oxygen demand). The model works on a cell basis only.

Application of AI Techniques for Identification of Unknown Ground Water Pollution Sources

A new methodology, based on the concept of statistical pattern recognition using BAYES' Optimal Decision Rule, has been developed for identification of unknown ground water pollution sources (Datta[14] et al., 1993). The function of the pattern recognition system is to statistically match an observed set of concentrations in the field with a comparable set obtained by simulating ground water transport for various disposal conditions. An expert decision support system has been attached to the pattern recognition system.[15]

Pesticide Run-off studies

The purpose of the studies held at *Rosemaund Experimental Centre (England)* was to study the risk posed to the aquatic environment by pesticides and to provide realistic field data on environmental concentrations of pesticides, so that model predictions can be checked and the model themselves improved, if necessary.[16] An essential part of the risk assessment concerns the need to predict likely environmental concentrations of products before they come into use. The most promising new approach to this problem involves the use of computerized models of the way pesticides behave in fields and their associate drainage water. In the years 1987-1990, the Rosemaund project monitored the run-off and leaching of 11 pesticides. The major finding was that a small proportion(<1%) of most pesticides was able to transfer from the fields to the stream within a few hours of rainfall, but that these levels peaked rapidly and then declined below detection limits once more. Computer modeling was only partially successful in predicting these events.

Health Risk from Radon

Recently, in Switzerland, attention has been brought to the health risk presented by *radon gas* inside houses and buildings, which might be located in the Alps region (presence granite rocks). Similarly the issue has also been investigated for the case when the construction material is made of granite. Some reports on this topic have been published.

Modern constructions are nowadays made to a better standard regarding insulation and the radon gas remains trapped inside the building for a much longer time that it was the case before. Especially in winter time there may be a radon problem when the houses are not ventilated properly.

1.4. Population and Meteorological Data

Within the framework of risk analysis, the effects consequent to the escape of hazardous materials in the surroundings are translated into the damage caused by them. In order to determine the number of persons who might be affected, knowledge about the population density in the surrounding is required as well as the weather conditions prevailing in the region.

1.4.1. POPULATION AND HOUSING DATA

The estimation of the societal risk from continuous releases or from a major hazard necessitates the determination of the density of the population around the production site or town area in case of transportation accident.

Collection of Data on Populated Areas
Experts dealing with risk assessment should have ready on hand information on the *population density distribution.* Other information required about the population includes the proportion present at different times of day, the percentage more vulnerable to the hazard (sick people, old people, and children), and the distribution between indoors' and outdoors' populations. *There exists several computer programs which process population data: TRIP*[17] (from SRD), *INTRAM*[18] etc. A methodology is presented in this later paper for estimating the population density and other population characteristics.

Population Density Distribution (urban, suburban, rural). Special geographical maps should show the *distribution of the population density* according to three categories: urban, suburban, and rural areas. For a risk estimate with regard to the presence of people, difference is made between the function of the area, such as:

- Residential areas, sub-divided into sparsely and densely populated areas
- Industrial areas, sub-divided into industrial installations and offices
- Recreational areas and others: sport-grounds, stadium, public swimming pool, camping, hospitals, etc.

Residential Areas. In order to determine the number of people which might be affected within a calculated damage distance, reference is made, most of the time, to population data per type of residential area or buildings. The accuracy of such a determination depends on the detail of the population data. If the damage area is of restricted nature, then local conditions play an important role, and consequently, sufficient accuracy can only be obtained from properly detailed data. If the damage affects a larger area, then data of a more global nature can suffice. Over- and under-estimations concerning an area should be averaged out. In conclusion, an acceptable reliability in risk estimations (for the case of accidental releases) can be obtained if, within a 400 m distance, detailed population data, such as from census, will be used. For distances larger than 400 m, global data relative to types of residential areas can be used.

Detailed population data are often available in municipalities, governmental and cantonal planning services. When detailed and up-to-date population data are available, it can be recommended to make use of it, for small as well as large damage areas.

Population Distribution. Especially for the assessment of accidental gas releases and the danger resulting from large fires, information on the *distribution of the population at day and night time* according to the age and sex of the people and their professional occupation is needed. The classification should show the percentage belonging to the following categories:

- Children below 6 years of age
- Children at school (6-19 years)
- Housewives (staying primarily at home)
- Workers and employees (going to work outside their house).

Such distinctions are for two reason necessary for the estimation of the population density and distribution at the time of an accident: (a) it may be necessary to *mitigate the effects* resulting from dense gas release or from other potential danger by evacuating part of the population concerned in the area in question; (b) for *dose-effects calculation and estimation* it is important to know the percentage of the population more vulnerable to the hazard (sick people, old people, and children).

The number of people present can be established by counting the persons who, at a given moment, are present in a given area. If, in the area under consideration, only housing is located, then the number of houses (or the number of apartments in a larger building) can be counted. Thereafter this number is multiplied by the average number of people per private house or apartment (according to domestic statistics, approx. 2.3 to 2.6). Far more people, as is usually assumed, are *staying indoors*. At certain hours of the day this proportion is changing due to people leaving their homes to go to work (rush hours), or at lunch time, or when housewives are leaving the house for going shopping. For calculation assume 1 or 2 people per house. The distribution between people present in the house is assumed to be 30-70% during the day or 100% at night.

Using mailing addresses information for estimating the number of people living in an area may be useful as well.

Nowadays the personal mobility has increased and people are traveling using public transportation or private cars and are partially protected against accidental toxic vapors releases, or could escape the danger area more easily. In *railway stations* the population density is fluctuating as a function of train arrival and departure. *The population distribution is also influenced by seasonal and weather conditions, which play a role especially during the weekend.*

Global data
It appears that the population density in the cities, at different distances from the center, satisfies reasonably well the following exponential function:

$$D(x) = D_0 \exp(-D_1 x) \tag{1.3}$$

in which

$D(x)$ = population density at distance x [persons/ha]
D_0 = population density in the center of the city [persons/ha]
D_1 = density gradient [km^{-1}]
x = distance from the center [km]

An inventory of population densities[19] in cities in the united-Kingdom, West-Germany and the USA leads to the following average values for D_0 and D_1:

D_0 = ± 100 persons/ha
D_1 = ± 0.25 km^{-1}
These values are higher in cities in Japan.

For larger damage area, it is sufficient to classify the area according to its type of use, whereby the following global inventory of population densities is provided:

TABLE 1.4. Population density as a function of the type of area

Type of Area		Population Density
Residential area:	nature area	0
	remote area	1
	scattered housing	5
	quiet residential area	25
	busy residential area	70
	urban area	120
Industrial areas:	low density of personnel	5
	medium	40
	high	80
Recreational area:	camp site	130
(only during summer)	tourist spot	200

Collection of Data Regarding Housing
In order to estimate the risk from explosions and from accidental dense gas releases, information is also needed on the *type of housing*. Information pertaining to the type of habitations in the country should also be recorded to be readily available for planning and emergency cases. Part of the information could be annotated or coded on the digitized maps showing urban areas and residential areas. The construction standards for Swiss habitations has changed over the years, as well as the comfort to be found (air-conditioning, ventilation, insulated windows with double glass layers, etc.).

Housing Type. In case of fire and explosion it is important to know the *type of house construction* (i.e. height and size of the building) and the type of *construction material* used (timber, bricks, concrete, etc.).

Houses and buildings should be classified according to their use:

- residential habitation
- industrial building, plant, manufacture
- stores and warehouses, etc.,

and according to the *type of ventilation:*

- building with no ventilation (only natural aeration)
- with low ventilation rate (only in bath and kitchen)
- with high ventilation rate (air-conditioning)

Housing type in working area. The same classification would apply to buildings in the working area. An additional code could give some indication on the class of danger resulting from an industrial activity (classified according to some scheme) and a fire index could give some indication on the fire protection encountered.

Housing type in rural areas. The same kind of classification could apply to rural areas, with an index indicating the type of activity (agriculture, cattle breeding, etc.)

Ventilation Rate. In case of accidental gas release it is important to know the ventilation rate of the different buildings due to the presence of air-conditioning systems and ventilation installations in kitchen and bathrooms. Knowing such facts can help estimating the time left before an eventual evacuation should take place.

The *season* during which the incident is happening is important, since in frosty winters most of the windows are closed over longer period of time than during summer. Air-conditioning may not be in action and ventilation may be reduced.

1.4.2. METEOROLOGICAL DATA

Using gas dispersion models means that meteorological data must be collected and compiled into a given form to be used as input by some more complex models. Especially in a planning phase there is plenty of time for gathering data for regions of interest.

Collection of Meteorological Data
The collected data shall adequately represent local meteorological conditions. Data processing and presentation should be performed in a consistent manner to allow an easy comparison of the results.

Meteorological Data for Nuclear Power Plants. For *nuclear applications*, meteorological and radioactivity (doses) data, in Switzerland, are collected from several measuring stations located at the vicinity of nuclear power plants, and further transmitted to a centralized work-station located at the *HSK*.[20] The data are transformed and dispatched on a screen, using Windows4GL from *INGRES* (a data base supplier) as visualization aids and programs developed by Siemens, Germany, in the course of the *project MADUK-ANPA*. Further data on radioactivity are transmitted to the *Swiss National Alarm Center*.[21]

Data on radioactivity and doses are also transmitted through the *NABEL network* (see below).

Meteorological Data for Military Use. For military use the meteorological data are needed for assessing fallout of radioactive dust particles after a nuclear explosion at low altitude or close to the ground. These information concern wind speeds and wind directions as a function of altitude, and can be obtained from the *Military Alarm Center* (Dübendorf, Switzerland).

Meteorological Data for Civil and Environmental Applications. The *Swiss Meteorological Institute* in Zürich, is coordinating the data acquisition from about 800 meteorological measuring/observing stations, which are dispersed on the whole Swiss territory.

> ***U.S. National Weather Service***
> *National Weather Service (NWS) meteorological data are, in the U.S.A., routinely available, and these data are familiar to most model users. Although the NWS does not provide direct measurement of all the data needed as model input, methods have been developed and successfully used to translate the basic NWS data into the needed model input (this is especially applicable to the regulatory codes/models used in the US).*

Organization of the Swiss Meteorological Networks
The meteorological stations are distributed into 10 different networks types with different programs of observation and are listed in TABLE 1.5. Part of the synoptic, climatological and agricultural-meteorological networks are automated and belong to the ANETZ (SMA-ANETZ automated network).

ANETZ Network (Automated Meteorological Network). The *ANETZ includes 72 automated weather stations,* transmitting measured weather data every ten minutes (some of them every hour) to AZEN, a central computer located in Zürich. From there the data are further transmitted to the mainframe computer METEOR for further processing.

ENET Network (Automated Complementary Network). The automated complementary network ENET has been developed for early warning in case of storms, weather changes and frost. 30 stations are located in the Swiss plateau (at an altitude below 700 m above sea level), ten are in the mountains (at an altitude above 2000 m), and 3 are used as ground-layer (convection room / thermal inversion) stations (below 1200 m). The data transmitted by ENET are shown in TABLE 1.5 and TABLE 1.6.

TABLE 1.5. Swiss meteorological networks

Network	**Number of stations**
- ANETZ (Automated measuring net)	
Synoptic stations	29
Climatological stations	38
Agricultural-meteorological stations	29
Nuclear power stations	5
- ENET (Automated supplementary net)	
ENET stations	4
ENET ground-layer stations	2
- Conventional stations' network	
Climatological stations (normal)	44
Climatological stations (with agro-meteorological data)	7
Rain precipitation stations (normal)	347
Rain precipitation stations (with agro-meteorological data)	1
Basic stations equipped with totalizers	74
Frost observation stations	22
Flight weather forecast	22
Airport weather forecast METAR (incl. SYNOP)	7(4)
AMETIS at Zürich Airport	9
- Storm-wind stations	
Storm warning - Gfeller net (fully automatic)	27
Storm warning (half automatic)	2
Phenomenological network	141
- Aerologic station	
Variable height measuring station	1
TOTAL SMA-stations :	779

NABEL Network. The *NABEL network* (National Network for Air Pollutants' Observation) *is collecting information on pollutants and toxic present in the atmosphere.*

NADAM Network. There exists also a special net NADAM, for automated doses' measurement and alarming (Netz für automatische Dosis-Alarmierung und Messung).

Meteorological Data Availability

The raw meteorological data which have been collected are first verified (error detection, plausibility tests, consistence, etc.) and, if necessary, corrected. Then they are compressed, and transformed for friendly customer use (daily averages, monthly values and yearly values tables are calculated). The finalized data are then stored on magnetic tapes as *"historical data"*. A daughter data base, *ENAD*, is stored at ETH computing center (RZ-ETHZ). *The ENAD data base contains historical hourly data, which are made available in the course of the month following the recording of the data. Climatic and precipitation data bases are updated on a yearly basis.* At the present time, the following *real-time meteorological data* are available for Switzerland (TABLE 1.7):

TABLE 1.6. Data transmitted by ENET stations

Swiss Plateau	• Wind • Barometric Pressure	every 10 min. every 10 min.
Mountains' stations	• Wind • Air temperature 2 m above ground • Snow height • Snow temperature at 5 different heights above ground	every 10 min. every hour every hour every hour
Ground-layer stations	• Wind at 150-250 m above ground • Temperature at 150-250 m above ground • Humidity at 150-250 m above ground • Temperature 2 m above ground • Humidity 2 m above ground	every 10 min. every 10 min. every 10 min. every 10 min. every 10 min.

When missing, wind data statistics should be collected for local conditions. These should include average wind direction and speed as a function of the day (or night) time. The seasonal variations should be known as well. The data collected should contain information on:

- Air temperature (at day, at night, seasonal variations)
- Air flow (wind direction fluctuations, wind speeds at different height, duration as a function of height)
- Thickness of the mixing layer
- Precipitation
- Humidity
- Solar radiation or sky cover.

Analysis and Presentation of Meteorological Data. Wind data are usually measured at 10 m above ground and reported on this basis. In some cases data should be obtained at the height representative of the effective point of release (to be evaluated on the basis of preliminary information). There are two basic steps involved in the analysis of meteorological data: determination of average values at regular intervals, and statistical analysis of these values.

Statistical Analysis. Two kind of statistical analysis are needed: those for evaluating the effects of *routine releases* (from chimneys) and of *accidental releases.* The first kind of data requires a *joint frequency distribution* of wind direction and wind speed for each stability class. For effluents subject to wash-out it will also be necessary to take into account precipitation classes. The second kind of analysis involves the *probability of occurrence* of different sets of meteorological conditions over different period of time.

TABLE 1.7. Availability of real-time meteorological data for Switzerland

Real-time meteorological data
a) Indirect Data Acquisition
The real-time meteorological data are made available to interested parties by the following media:

Standard messages from the automated networks (ANETZ/ENETZ). Standard coded data from the automated network (in SYNOP-code, KLIMA-code, or in ANETZ-code, and storm-wind messages, and radioactivity alarms). The data need to be decoded and further pre-processed before use.

Teletext. Reduced amount of data on actualized weather conditions and weather forecast are made available through the teletext system of the Swiss PTT.
Videotext. Standard actualized information regarding the weather conditions (wind, temperature) and weather forecast.

Telex for private customers. Specific data (hourly or daily).

Telepac, Telefax, Data bases. As from 1993/94 direct connection and services will be made available.

b) Direct connection to the RZ ETHZ data base
A direct connection is available for VAX/AEOLUS: climatic data base, precipitation data base, hourly ENAD data from the ANETZ network (measurement every 10 minutes); AERO- and SYNOP- barometric pressure and radioactivity data bases (containing data from the special networks).

1.4.3. METEOROLOGICAL ASPECTS AND DEFINITIONS

Atmospheric Turbulence
Fluctuations of wind direction are a direct indicator of turbulence. The diffusion characteristics of an atmospheric layer may also be partially determined by measuring the temperature difference between at least two levels. These levels should encompass the level at which the wind is measured. Depending on the particular characteristics of the area, additional investigations, for example measurement of temperature gradient, made a few kilometers away from the site under consideration may be necessary. In fact the turbulence indicator values introduced in the atmospheric model must be appropriately selected to avoid erroneous classifications, and therefore, inaccurate estimates of gas concentrations.

In cases were the atmospheric stability class are determined by visual observation of sky cover at various times of the day, the observation of the amount of sky cover and of the height of the clouds may be combined with data obtained from the reading of a actinograph or pyrheliograph. *The qualitative Pasquill classification scheme is used to determine the weather stability class and turbulence.* At night, net radiometers can be used to obtain the stability classification.

Some Meteorological Definitions

Advection
It is the horizontal heat-transfer caused by air currents in the atmosphere. With respect to gas dispersion in the atmosphere, advection is the main mechanism for downwind transportation.

Convection
Convection is responsible for the transport in the vertical direction. In the boundary layer, horizontal winds are typically 1 to 10 m/s. Compared to vertical winds, which are of the order of mm/s or cm/s, they provide very efficient transportation. Convection creates turbulence, an irregular swirly motion superimposed on the mean wind. The turbulence patterns are also called eddies. Convection creates turbulence, an irregular swirly motion superimposed on the mean wind. The turbulence patterns are also called eddies. Turbulence is usually a superposition of many eddies having sizes from the depth of the boundary layer (i.e., 100 to 3000 m in diameter) down to a few millimeters.

Turbulence
Turbulent diffusion of a pollutant is due to eddies in the atmosphere. Turbulence is usually a superposition of many eddies having sizes from the depth of the boundary layer (i.e., 100 to 3000 m in diameter) down to a few millimeters; these eddies also differ in intensity.

Large size eddies are produced by thermal gradients due to solar heating. Other large eddies are generated by the frictional drag on the air flowing over the ground: the resulting wind shear can become turbulent. Obstacles (buildings, trees, fences, and other obstacles) perturb the air flow, which causes turbulent wakes nearby and down wind of the obstacles.

The eddies most effective in dispersing a pollutant are those that have dimensions comparable to those of the cloud or of the width of the plume.

The intensity of turbulence depends on the atmospheric stability. This in turn is related to the atmospheric temperature lapse rate (or vertical temperature gradient. Although convection and turbulence in the atmosphere are very complicated processes, they can be included in dispersion models in a relatively simple way by the use of stability classes.

At large distance from the point of release, as the cloud or the plume grows, factors other than the intensity of turbulence come into play. These include:

- *distribution of the wind direction with height. Different layers of the cloud or the plume may move in different directions as a function of altitude.*
- *meandering of the wind and changes of atmospheric stability*
- *limited height of the lower mixing layer of the atmosphere.*

Stability Classes
Stability classes provide a scheme which relates the amount of turbulent energy in the atmosphere to general weather conditions such as wind speed or solar irradiation. One often used classification scheme is given by the "Pasquill Stability Classes", comprising six different classes (A-F) as a function of wind speed and day or night- time. This classification reflects many general features of the boundary layer. Dispersion model parameters (namely lateral and vertical dispersion coefficients) can then be defined on the basis of atmospheric stability.

Indicators of Turbulence

In using the models for atmospheric dispersion it is necessary to specify atmospheric stability in terms of or as a function meteorological parameters such as the following:

Temperature Laps Rate. This is the rate of decrease of temperature with height in the ambient atmosphere:

$$-\frac{dT}{dZ} \tag{1.4}$$

where

T = temperature (K)
Z = height above ground (m)

The bulk value is obtained by measuring the temperature at two levels, sufficiently separated in height. Sometimes the potential temperature lapse rate is used; it is effectively equal to:

$$-\left(\frac{dT}{dZ}+\Gamma\right) \tag{1.5}$$

where

Γ is the adiabatic lapse rate and is equal to 0.0098 Km^{-1}

Wind Direction Fluctuation. The magnitude and periods of the wind direction fluctuation are function of the intensity of turbulence (size of eddies, etc.). Consequently the wind direction fluctuation is used in practice to characterize atmospheric stability.

Insolation, Cloudiness, and Wind Speed. Thermal turbulence is related to heat flux. The cloud cover lessens heat loss or heat gain, tending to make stability neutral. Strong winds causing increased mixing, also lead to neutral stability. Clear skies at night with low winds make for a stable atmosphere, whereas similar conditions in daytime makes for an unstable atmosphere.

Richardson Number Ri. This reflects the imbalance between thermal or convective turbulence and mechanical turbulence. The Richardson number is a function of height, and is a parameter without dimension.

$$Ri=\frac{\frac{g}{T}\left(\frac{\partial T}{\partial Z}+\Gamma\right)}{\left(\frac{\partial U}{\partial Z}\right)^2} \tag{1.6}$$

where

g = acceleration due to gravity (ms^{-2})
T = temperature (K)
Γ = adiabatic lapse rate (Km^{-1})
U = mean wind speed ($m\ s^{-1}$)
Z = height above ground (m)

This is one of the basis parameters used in several atmospheric dispersion experiments. However, in general, the measurement of $\partial U/\partial Z$ needs accurate and sensitive instruments, and hence it may be difficult to evaluate *Ri* on a routine basis.

Bulk Richardson Number. This parameter can be measured more easily than the previous one, since it does not involve wind speed gradients. It is defined by:

$$Ri_B = \frac{\frac{g}{T}\left(\frac{\partial T}{\partial Z} + \Gamma\right) Z^2}{U_Z^2} \tag{1.7}$$

where

U_Z is the wind speed at the geometrical mean of the heights at which temperature is measured to obtain the temperature profile. The models described later in the next section may utilize one or more of these turbulence indicators together with parameters such as wind speed.

Mixing Layer

The mixing layer is defined as the layer of the atmosphere in which free vertical plume rise and growth is possible. Its thickness is called the mixing height H_m. An effluent plume growths vertically until the upper boundary is reached (mixing height). Quite often a stable layer occurring aloft (e.g., a temperature inversion) will present an upper boundary to free growth. The existence of such a lid is of great importance in any assessment of medium and long range dispersion within the friction layer. H_m may be estimated from temperature soundings.

Vertical mixing will stop at the mixing height H_m. The dispersion expression may be applied only up to a distance at which $H + 2\cdot\sigma_z = H_m$. Beyond about twice this distance, the vertical profile of concentration will be nearly uniform, and is given by: $\overline{\chi}$

$$\overline{\chi}_{H_m}(X,Y) = \frac{Q}{\sqrt{2\cdot\pi}\cdot\sigma_y\cdot U\cdot H_m}\cdot e^{\left[-\frac{Y^2}{2\sigma_Y^2}\right]} \tag{1.8}$$

For intermediate distance, the concentration may be obtained by interpolation. This equation is the same as for fumigation and implies only horizontal Gaussian dispersion, with uniform mixing of the vertical layers below the mixing height. *Information on mixing height has been published for some countries.*

The *existence of mixing layer* should be determined, and *its thickness evaluated,* concurrently with atmospheric stability. This information is needed to determine the upper limit of the vertical spread of the plume, when performing atmospheric dispersion calculation of positively buoyant gases.

TABLE 1.8 - TABLE 1.10 allow some one to determine easily the stability class and to estimate the wind speed from observation.

Stability Variation With Height. In different layers of the atmosphere, the temperature lapse rate may be different. Where the effluent encounters a stable layer, the vertical component of dispersion is reduced. On the other hand, a convective (i.e., unstable) regime would enhance the vertical dispersion. Omitting this fact for calculation will result with pollutant concentration being underestimated at large distances.

TABLE 1.8. Pasquill stability categories

Pasquill Category	Stability Classification	Temperature °C(standard) Ref. [22]	Temperature Gradient (°C / 100 m)
A	Extremely unstable	25.0	< -1.9
B	Moderately unstable	20.0	-1.9 to -1.7
C	Slightly unstable	15.0	-1.7 to -1.5
D	Neutral	10.0	-1.5 to -0.5
E	Slightly stable	5.0	-0.5 to 1.5
F	Moderately stable	2.5	1.5 to 4.0

TABLE 1.9. Equations for dispersion parameters.

(a) Urban Dispersion Parameters (for distances between 100 and 10,000 m)		
Pasquill Stability	σ_y**(m)**	σ_z**(m)**
A-B	$0.32\ x\ (1+0.0004x)^{-0.5}$	$0.24\ x\ (1+0.001\ x)^{-0.5}$
C	$0.22\ x\ (1+0.0004x)^{-0.5}$	$0.20\ x$
D	$0.16\ x\ (1+0.0004x)^{-0.5}$	$0.14\ x\ (1+0.0003\ x)^{-0.5}$
E-F	$0.11\ x\ (1+0.0004x)^{-0.5}$	$0.08\ x\ (1+0.00015\ x)^{-0.5}$
(b) Rural Dispersion Parameters (for distances between 100 and 10,000 (m)		
Pasquill Stability	σ_y **(m)**	σ_z **(m)**
A	$0.22\ x\ (1+0.0001\ x)^{-0.5}$	$0.20\ x$
B	$0.16\ x\ (1+0.0001\ x)^{-0.5}$	$0.12\ x$
C	$0.11\ x\ (1+0.0001\ x)^{-0.5}$	$0.08\ x\ (1+0.0002\ x)^{-0.5}$
D	$0.08\ x\ (1+0.0001\ x)^{-0.5}$	$0.06\ x\ (1+0.0015\ x)^{-0.5}$
E	$0.06\ x\ (1+0.0001\ x)^{-0.5}$	$0.03\ x\ (1+0.0003\ x)^{-1.0}$
F	$0.04\ x\ (1+0.0001\ x)^{-0.5}$	$0.016\ x\ (1+0.0003x)^{-1.0}$

Pasquill-Gifford Dispersion Coefficients.

In Gaussian plume modeling, the *rate of plume dilution is described by empirical dispersion coefficients*, which specify the horizontal extent and vertical extent (σ_z) of the plume versus downwind distance x from the source under different weather conditions, as characterized by *six atmospheric stability categories A through F* (TABLE 1.8). Several sigma schemes have been derived from measurements but probably the most common schemes currently in use are those due to *Pasquill and Gifford* and those due to *Briggs*.

For computer calculation, Briggs sigme curves are convenient to use since they are specified in an explicit functional form; σ_y, for example, is given by:

$$\sigma_y = a\,x\,(1 + b\,x)^{-1/2} \tag{1.9}$$

Thus, by inputting the above functional form and relatively few constants into a calculator, computer program or spreadsheet, (σ_z) and (σ_y) can be easily calculated at any x location of interest.

The Pasquill-Gifford sigma curves, on the other hand, were originally specified graphically. While graphs as presented by *Turner*, for example, are convenient for manual abstraction of sigma values, a functional representation comparable to the above equation is clearly more suitable for machine computations. Accordingly, various curve fit approximations to the Gifford sigma graphs have been presented.[23]

TABLE 1.10. Meteorological criteria for identification of Pasquill categories.

Surface Wind	**Stability Class: DAY**			**Stability Class: NIGHT**	
	Insolation			**Sky Conditions**	
Speed, U (ms^{-1})	**Strong**	**Moderate**	**Slight**	**Thinly overcast or ≥ 4/8 cloud cover**	**≤ 3/8 cloud cover**
U < 2	A	A-B	B	-	-
2 ≤ U < 3	A-B	B	C	E	F
3 ≤ U < 5	B	B-C	C	D	E
5 ≤ U < 6	C	C-D	D	D	D
> 6	C	D	D	D	D

Note: Wind speed, U (ms^{-1}) at 10 m above ground

1.4.4. EFFECT OF THE TERRAIN COMPLEXITY

For the assessment of continuous emissions and radioactive fallout, and especially in case of denser-than-air gas releases, the *topography* may play a non negligible role. Topographical features of interest include valleys, principal ridges and their heights, and land or water interfaces. The presence of isolated obstacles (e.g., hills), of wooded and forested areas, and of large artificial structures should be noted. Even shallow valleys (less than 100 m deep and 5-10 km wide) can affect the lower level winds considerably, and therefore reference to topographical maps is essential when a risk assessment has to be done.

In Switzerland digitized topographical data with equidistant elevation lines will be made available within the next five years. Such data could be used when calculations are made with codes requiring topographical data as input.

A methodology has been developed to use the advanced techniques of the *Complex Terrain Dispersion Model* in situations where on-site meteorological measurements are limited or unavailable. The approach, known as *CTSCREEN*, uses actual source and terrain characteristics to model plume impacts with an extensive array of predetermined meteorological conditions. CTSCREEN compares favorably with other complex terrain screening models (ones that require collection of on-site meteorological data) and is a great improvement over screening models that also use predetermined meteorology (Cimorelli et al.,[24] 1991).

1.5. Media and Modes of Environmental Transfers

To estimate *human exposure to hazardous substances*, one must establish:

- a credible source and mechanism of release to the environment;
- a medium of transport through the environment;
- a point of potential human contact with the contaminated medium; and
- an exposure route at the contact point.

For continuous emissions, the mechanisms of release and the receiving media are generally known or can readily be determined. The human activities and potential routes of exposure at each possible contact point (immersion, breathing, eating, drinking, etc.) define the pathways that must be evaluated between source and receptor.

1.5.1. SELECTING PATHWAYS

As a rule, pathways are selected to provide estimates of population-average exposures and maximum individual exposures. Each realistic pathway from source to receptor represents a unique mechanism of exposure. In selecting pathways, especially those following accidents, it is easy to become overwhelmed with considerations of "what-if" scenarios that postulate extreme combinations of unlikely pathways and events. *But little is accomplished by analyzing potential pathways of exceedingly low probability.*

The media and mode of environmental transfer of pollutants depend on:

- The medium into which they are initially released;
- The physical and chemical properties of the pollutants; and
- The pathway of transport and available opportunities for transferring from one medium to another.

The most extreme pathway scenario normally evaluated for continuous emissions is that of the "fence-post" or the "maximum individual," a person who, for example, lives his whole life at the boundary of a facility, drinks water from a well there, grows all of his food on a farm there, etc. Such an analysis is useful only to define the highest conceivable exposures to be considered for evaluation. It is not useful for estimating actual health risk, since no such person normally exists, and the results are only defining a hypothetical upper boundary of the exposures.

Some form of analysis involving realistic "maximum individuals," particularly *any classes of especially sensitive individuals* (i.e. children, old people, female, or male), is appropriate and useful. But care is required that the maximum scenario is quantitatively meaningful (i.e., its probability or its consequences are high enough to be worthy of attention). The selection of points to be actually evaluated has to be made in each case according to specific criteria. TABLE 1.11 shows some typical maximum exposure points that might be evaluated.

Media and Mode of Environmental Transfer of Pollutants
The first two factors can be determined directly from the *emission source inventory*, which is normally organized by receiving medium (air, water, land). The physical and chemical characteristics of the pollutants determine their *transferability* among media. The specific transfers depend on the presence of opportunities for transfer along transport pathways (points of direct contact between air and land, air and water, etc.) which are characteristics of the surrounding environment.

TABLE 1.11 Typical contact points for determining maximum exposure from continuous emissions.

Medium	Exposure Point	Exposure Route
Air	Nearest residence	Inhalation
	Nearest population magnet (school, shopping area, etc.)	Inhalation
	Occupied or non-occupied point of highest concentration	Inhalation
Surface water	Withdrawal point for drinking	Ingestion, dermal, inhalation
	Withdrawal point for agriculture	Inhalation, ingestion (food) dermal
	Nearest point for swimming / contact sports	Ingestion, dermal
	Nearest point for fishing	Ingestion, (food)
Ground water	Nearest potable well	Ingestion, dermal, inhalation
	Nearest agricultural well	Inhalation, ingestion (food)
	Nearest well for other uses	Inhalation, dermal
Soil	On-site	Dermal, ingestion
	Immediately adjacent to site (if restricted)	Dermal, ingestion
	Nearest cropland	Ingestion (food)

The *receiving medium* is often technology-specific; one technology may release a substance to the air and another competing technology may release the same substance or a transformation product (e.g., scrubber wastes) to water or land.

■ *Emissions to Air.* The heavy particles deposit rapidly to nearby surfaces on land or water. Lighter particles travel farther and deposit at lesser rates. Gases may be absorbed slowly or rapidly, depending on their reactivity with the surfaces they encounter. Many reactive gases change chemical and/or physical form in transit, which can change their absorption characteristics.

- *Emissions to Water.* They seldom reach the air, except for volatile substances like organic solvents. Mostly, these emissions change medium by direct deposition in bottom sediments, by uptake up and/or decomposition in the aquatic food chain, or by changing chemical and physical form during transport.

- *Materials Deposited on Land.* They routinely enter surface and ground waters by runoff and leaching, and enter the air through direct volatilization, chemical or biological transformation (fire, bacterial decomposition, etc.), or re-suspension. Rates are determined by the chemical and physical properties of the materials and the characteristics of their environment (e.g., rainfall, wind, permeability of soils, and cover).

The *physical and chemical characteristics* that are important in determining transfers among media are usually available in the environmental literature and are often included as part of the characterization of source terms or incorporated in standard environmental transport models. Expert judgment is helpful in selecting appropriate rate constants for less common pollutants. *Rate constants* are often complex functions of environmental conditions and *cannot necessarily be transferred from one environment to another* without careful evaluation.

1.5.2. CHEMICAL INDICATORS

Some care must be taken in cases where a *single chemical indicator* has been selected to represent a broader class of pollutants. Indicators are often developed for different purposes; sometimes they are just substances that are easy to measure.

Using a *particular indicator* may be useful for quantifying the presence of a class of pollutants in a source term, but the physical and chemical characteristics of the indicator may not provide a good representation of the transport and fate of that class of pollutants in the environment, or the health effects of exposure to them. Expert judgment is helpful in determining the usefulness of a particular indicator chemical in all stages of a risk assessment.

Transfers of materials from one medium to another are normally treated as a loss to the supplying medium and a source to the receiving medium. *Deposition* and *chemical transformation rates* are usually incorporated directly into environmental transport models, and models need only be appropriately linked at the loss-source term.

1.5.3. QUANTIFICATION OF EMISSIONS

In the absence of direct measurements of exposures produced by specific emissions, quantification of the pathway from emissions to effects must be made with models that simulate transport and transformation of materials in the environment. These models can range from the simplest of calculations done on hand calculators to state-of-the-art super-computer systems that solve coupled partial differential equations governing transport and transformation of pollutants.

Criteria for the Selection of a Dispersion Model
The characteristics that must be considered for estimating pollution dispersion include:
- *conditions under which materials are released*
- *chemical and physical characteristics of the materials released*
- *medium of transport*
- *geophysical characteristics of the pathway*
- *chemical and physical changes during transport*
- *matching of model output to information needed in the application*
- *availability and cost.*

Transport medium and materials released are major and obvious determining factors in the selection of appropriate models. Conditions of release include an important differentiation between *routine-continuous* and *short-term accidental releases*, which determine the time scale required for modeling and needs for probabilistic analysis.

Type of estimate for risk assessment
For risk assessment, model selection is driven by the type of calculation required for estimating the effects under consideration. For instance, assessing environmental damage and loads to a lake requires the estimation of *long-term average concentrations* (hours to seasons). In another case, determining whether or not an explosive limit might be exceeded, requires an estimate of *peak concentration over a short period of time,* perhaps of the order of seconds. In other cases, the assessing process may require an estimate of the surface area for which contamination limit has been exceeded. *In some cases, estimates of coexisting concentrations of more than one pollutant may be required.*

Conditions of Release
The conditions of release that must be considered cover a wide range, including physical and chemical form of the material, height of release and plume rise, smooth airflow or turbulence from nearby buildings or topography, still or flowing bodies of water, etc. *These source conditions determine the initial dilution of the materials, maximum impacts, and constrain possibilities for mitigating effects.*

Chemical transformations during transport can alter a toxic material from one form to another, to a harmless form, or from a harmless form to a toxic one. Analysts must determine whether this is important before models are selected. Similarly, removal mechanisms such as deposition or wash-out during transport can be significant and these mechanisms must be included in the capabilities of the selected models. Removal from one medium to another constitutes a source to the receiving medium and may expand needs for modeling to ensure comprehensive treatment.

Final and nontrivial considerations in selection of models for risk assessment are the practical ones of availability, costs of use, timeliness of results, etc.

1.6. Integration of Environmental Data

1.6.1. TRENDS AND DEVELOPMENT

In Germany, within the framework of the *"Technical Division of Computer Science in Environmental Protection"* of the Society for Computer Science (*Deutsche Gesellschaft für Informatik*), the working group "Integration of Environmental Data" was founded in 1992. The results of the first workshop were published by Denzer et al.[25] (1993). Some results concerning research and development as well as implementations related to an integrated approach of environmental data management are summarized shortly hereafter.

Many data concerning environmental protection are collected at the communal level by different departments (Dept. of Land Planning; Dept. of Housing and Construction; Dept. of Forestry, etc.), and appear also to some extent in national statistics. But these data are not collected and processed in a consistent manner and on a coordinated basis that would favor an inter-exchange with other communal or national Authorities and engineering offices dealing with environmental protection, and their use in conjunction with modern geographical information systems (GIS). Many data, that would be needed, are not to be found easily, or are obsolete, or are irrelevant for solving environmental problems. A first milestone for improving this situation has been set by the above mentioned workshop.

Geographical Information Systems and Cadastre

Topographical information in Switzerland is in the process of being digitized and can be displayed by means of geographical information's systems like *ARCINFO,* or *GIS/REGIS*. At any time it is also possible to add and project other information on geographical maps displayed, as required, with different levels of information and kinds of details. Such tools can even be combined with mathematical programs to make expert calculations. The use of such geographical management information systems is in Switzerland in an introductory phase and is not yet being used on a routinely basis as it is the case in the US, or in other countries.

Usually *cadastre* for a region should include the following information:

- Waste water canalization systems
- Communal repository
- Fruit trees and forests, biotopes, green surfaces
- Aquatic systems, aquatic systems potentially in danger
- Climatic data and energy consumption data; traffic data, etc.

The data are used for controlling and remedial purpose, for planning tasks and for documentation. In order to integrate such cadastres, new norms and standards must be developed and implemented on a nation wide basis or even better on an international basis in order to assure data consistency of a data base entry. The definition and use of meta-information describing data bases' contents and quality as well as models used for environmental assessment and their associated data, is a first step towards creating catalogues for modern information systems. The development of general purpose software tools for inter-linking and accessing the various data bases over network servers is not an easy problem. The software must be able to run on many different computer platforms and allow for easy navigation and extraction of data in tabular or graphical form. The search for information related to the content of data bases must be facilitated through the use of thesaurus (with synonyms), lexicons and reference knowledge. The way of aggregating the knowledge may vary and has also an influence on the portability: object-oriented representation of the knowledge versus conventional relational data bases are presently under discussion.

Since several years specialized information management systems for solving environmental protection tasks are emerging in various administrations. The development and implementing of such systems is quite often directed by the relevant department and is usually not coordinated with other cantonal administrations. This situation creates problems for integrating the different data bases into a communication network. One possible concepts for an open environmental data information system might involve the following tools[26]:

- Two interfaces are needed, one (a) between the local systems' services and the environmental network, and one interface (b) between the local system and the data bases of the different departments with their node-specific tasks:
    - At the local nodes (a), standardized programs will be used, which contribute to the good functioning of both the information system and the meta-information system, i.e., a server for the data and a server for the environmental data catalogues. This interface is needed for transforming the data before a transfer can take place. Here a global convention for an environmental data protocol must be elaborated.
    - The second interface (b) is inter-connecting the system services with the departments' data bases which run on different types of computers.
- The following protocols might be needed:
    - data structure protocol and method protocol
    - data, catalogue and meta-data protocols
    - basic protocol

Another concept for homogenizing and interfacing environmental data stored in relational data bases is based on the use of predicate logic (PROLOG language or similar).[27]

Current development on the needs and requirements for Environmental Information Systems (EIS) at the community level is reported below for Germany.

1.6.2.. ENVIRONMENTAL INFORMATION SYSTEMS (EIS)

Current EIS-Development

The *project AKASAAR* of the "Gemeinde Marpingen (Germany)" is dealing with the municipal waste water cadastre ("Kanal-Kataster"), for which minimal requirements have been defined and data are being collected.[28] Different computerized systems (dealing with cadastres for canalization' systems) were tested, such as:

SICAD-Kanal (WS-2000), GRIPS/Kandab-II (for a DEC station), PROCART (PC486), STAR-Carto (Apollo workstation), SIEMENS PCD-4T (GC-Kanal), GRAS. These systems make use of digitalized maps and of a GIS.

The *Ministry of Environment of Niedersachsen*, Germany, is implementing an environmental data base catalogue, *UDK*[29] *(Umwelt-Datenkatalog Niedersachsen)* which is a subsystem of *NUMIS*[30] *(Niedersächsisches Umweltinfomationssystem)*. UDK is a meta-information system that stores, manages, and retrieves information concerning regional decentralized data bases. It is similar to the catalog of a municipal library, which contains a description of the bibliographical format and content of books and reports. It is used inside NUMIS as an information management system. It brings an overview of data base contents and defines the quality and type of information stored therein, thus providing a contribution to the standardization process regarding the data collected. It is also used as a link between the owners of the different data bases and the users. The concept foresees a general purpose retrieval system, "*VISION-UMWELT*",[31] the UDK subsystem, and a GEOgraphic information system.

The *project INTEGRAL*,[32] an environmental information system with similar concept as NUMIS, is being implemented and tested on a small scale in Baden-Württenberg, together with *"Siemens-Nixdorf Informationssysteme AG"* (*GODOT* prototype project for geographical data).[33] In this system data objects are grouped in object classes; meta-data are then a description of these object classes. For each single class, methods are defined that allow for the manipulation of the instances, and the transformation thereof without loss of information. The concept is based on the use of UNIX-shell and X-Windows/OSF-MOTIF's user interface communicating over LAN and WAN (Internet). Client programs are called using remote procedure calls defined in an *"Object Request Broker"*[34].

An *expert system XUMA* ("Expertensystem Umweltgefährlichkeit von Altlasten") was developed for helping environmental engineers in choosing appropriate remedial actions for problematic repositories (with indiscriminate dumping). XUMA now runs on a UNIX platform and is coupled to a relational data base.[35, 36]

Environmental Information System Manager
An EIS-Manager (Environmental Information System Manager) helps organizing the information exchange and linking between different technical information systems. It allows to view information from the several offices of the Public Administration with different aggregation standards, and from distributed data bases comprising heterogeneous subsystems.[37, 38]

Integrated Data Bases for Water Management

Another example of environmental data integration is the *"Databases' system for water management"* located at the *Centre Informatique de l'Etat de Luxemburg* (CIE).[39] The main goal is to establish a general water balance, taking into account several hydrologic information and the results of geological, meteorological, and chemical analyses collected by different departments. The IBM mainframe computer of the CIE is used as central node for data and communication. All the data are stored there in a relational data base. The de-centralized data bases stored on PC use ORACLE Pro-C for MSDOS (with SQL as query language), and are interconnected in a local network over a UNIX-server. The program applications were developed using MS-WINDOWS.

CLAIR/CLEW System: Environmental Decision Support system for Air and Water Pollution Simulation and Control.

Features of the system make it useful in different applications related to integrated risk management, mainly for the case of continuous emissions (not accidents).

- Area stationary pollution sources data base;
- Technological measures data base (e.g. optimal pollution control systems, fuels emission sources liquidation, etc.);
- Industry and each particular emission source impacts on the atmosphere and aquatic systems;
- Ecological and economical multi-criteria effectiveness analysis of the industrial innovations;
- Optimized investments allocations for the air and water quality protection;
- Maximum resources calculation which are required for the atmospheric and water protection options;
- Multi-criteria air pollution minimization within the given value of expenditures;
- Effective set of the measures definition which could be applied to each source of emission;
- Surface water protection.

Environmental Decision Support System for Air and Water Pollution Simulation and Control

During recent years complex models have been developed for the simulation of the air or/and water pollution due to industrial activities. The *CLAIR/CLEW System* developed by IIASA is a software designed for decision support in problems of atmospheric and water protection for industrial risk management and planning on national, regional and enterprise levels. The model can be used within areas up to 400 x 400 km. The system has been developed and extended by IIASA; initial versions of it where validated in 30 case studies for the different climatic conditions and types of industries. The solution from the simulation process could be given for a set of pollutants simultaneously. The computer model requires minimal data input for its runs, taking into consideration the application environments in the developing countries. Default data are provided in the CLAIR/CLEW model.

Summary (Chapter 1)

This chapter summarizes the framework for the analysis and assessment of regional emissions when dealing with integrated area risk assessment and management. Continuous emissions generally lead to exposures that create chromic, long-term effects. A specific methodology for the analysis and risk assessment of continuous emissions is introduced; identification of sources of continuous emissions, characterizing the emission source inventory, selection of a pathway for analysis organized by a given receiving media (air, water, soil), use of models to calculate dispersion values in the receiving media, use of dose-response relationships to estimate the risk to population and environment. Information are summarized regarding the non-point source emissions, population and meteorological data, effect of the terrain complexity, media and models of environmental transfers, integration of environmental data by means of specialized data bases, and decision support systems, environmental information systems, multimedia technology, etc.

References (Chapter 1)

[1] WHO (1982). Rapid Assessment of Air, Water and Land Pollution, WHO offset publication No 62, Geneva, 1982.

[2] WHO (1983a). Compendium of Environmental Guidelines and Standards for Industrial Discharges, WHO, Geneva, 1983.

[3] WHO (1983b). Selected techniques for Environmental Managing - Training Manual, WHO Geneva, 1983.

[4] WHO (1988). Emissions, Environmental Transport and Dose-Response Models, Guidelines for Case Studies, WHO Geneva, 1988.

[5] WHO (1989). Management and Control of the Environment, WHO Geneva, 1989.

[6] UNEP (1985). The Environmental Impacts of the Production and Use of Energy Part IV. United Nations Environment Programme, Paris, 1985.

[7] OECD (1984). Emission Standards for Major Air Pollutants from Energy Facilities in OECD Member Countries. OECD Paris, 1991.

[8] OECD (1984). Emission Standards for Major Air Pollutants from Energy Facilities in OECD Member countries. OECD Paris, 1984.

[9] Environmental Data Services (Ends) Ltd., Dangerous Substances in Water: A Practical Guide, 1992. ENDS Ltd, Finsburry Business Centre, 40, Bowling Green Lane, London, EC1R 0NE.

[10] Hewlett, J.D.; Pienaar, L., Design and analysis of the catchment experiment,. In: E.M. White (ed.), Proc. Symp. Use of Small Watersheds in Determining Effects of Forest Land on Water Quality. Univ. of Kentucky, Lexington, (1973), pp. 88-106.

[11] Maas, R.P.; Smole, M.D., et al., Land Use Issues: Management and assessment. Practical guidelines for selecting critical areas for controlling nonpoint source pesticide contamination of aquatic systems. In: Perspective on Nonpoint Source Pollution, Proceedings of a National Conference, Kansas City, Missouri, May 19-22, (1985), pp. 363-367.

[12] Young, R.A.; Onstad, C.A.; et al., A method for prioritizing water quality problem areas. In: Perspective on Nonpoint Source Pollution, Proceedings of a National Conference, Kansas City, Missouri, May 19-22, (1985), pp. 368-373.

[13] Young, R.A.; Onstad, C.A.; Bosh, D.D.; Anderson, W.P., Agricultural Nonpoint Pollution Models (AGNPS) I and II - Model Documentation. Submitted to Minn. Pollution Control Agency in partial fulfillment of the Trust Fund Cooperative Agreement No. 12-14-3001-071. February 1985.

[14] Datta, B.; Levent Kavvas, M.; Orlob, G.T., Application of AI techniques for identification of unknown groundwater pollution. In: L. C. Wrobel, C.A. Brebbia (eds.), Water Pollution II, Modelling, Measuring and Prediction. Computational Mechanics Publications, Southhampton Boston. (1993), pp. 73-89.

[15] Hayes-Roth, F.; Waterman, D.A.; Lenat, D.B., Building Expert Systems, Addison-Wesley Publishing, Reading, Mass., 1983.

[16] Matthiessen, P.; Allchin, C.; Williams, R.J., et al., The translocation of some herbicides between soil and water in a small catchment , J. Inst. Wat. Environ. Manage., 6 (1992) 496-504.

[17] TRIP from SRD (Safety and Reliability Directorate), London, 1981)

[18] Petts, J.I.; Witherss, R.M.J.; Lees, F.P., The assessment of major hazards: the density and other characteristics of the exposed population around a hazard source, J. Hazard. Mater. 14 (1987) 337-363.

[19] Glickman, N.J.; White, M.J., Urban land-use patterns: an international comparison. Environment and planning A, 11 (1979) 35-49.

[20] HSK, Hauptabteilung für die Sicherheit Kerntechnischer Anlagen, Villigen, CH.

[21] NAZ, Nationale Alarmzentrale, Zürich.

[22] Standard deviation of horizontal wind direction fluctuation over a period of 15 minutes to 1 hour.

[23] Davidson, G.A., A modified power law representation of the Pasquill-Gifford dispersion coefficients. In: Journal of the Air and Waste Management Association (USA), 8 (1990) 1146-1147.

[24] Cimorelli, A.J.; Perry, S.G.; Burns, D.J., Advanced screening model for complex-terrain applications. Environmental Protection Agency, Research Triangle Park, NC (USA). Atmospheric Research and Exposure Assessment Lab., RN: PB--91-162693/XAB; EPA--600/D-91/005 (1991).

[25] Denzer, R.: Geiger, W.: Güttler, R., 1. Workshop Integration von Umweltdaten, am 9.-10-Februar 1993 auf Schloss Dagstuhl (Saarland), Kernforschungszentrum Karlsruhe, Bericht Nr. KfK 5187 September 1993.

[26] Denzer, R., Architektur offener Umweltinformationssysteme, Report 218/92, Universität Kaiserslautern, 1992.

[27] Beierle, C., Types, modules and databases in the logic programming language PROTOS-L. In: K.H. Bläsius, U. Hedstück and C.R. Rollinger (eds.) Sorts and Types for Artificial Intelligence, Vol. 418 of Lecture Notes in Artificial Intelligence. Springer Verlag, Berlin, 1990.

[28] Laub, W.; Osthof, M., Anforderungen an ein Umweltinformationssystem aus der Sicht einer Kommune. In: Denzer, R.; Geiger, W.; Güttler, R., 1. Workshop Integration von Umweltdaten, am 9.-10. Februar 1993 auf Schloss Dagstuhl (Saarland), Kernforschungszentrum Karlsruhe, Bericht Nr. KfK 5187 September 1993, pp. 1-19.

[29] Schütz, T.; Lessing, H., Metainformation von Umwelt-Daten-Objekten - Zum Datenmodell des Umwelt-Datenkataloges Niedersachsens, In: Jaeschke, A.; Kämpfe, T.; Page, B. and Radermacher, F.J., (eds.), Informatik für den Umweltschutz, 7. Symposium, Ulm, 1993.

[30] NUMIS, Grobkonzept für das Niedersächsische Umwelt-Informations-System INPLUS GmbH, 1989.

[31] Fachliches Feinkonzept für den Prototypen VISION-Umwelt des Führungsinformationssystems im Niedersächsischen Umweltinformationssystem, SEMA GROUP GmbH, 1992.

[32] Henning, I.; Schmidt, F., Integration von Daten und Methoden im Umweltinformationssystem des Landes Baden-Würtemberg. GI-Arbeitskreis "Integration von Umweltdaten", 1. Workshop, IBFI Schloss Dagstuhl / Saar, 9.-10. Februar 1993.

[33] Riekert, W.F., Das Forschungsprojekt GODOT: Geodatenhaltung mit objektorientierten Techniken. In: Geoinformationsysteme, 3. Internationales Anwenderforum Duisburg 93, Siemens Nixdorf Informations-systeme AG, München, 1993.

[34] Object Management Group. The Common Object Request Broker: Architecture and Specification. OMG, Framingham, MA, USA, 1992.

[35] Geiger, W.; Weidemann, R., The XUMA expert system for contaminated sites: functions, explanations and knowledge aquisition, 10th International Workshop "Expert Systems & their Applications", General Conference, Avignon 1990, Avignon, France, 28.5.-1.6.1990, pp. 951-961.

[36] Overlack, M., Integration des Altlastenexpertensystems XUMA in ein Umweltinformationssystem, Kern-forschungszentrum Karlsruhe, KfK 4933, November 1991.

[37] Denzer, R., Architektur offener Umweltinformationssysteme, Projektbericht UIS/1991-1, Universitaät Kaiserslläutern, 1991.

[38] Saarländisches Umweltinformationssystem - SUIS, Hauptuntersuchungsbericht: Ministerium für Umwelt, Saarbrücken, 1992.

[39] Klaus, R.; Schroeder & Associés, Luxemburg. Ein Wasserwirtschaftliches Datenbanksystem im Grossher-zogentum Luxemburg, 1. Workshop, IBFI Schloss Dagstuhl / Saar, 9.-10. Februar 1993.

CHAPTER 2

ATMOSPHERIC DISPERSION MODELS AND METHODS

2.1. Introduction

A considerable body of theoretical and experimental work is available on the dispersion of pollutants released into the atmosphere. On the basis of meteorological measurements, pollutant concentrations in the neighborhood of a nuclear power plant or a chemical plant or other type of industries can now be reasonably estimated.

Atmospheric transport models are used to predict the concentrations of airborne pollutants at various distances and directions from their points of release. They may be classified by any number of characteristics: treatment of topography, geographical scale, diffusion theories, source configuration, or relationship to regulatory requirements.

According to the size of the geographic region under consideration, we can roughly distinguish the following class of atmospheric models:

- *Global models:* These are used to simulate the dispersion of pollutants on a global scale. Box models are often used for this purpose.

- *Regional models:* They must be used when studying large regions, with dimensions of several hundred kilometers.

- *Mesoscale models:* Such models are optimal in the medium range for distances from ca. 10 to 100 km.

- *Local models:* This type of model is used for short range evaluation of concentrations within a radius of up to 10 km around the point source.

The problem of pollutant transport and dispersion at very large distances, for which different kinds of models are required (global dispersion models), is beyond the scope of this guide. As the distance increases, the vertical spread of the effluents becomes comparable in dimensions to the planetary boundary layer thickness and the usual assumptions of homogeneity make the model increasingly inapplicable. Also discussed in this section are the conditions under which the models are applicable, the modifications required for applying the models to specific practical situations, and the meteorological data required for input to the model. Models used for nuclear applications require some extra features described below and are nowadays quite often combined with consequence analysis (personal and collective doses calculations).

The goal of atmospheric dispersion modeling is to predict concentrations of pollutants as a function of time since release, and position with respect to the release point. The initial release and its characteristics are called the "source term". In most circumstances, the mass of substances released is small. Thus we can speak of diffusion and transport of week concentrations of various pollutants in air. For weak concentrations atmospheric dispersion mechanisms are not affected by the concentration of pollutants, and there is a proportionality between the amounts released and the concentrations.

The concentration of substances at some point downwind from the point of release into the atmosphere depends on the following:

■ *Type and amount of substance released.* These include chemicals, radionuclides, aerosols, dust particles.

■ *Wind direction and speed.* The horizontal transport of pollutants is dominated by advection, where the mean wind acts as the driving force.

■ *Atmospheric diffusion.* This depends on the characteristics of the atmospheric turbulence along the transport path. Convection is responsible for the transport in the vertical direction.

■ *Kinematics processes.* They affect air flow and turbulence due to terrain roughness, to the presence of lakes, hills, valleys, and forests. The presence of obstacles such as buildings have a non-neglecting effect. Due to friction, the horizontal wind can be drastically reduced near the ground.

■ *Conditions of release.* Such parameters like the release height, the thermal and kinetic energy of the carrier fluid play also a role.

■ *Various removal and transformation processes.* These include formation of aerosols, chemical transformation (due to air moisture), radioactive decay, wet or dry deposition, and resuspension.

> ***Volume source***
> *Where large volume sources are encountered, a virtual point source may be defined for the calculation of the horizontal plume spread, after shifting the origin upwind by a distance* x_1, *which defines the virtual source location, such that:* $\sqrt{2 p s_y}$ *corresponding to the travel distance* x_1 *equals the horizontal dimension of the volume source. Similarly for the calculation of the vertical plume spread, the origin is shifted by a distance* x_2 *such that* $\sqrt{2 p s_z}$ *corresponding to* x_2 *equals the vertical dimension of the volume source. The theoretical plume size then approximates the source size at the position of the source. For travel distances larger than a few kilometers, this correction factor is not significant.*

The different kinds of sources are usually classified in terms of spatial configuration and duration of release. Spatial configurations of interest are point sources, line sources, area sources, and volume sources. Line and area sources are not generally encountered in the analysis of radioactive releases from nuclear plants, but if encountered, they can be treated by subdivision into a number of effective point sources.

Actual sources of radioactive effluents are not point sources in the strict sense, but may be so considered for mathematical simplicity.

A distinction has to be made between *instantaneous releases or puffs* which are time dependent, and *continuous sources* assuming stationary meteorological and emission condition.

> - ***Puff:*** *An instantaneous release, or puff, forms a cloud that travels with the wind while expanding.*
> - ***Continuous release****: A continuous release may be considered to be composed of a series of puffs that form a continuous plume.*

Dispersion models for stationary meteorological conditions can be reduced to simple mathematical formulae. They can be lists of equations to be calculated by hand, single codes working on microcomputers, or more complex codes normally run on mainframe computers. Some even more complex codes require powerful super-computers, and considerable skill to run them. These complex codes are not normally available for routine risk assessment.

Most of the atmospheric dispersion models are deterministic. *Risk assessment assumes a probabilistic approach. This is achieved by using sets of emission values and dispersion conditions, each of which are affected by probabilities*. The model output is consequently also probabilistic.

Selection of appropriate meteorological dispersion models depends in part on the relative proportion of large point sources of emissions to the total regional emissions. Dispersion within 50-80 km of one to a few large point sources is normally simulated with some form of plume model and results for these few sources are added. If there are many large point sources or widely distributed smaller sources, plume models are not appropriate and regional average dispersion models must be used.

> ***Guide for Selecting Air Quality Models***
> *The U.S. Environmental Protection Agency has an especially useful guide for selecting appropriate air quality models from among a broad range of models that are in the public domain and available free as down-loadable code on a computer bulletin board system (SCRAMS) or at modest cost from the U.S. National Technical Information Service.*

The final outputs of dispersion models are derived from emission rates and atmospheric dispersion conditions, but they can be quite different according to the nature of the problem.

2.2. Environmental Dispersion Models

A review of the physical processes taking place during the dispersion of releases of pollutants into the atmosphere and the hydrosphere (surface as well as ground waters) has been prepared by Yadigaroglu and Munera[1] (1987). The analytical methods available for tracking pollutants in the atmosphere include local and mesoscale models (mostly based on Gaussian-plume dispersion), as well as regional and global models, where either more sophisticated numerical techniques or box modeling is used. Various removal processes such as physico-chemical transformations, wet and dry deposition, re-suspension, and plume rise affect aerial dispersion are discussed in above mentioned paper.

The mechanisms of transport in surface waters include mass transport by the waters themselves, dispersion, sedimentation, boundary exchange processes, and various forms of depletion. The models vary according to the type of surface waters considered: rivers, estuaries and tidal rivers, small lakes, open-coast water bodies, etc.

Once an aerosol or radioactive gas becomes airborne it travels and disperses in a manner governed by its own physical properties and those of the ambient atmosphere into which it is released. The effluent enters the atmosphere with a certain velocity and temperature which are generally different from those of the ambient atmosphere. The effluent motion has a vertical component due to the effects of vertical velocity and difference of temperature until they are dissipated. This upward rise of the effluents is called *plume rise* and changes the effective height of the release point. The path of the effluent is also affected by flow modifications near obstacles such as buildings and structures. The turbulent motion of the atmosphere causes random movement of the effluent resulting in a progressive lateral and vertical spreading by mixing with the air (*atmospheric diffusion*). The movements of the effluent with the wind after plume rise are called *transport*. The combination of transport and diffusion is called *atmospheric dispersion*.

In some cases, the effluent, while undergoing plume rise, transport and diffusion, may also be subject to processes such as:

- *Radioactive decay:* decay and/or buildup of daughter products
- *Wet deposition*
- *Rainout/snowout:* vapor or aerosol is scavenged by water droplets or snowflakes in cloud and falls out as precipitation
- *Dry deposition:* Sedimentation of aerosols or gravitational settling of particulate matter (with particle diameters larger than 10 μm. Impacting of aerosols and adsorption of vapors and gases on obstacles in the path of the wind.
- *Formation of aerosols and coalescence of aerosols*
- *Re-suspension* of materials deposited on surfaces.

Most of these effects can be expressed mathematically and are also included in modern computer codes.

2.2.1. GENERAL TYPES OF MODELS

Risk assessment requires careful selection of suitable models for description of natural phenomena and effects of pollution exposure. Choice of models sometimes depends heavily on available data and the purpose of the analysis. *Highly sophisticated models combined with inadequate data are surely the worst combination.*

Keeping in mind the main goal of risk management, the final product must be *a list of corrective measures* that are feasible, rational and in line with social and economic objectives. It is the spatial and technological harmony of solutions within an all-encompassing rational plan that must be the base for efficient risk reduction.

In selecting risk assessment models, analysts should include evaluation of their ability to address key problems, such as:

- Assessment of routine effects
- Establishing relationships between local, short-term effects and long-term goals
- Simultaneous evaluation of several different sources of risk according to different attributes
- Assessment of risks over time and the problem of discounting future risks and benefits
- Uncertainty analysis, and
- Synergetic effects.

Various environmental dispersion or transport models are available. Some are based on purely ecological principles; others favor a balance between economic parameters and the corresponding ecological risk. They fall into three general categories based on transport medium:

- Short and long-range air quality models
- Water quality models of various types and scales, and
- Terrestrial and aquatic food-chain models.

Selection of a suitable code must depend on the aims of the case study in which it will be used, and on an in-depth evaluation of the code's models, parameters, and implementation requirements, as well as verification of the adequacy of predicted results.

Long-range Atmosphere Models

The American National Standard Institute (ANSI) and the American Institute of Aeronautics and Astronautics (AIAA) have published a guide to reference and standard atmosphere models, which provides guidelines for selecting atmospheric models for engineering design or scientific research, especially in the field of meteorology, climatology, and aerospace sciences.[2] The ANSI/AIAA guide describes the content of the models, uncertainties and limitations, technical basis, data bases from which the model is formed, publication references, and sources of computer code for thirty atmospheric models for altitude from earth surface to 2500 kilometers. These models are general and describe the physical properties and chemical composition of the atmosphere as a function of altitude. Not included in this ANSI-guide are models describing exclusively low altitude phenomena, such as particulate aerosols or pollutants in the boundary layer and cloud properties as a function of altitude in the troposphere. Dynamic models of the atmosphere are also not included in this ANSI-guide. This reference is mentioned for the sake of completeness and will not be handled in this book.

Many dispersion models have been developed which give a description of the atmospheric dispersion at long distances, but a real time estimate of the concentration and ground deposition patterns requires a model with special structure and characteristics. It must be able to calculate the dispersion in a physically consistent way based on meteorological data that could reasonably be available in real-time; it must be flexible with respect to the source and pollutant characteristics and to the domain size and position. Although puff models can provide a useful insight into the pattern of long range dispersion, they are generally limited in their three-dimensional treatment of the windfield. Eulerian grid models which integrate the equation for advection and diffusion of material across a three-dimensional grid can incorporate the three-dimensional windfield, but are not well suited for the treatment of time-dependent release from a point source due to the problem of numerical diffusion and the large computer time required.

Based on above considerations, the *APOLLO model* has been designed to estimate ground level concentrations and deposition at distances up to some thousand of kilometers from one point source.[3] It is a *Lagrangian particle model*, representing the release as a sequence of particles, each advected across the domain according to the evolving three-dimensional windfield, and with a suitable parametrization of turbulent motions. It is considered as the most appropriate type of model for that kind of problem. It retains some basic concepts and features of puff models developed in recent years (ApSimon[4] et al., 1985; van Rheineck Leyssius[5] et al., 1988; Davis[6] et al., 1988).

Medium-and Short-range Atmosphere Models

The models used for regional risk assessment are of the short- or medium-range type and have been summarized, as far as possible, and descriptions thereof included in this book. *Reviews of the regional scale models* are given by Buckner[7] (1981) and Eliassen[8] (1980). A typical example of a long-range model is the Lagrangian code MESOS.[9, 10] Another review of the physical processes taking place during the dispersion of releases of pollutants into the atmosphere and the hydrosphere (surface as well as groundwater) has been prepared by Yadigaroglu and Munera[1] (1987). The analytical methods available for tracking pollutants in the atmosphere include local and mesoscale models (mostly based on Gaussian plume dispersion), as well as regional and global models, where either more sophisticated numerical techniques or box modeling is used. Various removal processes such as physicochemical transformations, wet and dry deposition, resuspension, and plume rise affect aerial dispersion are discussed.

2.2.2. APPLICATION RANGE OF ATMOSPHERIC MODELS

The formula and discussion presented here concerns primarily atmospheric dispersion models for assessing the risk from continuous emissions from a stack, or accidental releases of neutral and positively buoyant gases released continuously (Hanna[11], 1981). McCarthy et al.[12] (1990) have reported that most computational programs for atmospheric dispersion of pollutant materials are based on models that incorporate well-known Gaussian plume and puff solutions. The final outputs of dispersion models are derived from emission rates and atmospheric dispersion conditions, but they can be quite different according to the nature of the problem.

The problem of atmospheric dispersion has been treated theoretically by three approaches:

- K-theory approach or gradient transfer
- Statistical theory approach

- Dimensional analysis approach.

Each approach has certain limitations. Since the introduction of powerful computers and the development of adequate numerical methods, mainly based on the K-theory, more exact simulation could be obtained. In practice a double Gaussian dispersion equation has been assumed, the dispersion parameters of which are obtained by conducting field diffusion experiments and relating them to observed meteorological parameters, e.g., wind speed, turbulence, lapse rate. The simple models proposed in this guide belong to this latter approach.

Gaussian Plume Models. This type of models is widely used for estimating downwind concentrations of airborne material released to the atmosphere, and proved to be satisfactory in order to obtain a general idea of the plume behavior and the order of magnitude of the pollutant dilution factors. This class of models is based on the assumption that the plume will spread both laterally and vertically in accordance with a Gaussian statistical distribution. These models have been successfully used for many years to predict ground-level concentrations where reasonably flat terrain exists, and where an average, fairly uniform release rate can be assumed. Using such type of model for short-term predictions under conditions of complex meteorology or terrain may deliver results that may be uncertain by one or two orders of magnitude. It is also generally believed that near-filed Gaussian plume models are able to predict annual-average atmospheric concentrations of non-reactive pollutants to within a factor of two to four over flat terrain (Miller[13] et al., 1982).

The quality of the results may be significantly improved by calibrating the dispersion parameters with field tests. Nevertheless Gaussian models are neglecting actual topography, terrain roughness buildings and constructions which strongly influence the plume behavior. Such effects can be taken into account in more sophisticated mathematical models or in physical models.

The range of validity of the Gaussian plume formulation can however be extended for *mesoscale models*[14] by estimating average concentrations over larger distances through appropriate averaging procedures, which account for such factors as wind direction, stability category, duration of wet deposition, atmospheric mixing layer height, etc.

Simple Gaussian Plume Models. Over distances between one and about 50-80 kilometers on a smooth and flat terrain, dispersion from a point instantaneous or continuous sources can be described by Gaussian plume models. These models are derived from mathematical descriptions of the physical characteristics of dispersion in drifting plumes, and they produce an estimate of the concentration distribution throughout a plume as a function of a few source and meteorological characteristics. They require as inputs only a source term, an atmospheric stability category, and wind speed and direction. Estimated dispersion is governed by an increase in standard deviation with distance or by an increase in transfer time. Gaussian plume models assume the pollutant to be passive; they do not account for topography or changes in meteorological conditions. But they can accept exogenous variables such as release height, deposition parameters, and transformation kinetics. The location and magnitude of maximum ground level concentration can be derived from the Gaussian equation, which is useful for estimating potential maximum exposure from a single source. These models can be reduced to simplified nomograms giving dilution factors at various distances and they are available in easy-to-use microcomputer software packages.

If the source of pollutant material is instantaneous at a point, then a *Gaussian puff model* may be used. Gaussian plume models for elevated continuous and instantaneous (puff) point sources are described hereafter.

Gaussian Plume Models for Elevated Continuous Point Sources

The Gaussian Plume Model is used in more regulatory applications than any other model. It evaluates the dispersion of emissions from an elevated continuous point source (*Figure 2.1*).

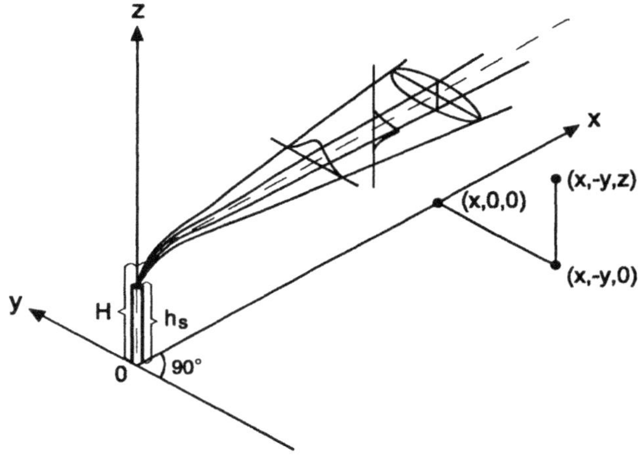

Figure 2.1. Coordinate system showing Gaussian distribution of a plume dispersion in the horizontal and vertical axis

The basis of the model is a simple formula which assumes constant wind speed *u* and complete reflection from the ground surface. The dispersion takes place along both horizontal and vertical axes as described by the following formula:

$$C=\left(\frac{Q_p}{\left((2\pi)^{3/2}\sigma_x\sigma_y\sigma_z\right)}\right)\cdot\exp\left(-\frac{(x-x_o)^2}{2\sigma_x^2}\right)\cdot\exp\left(-\frac{(y-y_o)^2}{2\sigma_z^2}\right)\cdot\left(\exp\left(-\frac{(z-H)^2}{2\sigma_z^2}\right)+\exp\left(-\frac{(z+H)^2}{2\sigma_z^2}\right)\right) \tag{2.1}$$

where

C	=	ground level concentration on the centerline of the plume (g/m³), at location of coordinates *x, y, z*
Q_p	=	emission release rate (g/s)
H	=	height of source above ground level including plume rise (m)
σ_y, σ_z	=	horizontal and vertical dispersion parameters (m) (functions of x).

x, y, z = coordinates of the bottom of the release stack (x - downwind, y - crosswind, z - vertical) (m).

where σ_x is the diffusion parameter in the x direction, and x_o and y_o represent the coordinates of the center of the puff in a horizontal plane.

For urban applications, consider *Pasquill stability class D* conditions, where the parameters σ_y, σ_z as a function of x (in meters) are calculated with the following relations:

$$\begin{aligned} \sigma_y(x) &= 0.16\,(1+0.004\,x)^{-0.5} \\ \sigma_z(x) &= 0.14\,(1+0.0003\,x)^{-0.5} \end{aligned} \tag{2.2}$$

For the case where different stability categories are prevailing, one can use the nomograms given in APPENDIX C for estimating the horizontal and vertical dispersion coefficients σ_Y and σ_Z as a function of the distance from the source. The characteristics of the different Pasquill stability categories are mentioned in Chapter 1 (TABLE 1.8 and TABLE 1.10). Additional graphs for estimating the values of the dispersion coefficients σ_Y and σ_Z are also presented in APPENDIX C.

For any x, the maximum concentration of pollutants will be reached on the x-axis. There is a distance x_{max} from the bottom of the stack for which this concentration will reach its higher value C_{max}, where u is the wind velocity (m/s), and e is the base of the natural logarithm equal to 2.718:

$$C_{max} = \frac{2Q\sigma_z}{\pi u e H^2 \sigma_y} \tag{2.3}$$

and this maximum concentration is achieved at the distance x_{max} where

$$\sigma_z = \frac{H}{\sqrt{2}} \tag{2.4}$$

This calculation is strictly correct only if the σ_z/σ_Y ratio remains constant for any x.

In most Western countries, Briggs' plume-rise equations are used. It is very important to account for plume rise since a factor of two increase in effective source height (stack height plus plume rise) will cause a factor of four decrease in maximum ground-level concentration. Plume rise at H_s is determined by emission exit velocity and temperature and distance from the stack. For *bent-over buoyancy plumes,* the plume trajectory takes the form:

$$z = 1.6\,F_o^{1/3}\,u^{-1}\,x^{2/3} \tag{2.5}$$

where the initial buoyancy flux F_0 is defined as:

$$F_o = W_o R_o^2 \left(\frac{g}{T_{Po}}\right) \cdot \left(T_{Po} - T_{eo}\right) \tag{2.6}$$

where

W_o = initial plume vertical speed (ms^{-1})
R_o = inside stack radius (m)
g = acceleration of gravity (ms^{-1})

Tp_o = initial plume temperature (K)
Te_o = ambient temperature at stack height.

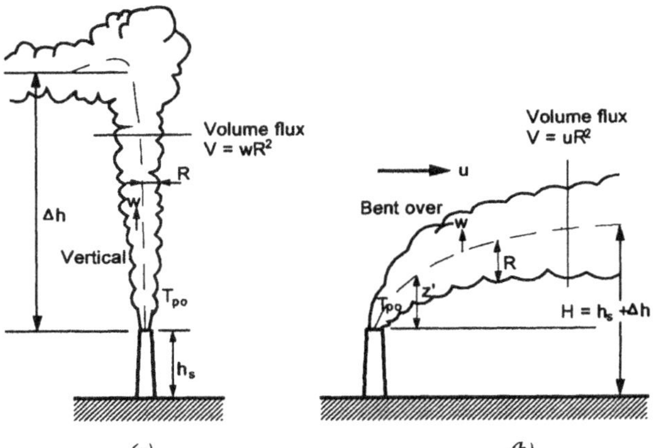

Figure 2.2. Important parameters and variables for plume rise calculations: (a) vertical, (b) bent-over plumes

$$\text{Environment stability: } s = \frac{g}{T_e}(\partial\, T_e\,/\,\partial\, z + 0.01 \quad {}^\circ\mathrm{C}\,/\,\mathrm{m}) \tag{2.7}$$

$$\text{Initial buoyancy flux: } F_0 = \frac{g}{T_{po}}\left(T_{po} - T_{eo}\right) w_o R_o^2 \tag{2.8}$$

Final plume rise for a bent-over plume in stable conditions is given by:

$$\Delta h = 2.6\left(\frac{F_o}{u\,s}\right)^{1/3} \tag{2.9}$$

where s is the ambient stability parameter:

$$s = \left(\frac{g}{Te}\right)\left(\frac{dTe}{dz} + 0.01^o\ K\,/\,m\right)$$

The ambient temperature gradient and, $\delta T_e/\delta z$, and wind speed, u, should be evaluated over the expected height range of the plume. The *wind speed at the plume* can be estimated from the wind speed, at a measuring height of 10 m, using the formula:

$$u(z) = u_{10}(z/10\ m)^p \tag{2.10}$$

where Irwin's table of the parameter p, as a function of the stability class, is given below:

TABLE 2.1. Irwin's Parameter, p, as a function of Pasquill stability class, and area (rural/urban).

Pasquill Stability Class	Urban Area parameter, p	Rural Area parameter, p
A	0.15	0.07
B	0.15	0.07
C	0.20	0.10
D	0.25	0.15
E	0.40	0.35
F	0.60	0.55

At heights above 200 m a constant wind speed equal to $u_{(200\,m)}$ should be used.
In the case of *calm wind*, final plume rise is given by:

$$\Delta h = 5.3 F_o^{1/4} \cdot s^{3/8} - R_o \tag{2.11}$$

For *nearly neutral conditions*, when the stability parameter, s, approaches zero, Briggs recommends to use the following formula:

$$\Delta h = 154 \left(\frac{F_o}{u u^{*2}} \right)^{2/3} \cdot h_s^{1/3} \tag{2.12}$$

where u_* is friction velocity and h_s is the physical stack height. Generally, there are other similar formula to be found in literature.

Influence of Buildings and Other Obstacles on the Plume Rise
A major problem in calculating plume rise is whether the plume will be influenced by a nearby building or other obstacle.

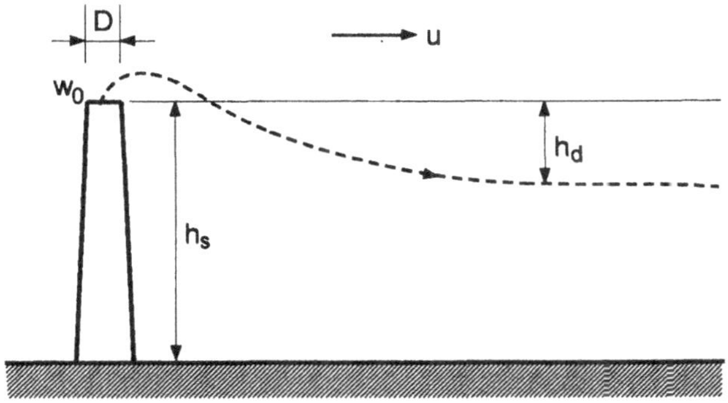

Figure 2.3. Schematic diagram of a stack

The important physical parameters related to a stack and to the plume are shown in Figure 2.3. These are:

w_0	=	Outlet gas velocity (m/s)
u	=	Wind speed (m/s)
D	=	Stack diameter (m)
h_s	=	Stack height (m)
h_d	=	Downwash distance (m)

First, if $w_0/u < 1.5$, stack downwash h_d must be accounted for in calculating the effective stack height, h_s':

$$h_s' = h_s - h_d = h_s - 4\left(\left(w_0 / u\right) - 1.5\right)R_0 \tag{2.13}$$

Next, according to Briggs (1973), let the cross-sectional building height be H' and its width be W, and let D^* represent the smaller value of the parameters H' and W. If the stack height is within $1.0\,D^*$ of the building parameters, the following revised effective stack heights will be appropriate:

$$h_e = h_s' \qquad \text{for } h_s' > H' + 1.5\,D^* \tag{2.14}$$

$$h_e = 2h_s' - \left(H' + 1.5\,D^*\right) \quad , \; H' < h_s' < H' + 1.5\,D^* \tag{2.15}$$

$$h_e = h_s' - 1.5\,D^* \qquad \text{for } h_s' < H' \tag{2.16}$$

If $h_e < 0.5\,D^*$, the plume could possibly be trapped in the cavity; h_e , in this case can be assumed to be zero, and the plume will have an initial area equal to D^{*2}.

Plume Rise

A plume emitted vertically from a stack will rise because of momentum and buoyancy, until turbulent mixing with the surrounding atmosphere terminates the rise. Analysts normally assume that all dispersion begins at a particular height above the actual stack exit. This height called the *effective stack height,* is the sum of the actual stack height and the plume rise. Plume rise is defined by the height of the plume centerline at the end of the emission phase before plume dispersion becomes important. The calculation of plume rise depends on the characteristics of the source and the dominant rise mechanism.

Plume Rise From Tall Stacks. For small volume sources with high exit speed (> 10 m/s) but little temperature excess (< 50°C over ambient), momentum dominates plume buoyancy. In this case plume rise H_p can be found from:

$$H_p = D_i\left(\frac{W_0}{U_s}\right)^{1.4} \tag{2.17}$$

where

H_p	=	height of plume centerline above emission source (m)
D_i	=	internal stack diameter (m)
W_0	=	vertical exit velocity (m/s)
U_s	=	mean wind speed at stack height (m/s)

Buoyancy becomes important for plumes with high temperature excess (> 50°C over ambient), and large volume (> 50 m^3/s). Then, plume rise depends on atmospheric stability, and momentum effects can usually be ignored. Briggs (1975) developed equations to provide a reasonable description of plume rise under different stability conditions.

Plume Rise in Stable Conditions. Final plume rise for a bent-over plume in stable conditions (stability classes E and F) can be calculated from:

$$H_p = 2.6\left(\frac{F_o}{SU_s}\right)^{1/3} \tag{2.18}$$

where

S = ambient stability parameter (s^{-2})
F_o = buoyancy flux (m^4/s^3)
U_s = mean wind speed (m/s)

The buoyancy flux and stability parameters can be determined from:

$$F_0 = gW_s\left(\frac{D_s}{2}\right)^2 \frac{\rho_{a_s} - \rho_s}{\rho_a} \tag{2.19}$$

$$S = \frac{g}{T}\left(\frac{\partial\theta}{\partial z}\right) \tag{2.20}$$

where

g = acceleration of gravity (m/s^2)
ρ_a = density of ambient air (g/m3)
ρ_s = density of stack gas (g/m3)
T = ambient air temperature (°K)
$\partial\theta/\partial z$ = vertical potential temperature gradient (°K/m).

The vertical potential temperature gradient $\partial\theta/\partial z$ has the value 0.02 (°K/m) for stability category E and 0.04 (°K/m) for category F.

For neutral or unstable conditions (stability categories A through D), the plume rise H_p can be calculated from the approximate formula:

$$H_p = 1.6\frac{F^{1/3}}{U_s}(3.5x^*)^{2/3} \tag{2.21}$$

where $3.5x^*$ equals the distance downwind at which plume rise terminates (m^4/s^3).

This distance can be estimated from:

$$x^* = 14\, F^{5/8} \quad (F < 55) \tag{2.22}$$

$$x^* = 34\, F^{2/5} \quad (F > 55) \tag{2.23}$$

Under light wind and calm conditions, normally associated with more stable circumstances, the above equations become invalid as the wind speed, U_s, approaches zero. In this case the plume rise can be best calculated from:

$$H_p = 5.0\frac{F^{1/4}}{S^{3/8}}$$

No matter which of the above equations is used, the resulting plume rise must be added to the actual stack height to determine the effective stack height for the dispersion calculations.

For effluents released from stacks which are at least $2\text{-}2^{1/2}$ times the height of adjacent solid structures, H_p can be reasonably estimated for neutral or unstable conditions by using the following equation:

$$H_p = 1.44 D_i \left(\frac{W_o}{U_s}\right)^{2/3} \cdot \left(\frac{x}{D_i}\right)^{1/3} - C \tag{2.24}$$

where

W_o	=	vertical exit velocity (m/s)
x	=	downwind distance (m)
U_s	=	wind speed at stack height (m/s)
D_i	=	internal stack diameter at exit (m)

and

$$C = 3\left(1.5 - \frac{W_o}{U_s}\right) D_e \tag{2.25}$$

where, C is a down-wash correction factor for $W_o < 1.5\ U_s$; here D_e is the external stack diameter (m).

When estimating H_p from equation (2.24), one should also estimate it by using the following expression:

$$H_p = 3 \frac{W_o}{U_s} \cdot D_i \tag{2.26}$$

The lower value calculated from the two equations (2.8 and 2.10) should be used as the more suitable estimate of plume rise. Caution should be exercised if the downwash correction brings the plume to less than 2-$2^{1/2}$ times the height of any adjacent building.

For *stable conditions* the result from equation (2.24) is compared with the results from the following two equations:

$$H_p = 4\left(\frac{F_m}{S}\right)^{1/4} \quad \text{and}\ H_p = 1.5\, S^{-1/6}\left(\frac{F_m}{U_s}\right)^{1/3} \tag{2.27}$$

and the smallest of the three values of H_p is used. In these equations F_m is a *momentum flux parameter* and S is a *stability parameter*, defined as:

$$F_m = W_0^2 \cdot \left(\frac{D}{2}\right)^2, \qquad S \approx \frac{g}{T} \cdot \left(-\frac{\partial \theta}{\partial z}\right) \tag{2.28}$$

where

g	=	acceleration due to gravity (ms^{-2})
T	=	ambient air temperature (°K)
$\partial\theta/\partial z$	=	potential temperature lapse rate (°K/m),

which may be approximated by $-(\delta T/\delta z + \Gamma)$, where Γ is the adiabatic lapse rate and is equal to 0.0098 °K/m. The stability classes have been defined previously. For *E stability class*, $S = 8.7 \times 10^{-4}$, and for *F stability class*, $S = 1.75 \times 10^{-3}$.

Plume Rise H_p From Short Stacks. Short stacks are defined as those of height h_{stack} less than 2-$2^{1/2}$ times the height of adjacent solid structures. For effluents released from short stacks the effluent plume can be considered as an elevated release whenever the vertical exit velocity of the plume, W_o, is at least five times the horizontal wind speed U at the height of release. In this case, the release should be evaluated as described above using equation. If W_o/U is less than one or unknown, a ground-level release should be assumed in order to obtain a conservative estimate of concentration. Where W_o/U is between one and five, a mixed mode should be assumed, in which the plume is considered to be an elevated release during a part of the time and a ground level release for the remainder of the time. In this case an entrainment coefficient E_t should be determined as follows:

$$E_t = 2.58 - 1.58\left(W_o / U\right) \quad for\ 1 \le W_o / U \le 1.5$$
$$E_t = 0.3 - 0.06\left(W_o / U\right) \quad for\ 1.5 \le W_o / U \le 5.0 \qquad (2.29)$$

The constants above are valid for a single building/stack situation.

The release should be considered elevated for 100(1-E_t) percent of the time. For each of these cases, the concentration should be weighted according to the fraction of time that each type of release occurs, and these weighted concentrations should be added together.

Plume rise H_s is determined by emission exit velocity and temperature and distance from the stack. There are two stages of behavior if the plume is separated by an artificial downwind distance V* (see also ref. [15]) defined by:

$$V^* = 2.16\, F^{2/5}\, H^{3/5}\ for\ H < 305\, m$$
$$V^* = 67\, F^{2/5}\ for\ H > 305\, m \qquad (2.30)$$

For $V \le x^*$, the plume rise is:

$$\Delta H = K \frac{F^{1/3}}{u} x^{2/3} \qquad (2.31)$$

with K = 1.6 - 1.8 (a value of K equal to 1.6 is suggested).

$$F = g\, v_s\, \delta_s^2 \frac{T_s - T_a}{T_s} \qquad (2.32)$$

where

U = wind velocity (m/s)
g = acceleration of gravity (m/s^2)
v_s = gas stack exit velocity (m/s)
δ_s = stack exit radius (m)
T_s = gas stack exit temperature absolute (K)
T_a = ambient temperature temperature (K)
F = momentum flux parameter

Gaussian Plume Models for Elevated Instantaneous (Puff) Point Sources

For Puff emissions, the Gaussian formula for ground level concentration has the following form:

$$C = \frac{\Delta M}{(2\pi)^{3/2} \cdot \sigma_x \sigma_y \sigma_z} \left\{ \exp\left[-\frac{(x-ut)^2}{2\sigma_x^2} - \frac{y^2}{2\sigma_y^2} \right] \right\} \left[\exp\frac{-(z-H)^2}{2\sigma_y^2} + \exp\frac{-(z+H)^2}{2\sigma_z^2} \right] \qquad (2.33)$$

where:

M = amount released (kg)
t = time elapsed after release (s)
Other symbols as in the previous case.

It is important to note that σ_y and σ_z in above equation for a puff are not the same as for a continuous plume. *Puff emission calculations are useful to calculate instantaneous or very short time emission dispersion conditions.* Ground level emissions are treated by setting $H = 0$. In that case, the influence of obstacles such as constructions, buildings, soil roughness may strongly affect the dispersion conditions and the above formula has to be used with great caution. It must be kept in mind that puff emissions could develop different spreading characteristics, and that continuous plume models have greater uncertainty. Because of lack of data, it is often assumed that $\sigma_x = \sigma_y$. The above equations are applied to ideal point sources from which the vapors are released. More complex relationships are available in the literature. Simplified relationships to calculate the ground level concentration from an elevated point and for puff emissions are presented next.

The *maximum ground-level concentration* can be estimated from the Gaussian plume model by differentiating the equation above with respect to x and setting the result equal to zero. When σ_y is proportional to σ_z, the maximum ground concentration occurs at a distance downwind, x_m, where 2 σ_z^2. The concentration at that distance is :

$$C_m = \frac{2Q}{\pi H^2 e u} \frac{\sigma_z}{\sigma_y} \tag{2.34}$$

Experience has shown that the distance x_m, is usually a few tens of stack height, h_s, downwind.

Because effective plume height, H, decreases as wind speed increases, there is a critical value of the wind speed for which the concentration C_m is maximized for all conditions. The corresponding formula is then:

$$Maximum\ C_m = \frac{1.1\,Q}{\pi H_s^2 e u} \frac{\sigma_z}{\sigma_y}$$
$$\text{at } x_m = \left(8/9a^2\right)^{1/2b} h_s^{1/b} \tag{2.35}$$

The critical wind speed, u_c, occurs when plume rise, Δh, equals 1/3 times the stack height, h, if it can be assumed that Briggs' equation for final rise in neutral conditions is valid.

Simplified Relation for an Elevated Point Source. The ground level concentration from an elevated point source is given by:

$$C = Q\,F / \pi\, \sigma_y\, \sigma_z\, u \tag{2.36}$$

where

C = ground level concentration on the center line of the plume (m³/m³).

The value of F can be obtained from the following relation:

$$F = \exp\left(H^2/2\sigma_z^2\right) \tag{2.37}$$

where

Q = gas release rate (m^3/s)
F = stack correction factor
H = height of stack or height of source above ground (m)
σ_y, σ_z = horizontal / vertical dispersion coefficient (m)
u = wind speed (m/s)

Simplified relation for Puff emissions. The following simplified equation may be used, whereby, in this case, the concentration is independent of the wind speed:

$$C = 2\,Q/\pi^{1.5} \cdot a_y\, a_z D^{(3-1.5n)} \tag{2.38}$$

where

C = centerline concentration (m3/m3)
Q = volume of gas release (m3/m3)
a_y, a_z = diffusion coefficients of the gas (dependent on weather conditions)
n = turbulence parameter

The weather conditions for simplified puff calculation are given in TABLE 2.2 which shows the values of the parameters a_y, a_z, and n as a function of the weather conditions.

TABLE 2.2. Weather conditions for simplified puff calculation

Weather Conditions	A	B	C	D	E	F
a_y	0.40	0.37	0.21	0.21	0.133	0.105
a_z	0.25	0.21	0.12	0.12	0.076	0.060
n	0.15	0.20	0.25	0.25	0.333	0.500

Pasquill stability categories can be identified from simple meteorological observations. The dispersion coefficients σ_y and σ_z are related to the stability of meteorological conditions. These have been classified in the so-called Pasquill stability categories defined in Chapter 1 (TABLE 1.8 and TABLE 1.10).

Simplified relations for calculating the concentration at cloud center

The concentration, at cloud center, of instantaneous point emissions (comprising short bursts or puffs) can be estimated using the following simplified relations:

a) Lapse condition (valid for Pasquill stability category A-B):

$$C = 45\,Q/(u\,t)^{2.74} \tag{2.39}$$

b) Neutral condition (valid for Pasquill stability category C-D):

$$C = 131\,Q/(u\,t)^{2.82} \tag{2.40}$$

c) Inversion (valid for Pasquill stability category E-F):

$$C = 493\,Q/(u\,t)^{2.47} \tag{2.41}$$

where

t = the time following a puff (s)
Q = total emission quantity instantaneously released (m^3/m^3)

There exist a large number of atmospheric dispersion models for flat terrain, based on homogeneous wind fields, as straight or segmented Gaussian plume, which in some cases give correct answers for minor complex terrain effects. The influence of complex terrain on atmospheric dispersion processes of airborne material is important in the local and lower meso-scale range. Therefore, dispersion modeling, particularly of emergency situations in real time or for planing purposes, is still a challenge, because of the complexity of the involved phenomena, their different space and time scales and the large amount of topographical and meteorological data to be processed.

The Gaussian Plume Model Applied to Complex Terrain
It is possible to apply the Gaussian plume model for calculating plume dispersion in complex terrain, but it is necessary to account for increases in σ_y and σ_z due to the great surface roughness and changes in the trajectory of the plume center line as it passes near a hill (*Figure 2.4*). Egan[16] (1975) has found that in complex terrain, the observed σ_y and σ_z curves are displaced by about one Pasquill stability class (i.e., Class C becomes class D). In stable conditions it can be assumed that the elevation of the plume center line does not vary with the underlying topography, and that the plume can impact a hill in its path.

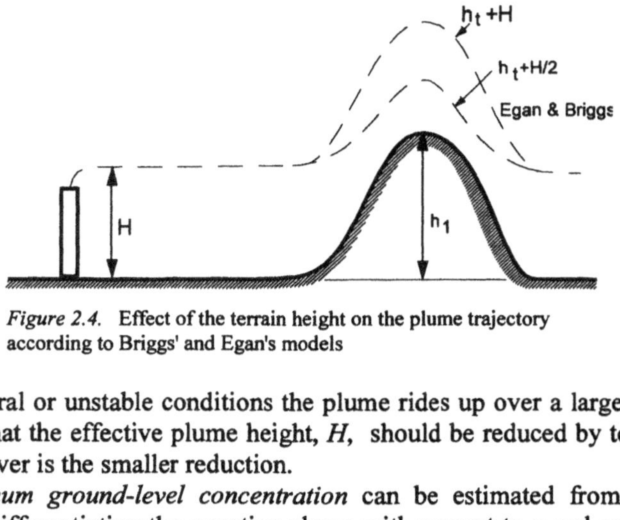

Figure 2.4. Effect of the terrain height on the plume trajectory according to Briggs' and Egan's models

But in neutral or unstable conditions the plume rides up over a larger hill. Briggs[17] recommends that the effective plume height, H, should be reduced by terrain height h_t or $H/2$, whichever is the smaller reduction.

The *maximum ground-level concentration* can be estimated from the Gaussian plume model differentiating the equation above with respect to x and setting the result equal to zero. When σ_y is proportional to σ_z, the maximum ground concentration occurs at a distance downwind, x_m, where 2 σ_z^2. The concentration at that distance is :

$$C_m = \frac{2Q}{\pi H^2 e u} \cdot \frac{\sigma_z}{\sigma_y} \tag{2.42}$$

Experience has shown that the distance x_m, is usually a few tens of stack height, h_s, downwind.

Because effective plume height, H, decreases as wind speed increases, there is a critical value of the wind speed for which the concentration C_m is maximized for all conditions. The corresponding formula is then:

$$Maximum\ C_m = \frac{1.1\,Q}{\pi H_s^2 e u} \cdot \frac{\sigma_z}{\sigma_y} \tag{2.43}$$

$$\text{at}\ \ x_m = \left(8/9\,a^2\right)^{1/2b} \left(h_s'\right)^{1/b} \tag{2.44}$$

The critical wind speed, u_c, occurs when plume rise, Δh, equals 1/3 times the stack height, h, if it can be assumed that Briggs' equation for final rise in neutral conditions is valid.

If the source of pollutant material is instantaneous at a point, then a *Gaussian puff model* may be used:

$$C = \left(\frac{Q_P}{\left[(2\pi)^{3/2}\sigma_x\sigma_y\sigma_z\right]}\right) \cdot \exp\left(-\frac{(x-x_o)^2}{2\sigma_x^2}\right) \cdot \exp\left(-\frac{(y-y_o)^2}{2\sigma_y^2}\right) \cdot \left(\exp\left(-\frac{(z-H)^2}{2\sigma_z^2}\right) + \exp\left(-\frac{(z+H)^2}{2\sigma_z^2}\right)\right) \tag{2.45}$$

where Q_P has units $g \cdot s^{-1}$, σ_x is the diffusion parameter in the x direction, and x_o and y_o represent the coordinates of the center of the puff in a horizontal plane. It is important to note that σ_y and σ_z in above equation for a puff are not the same as for a continuous plume.

McCarthy et al.[18], have reported that most computational programs for atmospheric dispersion of pollutant materials are based on models that incorporate well-known Gaussian plume and puff solutions. In these models, the spatial distributions of materials are considered to be Gaussian in one or more dimensions, and both the local sizes and the trajectories of the plumes or puffs are calculated independently according to the analysis of the particular forces acting on the material within the plume or puff.

Differences Among the Gaussian Models for Atmospheric Dispersion

The differences among the computational programs employing Gaussian models for atmospheric dispersion are primarily in the methods for calculating sizes and trajectories of the plumes or puffs, and in the methods for calculating the changes in the states of materials within the plumes or puffs. Growth and movement of a plume or puff are strongly dependent on local atmospheric conditions. Changes in state of the material within a plume or puff may be related to local atmospheric conditions or local concentrations within the plume or puff, among other factors. The effects of wind shear on the distribution of material are usually not included in Gaussian models because Gaussian plumes and puffs are symmetric about their centers. Eulerian or Monte Carlo methods (such as the random-walk technique), which can account for the effects of wind shear, were, in the past, employed in relatively few programs due to large computational times.

Remarks Concerning the General Use of Gaussian Plume Models

■ In the models reported above, the spatial distributions of materials are considered to be Gaussian in one or more dimensions, and both the local sizes and the trajectories of the plumes or puffs are calculated independently according to the analysis of the particular forces acting on the material within the plume or puff.

■ There exist a large number of *atmospheric dispersion models for flat terrain,* based on homogeneous wind fields, as straight or segmented Gaussian plume, which in some cases give correct answers for minor complex terrain effects. The influence of complex terrain on atmospheric dispersion processes of airborne material is important in the local and lower meso-scale range. Therefore, dispersion modeling, particularly of emergency situations in real time or for planing purposes, is still a challenge, because of the complexity of the involved phenomena, their different space and time scales and the large amount of topographical and meteorological data to be processed.

■ It is possible to use the Gaussian plume model in complex terrain, but it is necessary to account for increases in σ_y and σ_z due to the great surface roughness and changes in the trajectory of the plume center line as it passes near a hill. Egan[19] (1975) has found that in complex terrain, the observed σ_y and σ_z curves are displaced by about one Pasquill stability class (i.e., Class C becomes class D). In stable conditions it can be assumed that the elevation of the plume center line does not vary with the underlying topography, and that the plume can impact a hill in its path.. But in neutral or unstable conditions the plume rides up over a larger hill. Briggs[20] (1975) recommends that the effective plume height, H, should be reduced by terrain height h_t or H/2, whichever is the smaller reduction.

■ As already mentioned these models only apply to flat and smooth terrain. More sophisticated mathematical or physical models have to be used if topography, soil roughness, obstacles, (constructions, buildings) are to be accounted for. In areas of mountain ranges and deep valleys, wind and temperature fields produce complicated atmospheric dispersion patterns and distributions that are difficult to simulate. Complex terrain can result in several dispersion processes not encountered in flat terrain. These include: (1) channeling of flows along valleys, (2) trapping of pollutants within basins, (3) enhanced vertical dispersion through mechanically induced turbulence, (4) large vertical gradients in wind speed and direction, and (5) flow changes resulting from heating and cooling of sloping terrain. Finally the atmospheric conditions (wind, isobaric pressure, temperature, inversion conditions), and the period of the day/night have a non-negligible influence on the dispersion process.

■ The models only take into consideration the stationary meteorological conditions. Separate calculations have to be carried out if different meteorological conditions are occurring on a definite site as global results are obtained by summing up the partial results weighted by the probability of occurrence of the different conditions.

■ Puff dispersion calculations assume that pollutants ΔM are emitted with an instantaneous rate Q during a time span Δt so that $\Delta M = Q\,\Delta t$. If such puffs are being repeated indefinitely with Δt tending to zero, the emission becomes continuous and is described by the relevant formula presented earlier. On the other hand, if the instantaneous emission rate Q is time dependent, the resulting succession of puffs calculations can be used to simulate transient emission conditions.

Remarks Concerning Complex Gaussian Plume Models

■ The Gaussian plume models generally represent pollutants as puffs or a succession of puffs that are transported by wind trajectories within a varying meteorological field. The puffs expand in Gaussian fashion about a center, which is transported by local winds. Meteorological conditions and resulting puff expansion and transport are recomputed at fixed time intervals. Such models can account for orography, meteorological changes, and multiple sources (assuming additivity).

These approaches have given rise to a series of models, some of which are available as microcomputer software packages and others of which are run only on mainframe computers. The latter are generally more flexible, and "tailored-to-fit" runs can be made for specific conditions of release and local meteorology and topography. These models can also cope with complex, time-dependent release patterns. It is still impossible, however, to include highly detailed site characteristics and to reconstruct exact trajectories for specific puffs.

■ Plume models can be extended in order to take into consideration phenomena such as chemical or physical transformation during dispersion, radioactive decay, cloud depletion due to sedimentation (particles), washout by precipitation, impact on surfaces.

■ Whatever the model used, meteorological data have to be site and problem specific. It should be stressed especially that meteorological data collected on the nearby airport are of synoptic nature and especially applicable to air navigation. Data used in plume behavior models are different as far as the monitoring techniques are concerned, (type and location of sensors, frequency of measurements, averaging times etc.) Use of inadequate data may result in profoundly distorted or even false results.

Physically Based Models
Models of a region with precise representation of topography make it possible to simulate wind and temperature patterns and the exact behavior of one or several plumes. Such models are able to provide qualitative results that may be compared with sophisticated computer plume behavior models.

Regional Air Quality Models. The simplest air quality models *assume some linear relationship between regional average emissions and regional average concentrations.* The coefficients are estimated during a monitoring period and applied to some future period when emissions are different.

The first of these was the *Linear Roll-back Model* used in the early days of air quality assessment to estimate the effect on regional air quality of regulating specific sources. This model assumed that all sources in an area contributed to measured regional average pollution concentration in direct proportion to their relative contribution to total regional emissions. Information on emissions from all sources was used *to estimate an overall coefficient for concentration per unit emission from any source*, and this coefficient was then used to estimate the change in concentration that would be produced by a change (increase or decrease) in emissions from each source. *This modeling approach can be expanded to yield coefficients for seasons or for different meteorological conditions*, but the basic idea remains the same.

> *These models require simple data and are exceedingly easy to use, but are useful only over a small range of changes from the observations and only for pollutants without complex atmospheric chemistry that makes the emission-concentration relationship non-linear. They contain no causal mechanisms that can be adjusted for new conditions.*

The *regional air quality model* of the International Institute for Applied Systems Analysis (IIASA) is *more sophisticated than linear roll-back*, but is still easy to use. *This model consists of an array of coefficients to be entered into a dispersion equation that is a simple function of the size of the area modeled.* The coefficients were derived as generalizations of results from repeated runs of more complex mathematical models under a broad range of representative conditions.

The resulting equation yields ground-level concentration per ton emission for:

- Source height categories : Low, Medium, and High
- Meteorological conditions: Unstable, Neutral, and Stable,

at wind speeds:

- very low : < 2.0 m/s
- low : 2-5.0 m/s
- moderate : 5-7.5 m/s
- high : > 7.5 m/s

The *total emissions* are classified by their *relative frequencies*, according to emission height and meteorological category. The appropriate coefficients are applied to the resulting array, and the results are added.

Other coefficients are provided for ratios between regional average concentration and peak concentration, for a uniform emission density, for areas without pronounced centralization, and for areas with pronounced centralization of activities. These can be used to estimate maximum exposures. *Default coefficients* are provided for the three emission height categories to use for approximations in the absence of the necessary data on the frequency of meteorological conditions.

A broad range of more *complex Lagrangian* and *Eulerian regional wind-trajectory models* is available from various government agencies and packaged in user-friendly formats by private computer software firms. *These models are data-intensive, require considerable knowledge and experience to operate properly, and their relative applicability is problem-specific.* Analysts should seek advice on these from experienced meteorological modelers.

2.2.3. ATMOSPHERIC DISPERSION IN NUCLEAR POWER PLANT SITING

The IAEA has prepared a safety guide regarding atmospheric dispersion in nuclear power plant siting. This guide contains useful definitions and formulas for atmospheric dispersion calculations, for plume rise and effective stack height. A content summary of this IAEA guide is given below.

IAEA Safety Series No. 50-SG-S3, Vienna (1980)

Atmospheric Dispersion in Nuclear Power Plant Siting; - A Safety Guide.
Methods are proposed for evaluating time-integrated short and long-term release, and atmospheric dispersion, using simplified methods, Pasquill-Gifford type methods, temperature laps-rate method, wind fluctuation method, 'Split-Sigma' method as well as methods based on both temperature and lapse rate.

Meteorological aspects are considered and the precision, nature and scope of meteorological data to be collected are indicated. These include air temperature, air flow (wind direction, associated speed and duration, thickness of mixing layer, precipitation, and humidity). Specific data are also required for estimating the atmospheric turbulence and calculating the plume dispersion parameters in the x, y, and z direction, which will be used in the dispersion models.

Other parameters of influence are: influence of the terrain (valleys and hills), surface roughness, influence of urban location, water body, flat dessert area, wind direction, stability variation with height, building wake effects, fumigation, low wind speed dispersion, mixing height, and also the influence of cooling towers on atmospheric dispersion.

Different tables and graphs listed in this guide concern: the equivalent height of plume rise; several methods for determining the weather stability class (Pasquill method and modifications); a modified bulk Richardson number method for estimating the dispersion coefficients.

2.3. Computer Codes for Atmospheric Dispersion Modeling

Earlier atmospheric models were reviewed by Hanna[11] (1981); another review was recently prepared at PSI by Yadigaroglu[21] et al., (1987), and one by Herrnberger[22] et al., (1993), who reviewed 64 atmospheric dispersion models, capable of handling complex terrain effects. The models have been evaluated and ranked for nuclear application, using weighted factors. Since the nuclear requirements are set higher than for normal dispersion of pollutants and gases, the codes proposed below can also deal with atmospheric dispersion in complex terrain, and some do also include models for the dispersion of heavy gases. The outcome of this study is summarized below.

2.3.1. ATMOSPHERIC DISPERSION CODES FOR NUCLEAR APPLICATIONS

The most interesting atmospheric dispersion codes for nuclear applications were selected according to given criteria, on the basis of questionnaires sent to the codes developers. Models have been tested and evaluated at PSI, *using the field data obtained from the tracer experiment SIESTA*[23] *over the Swiss plateau.* Mean wind and turbulence data were measured in conjunction with atmospheric dispersion SF_6-tracer during weak wind situations over the Jura ridge and the hilly pre-alpine region. The ability of the codes to predict the dispersion of heavy gases has, however, not been tested.

Starting with a review of the state of the art of atmospheric dispersion modeling of airborne substances in complex terrain, five different types of dispersion models were found:

- *Analytical or Gaussian puff model:* the cloud is simulated by a series of puffs. Their growth and inner concentration pattern follow a Gaussian distribution. Nevertheless *Gaussian models are neglecting actual topography, terrain roughness buildings and constructions which strongly influence the plume behavior.*

- *Statistical / Monte Carlo simulation of advection / dispersion or Lagrangian particle models:* In the Lagrangian particle method the advection and dispersion is simulated by random walk (using a Monte Carlo technique). The physical process can be described in any details. Lagrangian solutions often proceed in two steps: first, the fluid trajectories are obtained by a solution of the mass conservation equation. In this step the vertical component of the wind is usually ignored. Second, the diffusion problem is solved. The results of the computation contain inherent statistical fluctuations

- *Deterministic advection/diffusion or Eulerian grid models:* In the Eulerian grid method the process is approximated by an advection/diffusion differential equation, which is integrated numerically by discrete ordinates in space and time (K-theory). In the gradient-transport approach, turbulent fluxes of pollutant concentration C are assumed to be proportional to the mean gradient of C:

$$\overline{-w'C'} = K\,\delta C/\delta z \tag{2.46}$$

The gradient transport model can be written:

$$\frac{\delta C}{\delta t}+u\frac{\delta C}{\delta x}+v\frac{\delta C}{\delta y}+w\frac{\delta C}{\delta z}=S+\frac{\delta K_x}{\delta x}\frac{\delta C}{\delta x}+\frac{\delta K_y}{\delta y}\frac{\delta C}{\delta y}+\frac{\delta K_z}{\delta z}\frac{\delta C}{\delta z} \tag{2.47}$$

where S represents internal sources or sinks, and K_x, K_y, K_z are eddy diffusivity coefficients in the x, y, and z directions, respectively. Analytical and numerical solutions to this equation are used in regulatory applications throughout the world.

The method suffer from deficiencies in the nearby region of point or line sources, or for small diffusion times. The results contain inherent numerical errors *called numerical diffusion.* Strictly speaking, use of the gradient transport model is valid only if the size of the plume is greater than the size of the dominant turbulent eddies, so that all the turbulence implicit in K is taking part in the diffusion. For this reason the gradient-transport model is most valid for vertical diffusion from near-ground releases at down-wind distances less than about 10 km.

It should be noted that knowledge of diffusion parameters σ_y and σ_z in the Gaussian model and K_y and K_z in the gradient transport model is required.

■ *Combination of statistical and deterministic or hybrid:* these simulate the advection by Lagrangian particle method and the dispersion by Eulerian grid, or concentration gradient, or diffusion method.

■ *Flow models with integrated statistical or deterministic dispersion part or fluid dynamic models:* Modern dynamic fluid processors (NHEL) contain either Lagrangian or Eulerian type dispersion models and have recently been applied to atmospheric dispersion problems.

More details of the theoretical basis involved in the different types of models can be found in a previous evaluation of flow and dispersion models.[24] Comparing the 4 main model types by their mean value performance measure, they all look similar and the differences between the model types are relatively small and within the accuracy of the performance measure of about 5-10%. Nevertheless a general trend appears, showing the highest performance measure for the NHEL, which are extensively developed engineering tools, and the lowest for the Puff models. The latter are more rudimentary regarding the incorporation of modifications and extensions of the straight Gaussian plumes. The model categories between the Eulerian and the Lagrangian types, are still in full development and the codes are getting more operational.

Recommendation for the Choice of a Model

A suitable model has to meet general scientific performances (consistent theory, theoretical and experimental validation, publication in review journals, etc.) as well as *project specific* performances. Description of complex terrain effects and heterogeneous stratigraphy by inhomogeneous windfields, drainage flow, weak wind situations, should be possible and models incorporating such features are highly recommended. In conjunction with pollutants it is quite often necessary to handle physical-chemical transformation of chemical compounds in contact with moist air. For nuclear applications, the model should also be capable of predicting radioactive decay.

The technical and computing performances of the model are also important. The following requirements should be met by codes dealing, among others, with dispersion of radioactive aerosols and particles:

- Standard models for deposition, sedimentation (and dose impact from inhalation), groundshine and cloudshine
- Modeling range of some 10 km downwind (local and meso-scale)
- Real time simulation and prognostics of typically 6 h, in less than 1 h CPU time (on a RISC 6000-500 computer)
- User friendliness, menu-driven input/output, graphics capabilities (GIS)
- Two and 3-dimensional iso-surface over site landscape and geography.

If the wind and turbulence data are not available from some meteorological monitoring networks, such as the Swiss meteorological network "ANETZ", or other private monitoring network, for instance "Safer System" used by some chemical companies (Ciba, Sandoz, etc.), or from nuclear power plants (Beznau-I,-II; Mühleberg, Leibstadt), then a flow model must be used to generate them for use by hydrodynamic and 3-D models.

Review of Gaussian Puff Models
The models selected as best Gaussian puff models are:

- TRAMES (with a Performance Index, P.I. = 1.5), and
- LINCOM-RIMPUFF (P.I.= 4.3).

Further candidates are:

- SIMACCS, and
- UK-ADMS.

As an example, the mesoscale dispersion model RIMPUFF, developed for nuclear applications, will be shortly described:

LINCOM-RIMPUFF is a combined flow/puff model, and was developed at Riso National Laboratory for the Vandenberg AFB Meteorology and Plume Dispersion Handbook and is suitable as is for real time response to emergency spills and vents of gases and radionuclides.

LINCOM is a linear, diagnostic, spectral, potential flow model which extends the Jackson-Hunt theory of non-hydrostatic, adiabatic wind flow over hills to the mesoscale domain. It is embedded in a weighted objective analysis (WOA) of real-time Vandenberg tower winds and may be used in ultra-high speed lookup table mode.

The RIMPUFF model is a fast and operational computer code suitable for real-time simulation of hazards from radioactivity released to the atmosphere. The model (using output from the LINC2W program) applies to non-homogeneous terrain and moderate topography on a horizontal scale of 0 to 50 km, and responds to changing (in-stationary) meteorological conditions. The Lagrangian puff-model simulates time-changing continuous releases by sequentially releasing a series of Gaussian shaped puffs at a fixed release rate on a specified grid. The amount of radioactive material allocated to individual puffs equals the release rate times the elapsed time between puff releases. RIMPUFF is equipped with computer-time effective features for terrain and stability-dependent dispersion parametrization, plume rise formulas, inversion and ground-level reflection capabilities and wet/dry (source) depletion. In addition, the code optionally provides local relative diffusion parametrization and a scheme for horizontal/vertical shear diffusion. It can treat plume bifurcation in complex terrain by using a puff-splitting scheme.

When applied to orographically (rain clouds) influenced dispersion scenarios, RIMPUFF can be interfaced with a flow-model, allowing the flow-model to compute the larger scale (e.g. beyond 500 m grid resolution) mean windfield. In diagnostic mode and for complex terrain with stationary meteorological conditions, radioactive exposure is well assessed via the release of a single initial puff. With optimization, processing time for a two hour forecast with RIMPUFF would be on the order of 2 CPU minutes or less on a PC/AT-286 system. The program is available on loan from *Risoe Library, DK-4000 Roskilde, Denmark.*

Review of Lagrangian Particle Models
In the case of Lagrangian particle models, the differences between the models are less noticeable. Five models were found to have a performance index above the mean:

- LASAT (P.I. = 2.3),
- EXPRESS-II (Hybrid model with P.I. = 3.3),
- PARTRAC (P.I. = 5.3),
- LAPMOD (P.I. = 7.2), and
- ADPIC (Hybrid model with P.I. = 8.4).

Further candidates are: MC31, SPRAY, TRAVELING, WCCCS/ARCO, and LASAT/G.

Review of Eulerian Grid Models
They can be grouped in two groups with high and low performance respectively:

- (EULERIAN)-HERMES (P.I. = 1.3),
- EXPRESS-II (P.I. = 3.3), and
- TRAPPA (P.I. = 7.1),
- ADPIC (P.I. = 8.4), and
- AIRVIRO (P.I. = 9.5).

The Hybrid models EXPRESS-II and ADPIC are the top models in this category. The model RAM cannot work with grid sizes smaller than 5 km. DREAMS-2 is not yet ready for use, and MARS can only disperse inert substances (no chemical reacting).

Review of Fluid Dynamic Processors
These models show similar distribution as the Lagrangian models, with many models having a performance index close to the mean and two top candidates:

- EXPRESS-II, and
- ADPIC.

Unfortunately most of these model types are still under development and not yet ready for use by third party. It has been reported that such models were already successfully applied in atmospheric dispersion, and that they showed very promising results in the 2-dimensional simulation of eddy fields behind ridges or in heavy gas dispersion scenarios.

The "Environmental Division of ENEA-DISP" [25] has applied for several years the *Atmospheric Diffusion Particle-in-Cell model ADPIC* (Lange[26], 1978) in conjunction with the *MATHEW mass consistent diagnostic wind field model* (Sherman[27], 1978) originally developed at the Lawrence Livermore National Laboratory, for both the real time and the forecasting of short term atmospheric dispersion in complex terrain. In ADPIC, Lagrangian "mass" particles are transported inside a fixed Eulerian grid, and the diffusive velocity of the particles is derived based on the K-theory.

Recently the option for Monte Carlo statistical diffusion has been introduced into ADPIC (Lange[28], 1990). Later on a new code, ARCO, was developed and is presently operational in the ENEA-DISP real time system for the assessment of the consequences of accidental releases into the atmosphere.[29]

Modeling Problems and Validation with Field Experiments

As mentioned above some of the most promising models have been tested and evaluated on the basis of the tracer field experiment SIESTA. *The models ADPIC2 and LASAT were taken as reference for this discussion, which is based on the comparison of the field test results with model predictions.*

In general the longitudinal plume size depends on the advection process determined by main wind speed. The cross wind spread or the lateral plume size depends on the treatment of the horizontal and vertical material exchange, or on the order of closure and type of parametrization in the turbulence modeling.

However in the case of Eulerian grid models, the cross wind spread can be heavily distorted due to non-physical "numerical diffusion", which is generated by the solver algorithm of the advection / diffusion equation. In this respect the model is particularly sensitive, if step concentration gradients occur near the point source region. Such gradients can be taken into account by using a finer grid mesh nested inside the coarse mesh.

The plume calculated by RIMPUFF and AIRVIRO are essentially shorter and remain within the modeling domain. The plume of TRAMES is only slightly shorter. The vertical treatment of the Puff model TRAMES is better, because the Puffs are transported along a 3-D trajectory derived from the 3-D windfield. From there on no more material exchange is possible and the simulation gets truly 1-dimensional. However the starting point of a trajectory depends on the effective emission height.

The plume size of the Eulerian model AIRVIVO is more similar to the Puff models than to the other Eulerian models. On the other hand AIRVIVO and RIMPUFF show the same good performance close to the source region, due to the Puff approximation applied in AIRVIVO and the smaller grid size than in TRAMES. *These models are less suitable for urgent emergency situations.*

Due to numerical diffusion, the Eulerian model HERMES shows large cross wind spreads. LASAT and TRAPPA, both use all the available stability information and their plume looks similar. TRAPPA has a fairly low numerical diffusion due to the use of advanced advection schemes with an accuracy of second order in space and time. HERMES seems to be not well suited for a straightforward and realistic simulation. *The models ADPIC2, EXPRESS and HERMES were capable of predicting a possible plume bifurcation into a side valley.* These models simulate the dispersion in Cartesian and pressure coordinates as well.

Conclusion About the Comparison of Dispersion Models

From the qualitative comparison made previously, it was concluded that:

- The Puff models RIMPUFF and TRAMES and the Eulerian model AIRVIRO cannot be recommended due to their generally reduced plume sizes caused by limitation in the vertical material exchange treatment. The Model HERMES shows too large cross wind spread and should not be used.

- A comparison of calculated data with predicted data (using scatter plots) show that the models PARTAC, LAPMOD, (TRAMES, HERMES) overestimate the measurements, and that the models ADPIC2, EXPRESS, (AIRVIVO) underestimate them.

- It was found that Lagrangian types models run faster.

> *Comparing also the performance of the models on a statistical basis it was found that ADPIC2, LAPMOD, EXPRESS, and TRAPPA were excellent.*

There is however a problem for *nuclear application*. Up to 110 radioactive mother and daughter isotopes can be relevant for calculating the dose impact on the environment. Through radioactive decay they can be transformed from a noble gas to an aerosol and vice versa. The sedimentation rates and the dispersion characteristics are both different. *None of the evaluated models can simulate the dispersion of this type of complex decay chains*. The simulation of the resuspension of particles is only possible in ADPIC2, but with great uncertainties.

All models are written in FORTRAN 77, except LASAT which is written in C, and they run on VAX, CONVEX or IMBM-RISC computers.

Considerations given by Taibi[30] on the use of the *generalized Gaussian plume model* was derived from a sensitivity analysis, whose aim was to study the influence of the dispersion parameters of the most quoted models on the estimated results. This study is a follow-up to previous analyses in which releases in the short, medium, and long-terms were taken into account. His conclusions has been reported regarding the influence on the prediction of the ground-level pollutant concentration of the adoption of the more common models describing physical phenomena such as: the thermal buoyancy of the plume, vertical profile of the wind speed, gravitational settling and dry deposition.

TABLE 2.1. Evaluation of atmospheric dispersion models

Model number	Model Name	Institute / Country	Merit Points	Weighted Merit Points	Comments	Performance Level Type: A/B/C
1.1	MC 31	CEA / F	96/80	245/202/19	L	L: 1/2/6
1.2	SPRAY	ADRIA / F	93/74	204/166/41	L,@,(OP)	L: 5/6/1
1.3	EULERIAN (HERM)	ADRIA / F	107/90	262/220/26	H,E	E: 1/1/3
1.4	SIMACC	ADRIA / F	105/85	253/199/37	P,IN	P: 3/2/2
1.5	TRAMES	ADRIA / F	114/92	262/212/41	P,IN	P: 1/1/1
1.6	TRIO VF	CEA / F	101/87	213/182/37	NH,E,@	NHEL: 6/6/3
2.1	TRAVELING	KfK/ D	91/80	204/183/16	L,@,(OP)	L: 5/3/7
2.2	DRAIS	KfK/ D	78/72	155/147/4	E,@,(OP)	E: 9/8/8
2.3	LASAT	Janicke / D	97/88	239/221/20	L	L: 2/1/5
2.4	HERMES	EDF / F	100/84	223/195/22	D,H,E,@,(OP)	NHEL: 4/4/5
2.5	GESIMA	GKSS / D	86/77	192/174/20	NH,E,@,(OP)	NHEL: 8/7/6
2.6	CANERN	EC / CdN	23/107	277/243/38	D,H,NH,E,L	NHEL: 2/2/2
3.1	PHYSICS (SPEEDI)	JAERI / J	48/45	109/103/6	H,@,(OP)	-
3.2	EXPRESS-I	JAERI / J	66/60	135/124/9	IN,@	-
3.3	EXPRESS-II	JAERI / J	89/75	208/170/10	LE	L: 4/5/9 E: 5/5/7
3.4	PHYS+ EXPRESS-II	JAERI / J	85/72	206/171/10	H,L,(OP)	NHEL: 7/8/8
3.5	FLOWSTAR	CERC / GB	90/75	194/164/24	(NH),@	-
3.6	UK-ADAMS	CERC / GB	78/89	219/197/20	P,NH,(@)	P: 4/3/5

TABLE 2.1. Evaluation of atmospheric dispersion models (continued)

Model number	**Model Name**	**Institute / Country**	**Merit Points**	**Weighted Merit Points**	**Comments**	**Performance Level Type: A/B/C**
4.1	AVACTA	JRC / CEC	54/52	124/120/4	P,D,@	-
4.2	MCLAGPAR	JRC / CEC	60/58	142/137/5	D,L,@	P: 6/5/7
4.3	LINCOM / RIMPUFF	Risoe / DK	118/102	255/212/29	P,IN,(NH)	L: 11/11/10
4.4	DIANE	CHAM / GB	127/110	282/245/35	NH,E,L,@,(OP)	P: 2/1/3
4.5	MESODYN	PSI / CH	37/35	92/88/0	NH,@	NHEL: 1/1/4
4.6	CONDOR	PSI / CH	33/29	76/70/20	I,D,@	-
5.1	DREAMS-1	PROCLIM / CH	76/65	174/151/19	H,(OP)	-
5.2	DREAM-2	PROCLIM / CH	60/69	167/144/19	L,E,@	L: 10/10/6 E: 7/9/4
5.3	PARTRAC	PSI / CH	80/66	187/156/27	D,L,@	L: 8/9/3
5.4	FLOW3/DR	UKAEA / GB	115/94	228/185/46	NH,E,L,@	NHEL: 3/5/1
5.5	WOCCS/ARCO	ENEA / I	80/63	192/157/30	D,L,@	L: 7/8/2
5.6	AVP	CIEMAT / E	57/51	110/94/16	P,@,D	P: 7/6/6
6.1	LASAT-G	GRS / D	81/70	196/174/26	L,@,(OP)	L: 6/4/4
6.2	RIMPUFF-G	GRS / D	79/70	190/169/25	P,@,(OP)	P: 5/4/4
6.3	NOABL	GRS / D	60/48	133/11/24	IN,D,(OP)	-
6.4	TR3	Uni Bonn / D	63/50	148/120/30	E,@,(OP)	E: 9/9/2
6.5	FOOT 3D	Uni Bonn / D	59/44	130/98/34	NH,(OP)	-
6.6	MESOSCOP	DLR / D	(46/46)	(106/106/0)	NH,E,@,(OP)	-

TABLE 2.1. Evaluation of atmospheric dispersion models (continued)

Model number	Model Name	Institute / Country	Merit Points	Weighted Merit Points	Comments	Performance Level Type: A/B/C
7.1	TRAPPA	Uni Karls-ruhe / D	73/69	162/150/16	E,@	-
7.2	LAPMOD	Uni Karls-ruhe / D	78/74	173/160/16	L,@	-
7.3	MARS	Uni Karls-ruhe / D	77/73	168/156/16	E,@	-
7.4	MEMO	Uni Karls-ruhe / D	73/69	160/148/16	NH,H	-
7.5	CONDOR	Uni Karls-ruhe / D	62/57	137/124/16	D	-
7.6	MIMO	Uni Karls-ruhe / D	59/55	136/124/16	NH, (OP)	-
8.1	TOPOG	LLNL USA	52/46	120/102/14	Pre-process	-
8.2	MEDIC	LLNL USA	62/58	142/130/8	Pre-process	-
8.3	MATHEW	LLNL USA	70/62	154/138/12	D	-
8.4	ADPIC	LLNL USA	100/86	236/202/12	LE	L: 3/2/8 E: 3/3/6
8.4	PLCNT	LLNL USA	67/53	154/120/12	Post-process	-
8.6	ADREA	Demokr Gk	86/80	218/210/10	NH,E,@	NHEL: 5/3/8
9.1	FITNAH	Uni Hano-ver / D	50/42	118/102/24	NH	-
9.2	RAM	SMIH / S	95/90	214/203/10	E,@,Grid too big	E: 4/2/7
9.3	METRICS	Uni Ham-burg / D	88/82	187/171/16	NH,H,E	NHEL: 9/8/7
9.4	METRAS	Uni Ham-burg / D	85/79	179/163/16	NH,H,@	-
9.5	AIRVIRO	INDIC / S - CH	100/81	237/196/44	D,H,E,@	E: 2/4/1
9.6	LADI / TRAMO	GEOS / CH	56/49	131/116/10	D,L,@	L: 12/12/9

TABLE 2.1. Evaluation of atmospheric dispersion models (continued)

Model number	Model Name	Institute / Country	Merit Points	Weighted Merit Points	Comments	Performance Level Type: A/B/C
10.1	RAMS	CSU / USA	NO	Response	NH	-
10.2	LPDM	CSU / USA	NO	Response	L	-
10.3	ARAMS	R*CAN / USA	NO	Response	NH	-
10.4	ALPDM	R*CAN / USA	NO	Response	L	-
10.5	HIRLAM	DMI / DK	NO	Response	H	-
10.6	FRACTAL	Uni Pavia / I	NO	Response	FRACTAL	-
10.7	LAGRANGE	SCS / D	NO	Response	L	-
10.8	MONARC / DIAMO	GEOS / CH	NO	Evaluation	NH,D	-
10.9	FLOW3D/ R2.4	UKAEA / GB	As 5.4 but	Older version	-	-
10.1	ASTEC	UKAEA / GB	NO	Older version	-	-

Legend for TABLE 2.1. Abbreviations / Model Types

Dispersion models	Flow models
P : Gaussian Puff	IN : Interpolating
E : Eulerian	T : Trajectory
LE : Hybrid	NH : Non-hydrostatic
L : Lagrangian	D : Diagnostic
NHEL : Fluid dynamic processor	PP : Pre/Post-processor
	H : Hydrostatic
	@ : No dose model

Model performance
A: Overall modell
B: Dispersion model
C: Input/output capabilities

Sum of merit points:	Model Type A / B
Weighted sum of merit points:	Model Type A / B / C

Statistics

Number of dispersion / flow models evaluated:	38 / 16
P (Gaussian Puff):	7
L (Lagrangian):	11
E (Eulerian):	8
LE (Hybrid):	2
NHEL (Fluid dynamic processors) :	10

There are 10 references related to above tables (Ref. [31 - 40]).

Summary
For dispersion modeling in complex terrain, the Puff models show weak performance. The plumes are in general too short. The Eulerian grid (diffusion) models can strongly suffer from numerical diffusion dominating the crosswind spread. The description of turbulence and of planetary boundary layer stratification differ much from model to model. Hybrid and Lagrangian models showed an acceptable overall performance.

2.3.2. ATMOSPHERIC DISPERSION CODES

Various regulations require the assessment of the potential impacts of chemical releases from industrial facilities.

In Switzerland, the Federal Authorities (see the legislation[41] entitled "Störfallverordnung", 1992) require that industry together with the Swiss Regional Authorities and the "Cantonal Chemists" address air-quality impacts of new or modified emission sources, potential impacts of accidental releases, and the residual risk associated with emissions of hazardous air pollutants after installation of required control technology. A handbook has been published, which contains indications on the way to prepare the risk assessment study and to report to the Federal Authorities.[42]

Some of the models and methods summarized below may help assessing risks from atmospheric dispersion.

Models from "ENERGY" Data Base
A short review of the models and papers stored in the "ENERGY" data base is given below.

Transport Model ATM. The ATM model can be used to estimate the transport of a pollutant from an emitting source to some receptor point.[43] It is based on a Gaussian plume model and is capable of calculating wet and dry deposition and air concentration of non-radioactive pollutants at various receptor points located within 50 kilometers of a source or sources. The model can treat point, area, and line sources, including wind blown sources. The use of either the Pasquill-Gifford or the Briggs-Smith dispersion parameters can be specified. The program is written in FORTRAN IV to run on an IBM360; OS-360/370.

Atmospheric Dispersion and Deposition

Air Quality Modeling. To quantitatively assess these impacts, one must relate the downwind concentrations of the chemicals of concern to their emission rates. Computer-based air-quality modeling (also called air-pollution modeling) is used to perform such calculations. Applications of air-pollution modeling in the chemical process industries (CPI) include: permitting of a new or modified source; estimating public health risks associated with emissions of toxic chemicals; and estimating the consequences of accidental releases of hazardous substances. Air-pollution modeling is most valuable to industry when it is used pro-actively as a management tool rather than in reaction to regulatory constraints. In order to use air-pollution modeling effectively as a management tool, one must understand the existing and pending air-quality regulations, as well as the fundamental of this discipline.[44]

2DPUF Model. A sequential Gaussian puff model, 2DPUF, has been developed by Westinghouse Savannah River Co. The Environmental Transport Section's (ETS) two-dimensional, sequential Gaussian puff transport and dispersion model for emergency response has been reported in detail.[45] The advantages of this new model over the ETS's PUFF/PLUME model are discussed and a flow diagram for the modules comprising the 2DPUF code and a description of each module is presented. Options are provided for calculating a two-dimensional wind field.

GALSON-ROOFTOP Model. It is a model for estimating pollutant impact in urban areas.[46] Regarding potential exposures to trace amounts of air toxics in urban areas, air quality analysts are often faced with a complex dispersion modeling problem. Unlike emissions of criteria pollutants, which are usually emitted from well designed stacks, air toxics are often emitted from short roof-based stacks and vents. Plumes emitted in these near-roof regions generally do not follow the conventional dispersion processes typical in the free atmosphere. The GALSON ROOFTOP dispersion Model has been developed to simulate near-roof dispersion surrounding a simple building. The ROOFTOP model is based on semi-empirical observations from wind tunnel studies of dispersion near structures. Simple structures can be analyzed directly. More complex structures must be approximated by reference to aerodynamically equivalent simple shapes. The model, which can be used as a screening tool to indicate if a potential impact problem exists, has been adapted to include multiple emission points, elevated sources, momentum plume rise and elevated receptors. It can provide sensitivity analysis of key assumptions, such a plume rise from roof-level stacks, as well as predict impacts. Results can be obtained as estimated concentrations, or as graphical representations of the building aerodynamics to allow for qualitative analysis.

METEC Code. The computer code METEC was set up for assessing the environmental impact of industrial releases.[47] It is based on the Gaussian plume model. The code is able to evaluate air concentration, ground deposition (dry and wet) and, for radioactive pollutants, the doses for external exposure and intake by inhalation. The code takes into account various phenomena such as the thermal buoyancy of the plume, vertical profile of the wind speed, gravitational settling, etc., as well as, various sets of empirically determined dispersion parameters.
A sensitivity analysis on some of the most used models was prepared in order to investigate the real influence on the predictions of taking into account the different physical phenomena and what is really implied when a particular set of dispersion parameters is adopted instead of another one.

Other Types of Models

Forecasting Model. A microcomputer-based *forecasting model with potential applications for emergency response plans and air quality studies* has been developed.[48] Most atmospheric transport and diffusion models within emergency response systems have very limited physics and are forced to rely on the assumption that wind and turbulence conditions at the time of the release will be representative over the period for which dispersion must be predicted. This model incorporates features for dealing with changes in wind and turbulence conditions. This model was tested in complex terrain with good experience.

K-Theory Model. Another model based on K-theory[49] has been developed at Louisiana State University for describing the *short range air dispersion from area sources of nonbuoyant toxics*. Model parameter estimation is via boundary layer theory. Lateral dispersion by plume meander is considered but all other sources of horizontal dispersion are neglected. The model can be applied on and near area sources and it can be adapted for predictions of downwind concentrations with a wide variety of meteorological inputs. The model has been evaluated by simulating the data obtained during atmospheric tracer studies and by comparison to vinyl chloride concentrations near the BKK landfill in southern California. The model appears to represent a useful and accurate tool for regulatory planning and risk assessment close to area sources of toxics.

Model Incorporating Reactions' Kinetics. For some applications it is necessary to use models incorporating kinetics' equations. Plume models for predicting the concentration due to a nonbuoyant release from a source, whose strength decreases exponentially with time, of a pollutant that reacts in the atmosphere at a first-order rate, are described.[50] Models were developed that combined the conventional Gaussian plume with one of three different one-dimensional models to describe advection and dispersion in the downwind direction: advection without downwind dispersion, advection and dispersion by a Fickian puff, and advection and dispersion by a puff of constant size. In addition to these one-dimensional dispersion models, a model based on three-dimensional Fickian puffs was also developed.

CTDMPlus code for Complex Terrain. The Complex Terrain Dispersion Model Plus (CTDMPlus), for use with microcomputers, is a refined air quality model for use in all stability conditions for complex terrain applications.[51] It contains the technology of the original Complex Terrain Dispersion Model (CTDM) for stable and neutral conditions, but also models daytime, unstable conditions. The model makes use of considerable detail in the terrain and meteorological data (as compared to current EPA regulatory models) and requires the parametrization of individual terrain features, thus considering the three-dimensional nature of the interaction of the plume and terrain.

OCD model. The Offshore and Coastal Dispersion Model[52] was adapted from the EPA guideline model MPTER to simulate the effect of offshore emissions from point sources in coastal regions. Modifications were made to incorporate plume transport and dispersion as the plume crosses the shoreline. Modifications included the use of turbulence intensities to define dispersion regimes, use of the RTDM Model to treat plume reflection in elevated terrain, use of the BLP Model to incorporate building downwash and plume rise to the model, partial plume penetration into elevated inversions, a virtual source to describe plume growth as the plume intercepts the overland internal boundary layer, and a continuous fumigation modular. Three-dimensional forecasting models on desk-top computer.

3-D Forecasting Model. The ability of using an engineering workstation to operate a *three-dimensional mesoscale modeling system to describe the transport and dispersion of atmospheric pollutants over complex terrain* surrounding Vandenberg Air Force Base (VAFB), Vandenberg, California, has been reported by Yamada[53] et al., (1990). The modeling system is composed of two numerical codes, HOTMAC, and RAPTAD. HOTMAC is a mesoscale forecast code that is able to model three-dimensional distributions of wind speed, wind direction, turbulence, temperature, and water vapor. RAPTAD is a Lagrangian puff code based on the Monte Carlo statistical diffusion process. RAPTAD can be used under extreme conditions where a conventional Gaussian plume model may fail.

ARCO Model for Atmospheric Dispersion in Complex (Terrain). Dispersion in complex terrain or under complex meteorological conditions requires the use of three-dimensional diffusion models coupled with objective or prognostic windfield models. With respect to Eulerian grid models, Lagrangian particle models have the advantage of being conceptually and computationally simpler by avoiding numerical diffusion. Particle models can be purely deterministic or possess statistical characteristics. In the first case particle motions are uniquely calculated based on some closure of the diffusion-advection equation; particle-in-cell models based on K-theory belong to this category. In the second case, Monte Carlo techniques are used to produce semi-random perturbations and the dynamic of each particle represents just one realization of an infinite set of possible solutions.

The atmospheric dispersion model ARCO (Atmospheric Release in Complex Terrain) has been developed by ENEA-DISP[25] for the study of accidental or routine emissions by point sources over a complex terrain. It is a Lagrangian particle model, in which the pollutant plume is simulated by a large number of particles with a velocity which is the sum of the advection and turbulent components. The advection field is provided by a diagnostic wind field model coupled with ARCO, while the turbulent motion is computed using a Monte Carlo statistical technique and on the basis of different parametrizations of the dynamic and thermal structure of the atmospheric boundary layer.

Fugitive Dust Model (FDM). A technical description and user's instructions for the Fugitive Dust Model (FDM) have been reported by Winges[54] (1990). FDM is a computerized Gaussian-plume air quality dispersion model, specifically designed for estimation of concentrations and deposition impacts from fugitive dust sources. The sources may be point, line, or area sources. FDM employs an advanced gradient-transfer particle deposition algorithm. Gravitational settling velocity and deposition velocity are calculated by FDM for each of up to 20 user-specified particle size classes. Descriptions of three performance evaluations of FDM and the EPA Industrial source complex model are given. Also included with the manual is a floppy diskette containing FORTRAN source codes, PC executable codes, and test data sets.

Models for Simulating Short-Term, Short-Range Air Quality Dispersion Phenomena
Basic information on deterministic mathematical techniques for the simulation of short-term, short range air quality dispersion phenomena can be found in the Encyclopedia of Environmental Control Technology.[55]

Encyclopedia of Environmental Control Technology
The authors describe the different components, or models, that can be used in this simulation. First, meteorological models, which provide the input to dispersion models, are discussed. Plume rise phenomena are presented in "Plume Rise Formulations", and the major categories of dispersion models are analyzed: the Eulerian models, in which the computational domain is divided into grid cells; the Gaussian models, which are based on the explicit Gaussian dispersion formula; and the Lagrangian models, in which the reference system moves with the average wind speed and direction. A few special applications of dispersion models are also presented.

PAL model. PAL is a Method Of Estimating Short-Term Dispersion Using Gaussian - Plume Steady-State Assumptions.[56] The algorithm can be used for estimating concentrations of non-reactive pollutants at 99 receptors for averaging times of from 1 to 24 hours and for a limited number of point, area, and line sources (99 of each type). Calculations are performed for each hour. The hourly meteorological data required are wind direction, wind speed, stability class, and mixing height. Single values of each of these four parameters are assumed representative for the area modeled. *The PAL model can treat deposition of both gaseous and suspended particulate pollutants in the plume* since gravitational settling and dry deposition of the particles are explicitly accounted for. The software is written in FORTRAN for implementation on an IBM-PC microcomputer.

PEM Model. The Pollution Episodic Model (PEM)[57] is an urban-scale (up to 50 km distances) air pollution model capable of predicting short-term (1 to 24-hour) average surface concentrations and deposition fluxes of two gaseous or particulate pollutants at up to a maximum of 2500 ground-level receptors located on a 50 km by 50 km square receptor grid.[58] Predictions are based on steady-state Gaussian plume assumptions, Briggs' plume rise formulations, and Pasquill-Gifford (P-G) dispersion parameters. The surface concentration and deposition flux estimates of two independent non-reactive (gaseous or particulate) pollutants or one pollutant with first-order chemical decay can be obtained as special cases of the model. Up to 300 point sources and up to 50 area sources may be included in the model inputs.

A Desk-Top Computer Forecast for Emergency Response Planning. Bice[59] et al., (1989) have demonstrated the feasibility of running a three-dimensional weather forecast model on a desk-top computer.

For emergency response planning, the ability to forecast wind temperature, and atmospheric turbulence is essential in order to predict the transport and diffusion of accidental release of toxic materials. Most models for emergency response are based on the assumption that wind and turbulence condition at the time of the release will be representative over the period for which dispersion must be predicted. For small releases where the principal concern is within the first few kilometers of travel, such an assumption is appropriate.

However, for large releases during stable conditions, the plume may travel for several hours and the assumption of persistence is invalid. Under these circumstances a model which can forecast change in wind and turbulence conditions is required. Persistence is a particularly poor assumption to make in mountain valleys and coastal locations. In coastal locations, we have land-sea breeze phenomena, whereas the mountain valleys we have the mountain valley wind system. In both instances abrupt changes in wind direction can be expected during the diurnal cycle, and the wind direction at a given time is very sensitive to the location where the winds are measured. A diagnostic model needs a large number of measurements in time and space to represent accurately the airflows in complex terrain. Such extensive measurements could become very expensive and are not practical.

Diagnostic models, by themselves, are not adequate for emergency response management since they have little predictive capability. A forecast model, on the other hand, can forecast locally generated diurnal circulation driven by terrain features. Three-dimensional forecast models for practical applications and results are discussed.

ISCST Code. The ISCST code has been developed for improved dry deposition estimates for air toxics risk assessment. In recent health risk assessments, EPA-guidelines dispersion models have been employed to estimate the amount of particulate and gaseous material deposited on the earth's surface. These estimates of dry atmospheric deposition are used in the health risk assessment. *A dispersion model frequently employed to estimate dry deposition is the Industrial Source Complex (ISC) code.* The theoretical basis for the Gaussian plume model concept upon which ISC is based are briefly summarized. Certain of the fundamental assumptions and limitations underlying the Gaussian model are outlined. Alternative conceptual approaches in treating dry deposition in Gaussian plume calculations are presented. Information on the relative accuracy of different methods for simulating dry deposition is given. The improved ISCST code incorporates a refined dry deposition estimation procedure.[60]

CDM2 Climatological Dispersion Model. The Climatological Dispersion Model[61] - Version 2.0 (CDM-2.0) determines long-term (seasonal or annual) quasi-stable pollutant concentrations in rural or urban settings using average emission rates from point and area sources and a joint frequency distribution of wind direction, wind speed, and stability. The Gaussian plume hypothesis forms the basis for the calculations. Computations can be made for up to 200 point sources and 2500 area sources at an unlimited number of receptor locations. CDM-2.0 includes the following options: 16 or 36 wind-direction sectors, initial plume dispersion, buoyancy-induced dispersion, stack-top down wash, and gradual (transitional) plume rise. The user has a choice of seven dispersion parameter schemes. Software Description: The software is written in the FORTRAN programming language for implementation on an IBM-PC compatible microcomputer. Data are in compressed format but a dearchive utility is included.

AMETISTA Model for the Evaluation of Long-term Atmospheric Pollution Caused by Stack Emissions. AMETISTA is an air quality model based on meteorological time series and turbulence analysis for evaluating the long-term air pollution caused by emissions from more co-located, elevated point sources.[62] It can simulate the dispersion of plumes emitted by up to twenty release points through the iteration of a straight line, hybrid plume dispersion model, mostly based on the equations of HPDM model (Hanna and Paine[63], 1989). The model is valid for tall stacks in relatively flat terrain and has been extensively validated against field data from the Kindcaid and Bull Run power plants in USA. Hourly air concentrations are calculated on a rectangular grid and at a number of receptor points whose coordinates can be specified as input.

AMETISTA is written in FORTRAN 77 language, and is one module of ARIES-W (Atmospheric Release Impact Evaluation System - Workstation), an integrated hardware/software system for the study of the atmospheric dispersion at different space and time scales developed at ENEA-DISP.

Two categories of meteorological data can alternatively be used for the estimate of the atmospheric dispersion over long time periods:

- *Joint Frequency Functions (JFF*i.e., the statistical frequencies of the occurrence, at a certain site, of a meteorological situation generally defined by some stability parameters (usually Pasquill-Gifford categories), wind speed and a sector of wind direction. Such data are officially published in some countries, and
- *Time Series of Meteorological Observations (OTS)* at time intervals of the order of one hour.

The model based on JFF are relatively simple, but have on the other hand several drawbacks regarding the characterization of the diffusion properties of the atmospheric boundary layer based on the stability categories (which is not quite appropriate for unstable conditions), and the calculated concentration values, and the statistical indexes that can be derived from them (e.g., average, median, percentiles of the distribution).

The model based on OTS consists of an iteration of a short-term, straight line plume model for the number of cycles needed to cover the period of interest, for example one year. From the primary variables contained in OTS it is possible to derive the meteorological scaling parameters, such as the sensible heat flux, the Monin-Obukhov length, and the mixing height, thus it is suitable to apply as a short-term model an *advanced hybrid model* (i.e., non-Gaussian in the vertical). By means of time interpolating or averaging algorithms it is possible to reduce original OTS to hourly OTS, and from this to obtain the statistics on the concentrations for sampling times multiple of one hour.

The advanced diffusion model AMETISTA was developed on the basis of above considerations, and makes use of meteorological scaling parameters.

PTPLU Model. PTPLU is a point-source dispersion Gaussian screening model for estimating maximum surface concentrations for 1-hour concentrations.[64] PTPLU is based upon Briggs' plume-rise methods and can use either Pasquill-Gifford or Briggs' urban dispersion coefficients. PTPLU is an adaptation and improvement of PTMAX which allows for wind-profile exponents and other optional calculations such as buoyancy-induced dispersion, stack downwash, and gradual plume rise. PTPLU produces an analysis of concentration as a function of wind speed and stability class for both wind speeds constant with height and wind speeds increasing with height. Using the extrapolated wind speeds and the options allows the model user a more accurate selection of distances to maximum concentration. PTPLU1 is the interactive version of the model. Software Description: The model is written in the FORTRAN programming language for implementation on an IBM 3090 computer.

Calculation of Initial Conditions for Material Transport in Mesoscale Circulation Models. A numerical model has been developed for predicting the dispersion of *pollutants emitted from a stack into the atmosphere.*[65] The model consists of two parts: (1) an integral model utilized to simulate the jet-like near field of the dispersing plume, and (2) a diffusion model for the subsequent phase which is dominated by atmospheric turbulence. The two submodels are linked to each other. The coupled model provides initial conditions for the simulation of pollutant transport in meso-scale grid models.

Meso-Scale Models
Other kinds of models are:

- The KNMI mesoscale air pollution model (Dop[66] et al., 1982)
- A surface depletion model for deposition from a Gaussian plume (Hort[67], 1977).
- METRAS is a transport and flow model (Schlünzen[68], 1988).

Comparative calculations and validation studies with atmospheric dispersion models were per formed at KfK (Päsler-Saur[69], 1986). Another review on particle and gas dry deposition was prepared by Sehmel[70] (1980).

2.4. An Environmental Decision Support System "Inter-CLAIR"

Risk assessment of complex technological systems operating in a given environment has to be made usually by means of computer simulation tools. This allows the forecasting and analysis of cases which have not yet taken place or are most improbable in reality. The *Environmental Decision Support System "Inter-CLAIR"*[71] provides potential users with a the set of practical tools for performing such applications. It helps to identify solutions to problems of air pollution protection, air and surface water quality management in large industrial complexes, including energy production systems. The decisions are supported for environmental, economics and innovation planning at the national, sectorial and enterprise levels, taking into account the best available technology, national pollution standards, and the great variety of emission sources related to air and surface waters.

Inter-CLAIR integrates basic knowledge from environmental engineering, mathematics, natural sciences and economical modeling as applied to the field of integrated environmental and health risk assessment and management.[72] It simulates specific processes for ecological, economical and risk assessment analysis in regions with dimensions up to 400 x 400 km (about 300 x 300 miles). The system can be applied to a variety of industrial areas, cities and relatively small size states. The different versions of the computerized simulation system "Inter-CLAIR" have been used in a number of case-studies for several European and Asian countries. Inter-CLAIR could provide a common base for solving potentially conflicting environmental risky situations in a given region, including transboundary pollution evaluation and mitigation.

The following system's features are considered to be useful for a number of other various applications:

- data base of regional stationary pollution sources
- data base of technological measures (e.g. new clean technologies, optional filters, fuels, liquidation of emission source, etc.)
- data on industrial and source-specific impacts on the atmosphere
- ecological and economical effectiveness analysis of overall industrial innovations
- optimized investment allocations for air quality management
- maximum resource calculation required for atmospheric protection options
- air pollution optimal reduction within specified expenditure constraints
- the effective set of technological measures that could be applied to any emission source
- critical load estimation for various environments and receptors.

Optimal simulation solutions for pollution abatement can be calculated simultaneously in Inter-CLAIR for any given set of pollutants (e.g. SO_x , NO_x etc.) and criteria. The system requires a minimal input data set. Some of these data are available from statistical books and specialized publications. Wherever possible, default data have been provided within the Inter-CLAIR system. The required data to run this system include: i) emissions' sources data; ii) technological measures data; iii) meteorological data for the region, averaged over a period of ten or more years.

Data input handling is facilitated by a set of friendly menus and on-line cross-reference help procedures. The interface is user-friendly and easy to be used by an inexperienced risk assessment analyst. The interactive work with the system is supported by on-line explanations and guiding illustrations. After running the simulation model, the following resulting data and maps are available for a potential decision maker:

- maps of emission distribution for a given number of pollutants
- maps of economical and environmental damage distribution
- maps of pollutant concentration distribution in the region

- maps of wet and dry deposition for sulfates and nitrates
- sulfate and nitrate transportation diagrams out of the area considered
- maps evaluating danger (due to acid rain) to coniferous forests
- graphs of optimal strategy as a function of expenditures for atmospheric pollution reduction by each selected criteria (emissions, concentrations, depositions, transportation out of the region, danger to coniferous forests) and any combination thereof
- compromised optimized solution sets, for a final decision making process.

A number of new models and subsystems are expanding the capabilities of Inter-CLAIR:

- car and city traffic environmental impact analysis
- simultaneous air and surface water pollution and risk estimation
- monetary resource distribution for air and water quality protection.

Hardware Requirements. The Inter-CLAIR system currently runs on IBM-PC/AT 386, 486, or compatible with EGA or VGA/SVGA/XGA graphics adapter, co-processor and a hard disk under DOS ver.3.xx or higher. The system requires about 5 MB of disk space.

Running Case Study with Inter-CLAIR

There are several stages of "on-site case study" using the Inter-CLAIR system. All of them are performed by a small team of risk experts and engineers, including also experts on emission sources and technological systems. Relatively small parts of the statistical data and measured information can be extracted from files of the environmental monitoring data normally stored by official meteorological services. The rest should be provided by the local experts (sources of emission, technological measures). The following stages have to be considered for running a case study:

- Site selection and team formation
- Identification of the emission sources
- Definition of the time span or time prognoses variations
- Definition of pollutant sets (by the most pressing indicators)
- Definition of possible and available technological sets of measures
- System runs and calculations using several air pollution criteria
- Optimized strategies identification by given criteria
- Results analysis
- Initial data correction and model adaptation
- Final system runs and calculations
- Report preparation and recommendations validation
- Case study task force group meeting with final results discussions (with possible invitations of authorities or decision makers and/or mass media representatives, if necessary)
- On-site system installation, local personnel education and training for support to local decision makers.

Meteorological Data for Cities and Regions. All meteorological data are defined as the average arithmetic mean over a period of at least 10 years. This should average possible weather deviations that could occur during some years. Meteorological data necessary for Inter-CLAIR system simulation runs can be found in climatic reference literature and regional hydro-meteorological services. While preparing the data, the following should be kept in mind:

- average surface temperature (in degrees Centigrade)
- average surface wind speed (in m/sec)
- average surface wind speed re-iteration (should be defined by 8-directions, with data being represented without dimension or as percentage).

For the wind directions considered in the Inter-CLAIR system, it is assumed that the first direction is the "northern wind" from the South to the North. Chemical pollutants and the products of their chemical reactions are transported for several hundred kilometers by winds. This usually occurs in the so-called "boundary" or "mixing" layer - the layer of atmosphere extending from the earth's surface to a height of about 1000 m. Pollutants are transported according to the model up to 400 km for several dozen hours. During this process they interact with the earth's surface (dry depositions) and are washed out. Therefore it is important to have information available on precipitations and the boundary layer.

City/region Pollutant Data. Pollutant data consist of the names and aggressiveness coefficients of pollutants for which calculations will be made. In the city data entry forms, values of maximum occasional permissible concentrations (MOPC) are also included. The maximum number of the selected pollutants for Inter-CLAIR run is currently seven. The Inter-CLAIR system allows the detailed study of sulfur and nitrogen compounds in the atmosphere. For the other four pollutants, one can define only emission and economic damage estimations if they are included in the simulation.

Emission Source Parameters. Chimney emission parameters are directly related to the technological processes, in particular industries. The necessary information can be gathered in a number of ways, some of which are simpler and more cost-efficient than others. For example, in one case the chimney geometry might be calculated using its shadow. The estimation of emission parameters, however, is somewhat more complicated. There are two ways of finding a solution - through relatively simple measurements, or using engineering methods based on the technologies used.

Technological and Economic Characteristics of Atmospheric Protection Activities
If the set of sources and the corresponding set of parameters for a city/region can be identified, these data are then used in the Inter-CLAIR system to estimate atmospheric air pollution. Thus data prepared according to the specifications of the section "Emission Sources Data" describe the regional air pollution situation. Any new technological, economic, or geographical innovation in the city/region leads to changes in some emission source parameters. If such changes are defined quantitatively and are technically feasible, it is considered by the Inter-CLAIR system to be a "technological measure". *"Quantitatively"*

means that the expenditure estimations for the measure's implementation as well as new emission source characteristics are known. *"Technically feasible"* means that the possibilities for a measure's implementation for the given sources can be determined only by experts, and only for measures defined quantitatively. The general scheme for defining measures in the Inter-CLAIR system allows for analyzing any such measure effectiveness, regardless of the number of pollution sources and/or pollutants, and ecological consequences of sources dislocation and/or liquidation. It provides a universal technique for analyzing new measures in regional or city industries. The following data are necessary for each measure:

- investments and annual costs
- numbers of the sources whose characteristics changed (modified sources)
- characteristic variations by each source to be modified and defined.

The Inter-CLAIR system provides possibilities to simulate any number of situations in a city/region while undertaking a variety of possible technological atmospheric protection measures. The price paid for achieving these benefits is the necessity to arrange the complete set of possible measures into groups to simplify the interpretation of calculation results.

Models Included in Inter-CLAIR

There are many mathematical, physical, chemical and economical models developed and integrated into the Inter-CLAIR system.

City Pollution Estimates. The state of the city air basin cannot be estimated properly on the basis of a single criterion. Therefore the Inter-CLAIR system makes use of the set of relatively independent estimations. This means that different models are used to get the set of estimates. Emissions' distribution, economic damage and maximal pollutant concentrations fields are calculated during this computational stage. The calculation runs should be performed only after all data preparation has been carefully accomplished. Several means of validating data are incorporated into the system.

Choosing Efficient Strategies. Several hundred, even a thousand, pollutant emission sources are present in the territory of a real city/region. Five to fifteen air protection measures can be implemented at each source. Thus, a multitude of alternative measures can be foreseen for protecting the air basin from pollution. Special optimization algorithms are required for a comparative analysis of these variants. Optimal strategies can be selected on the basis of an algorithm that includes the following four stages.

Stage 1: Simulating a bank of initial measures.
Stage 2: Forming the permissible sets of measures
Stage 3: Optimizing the distribution of costs of air protection measures
Stage 4: Multi-criteria analysis of optimal measures' sets.

Concentration Indices. Concentration indices are formed on the basis of calculated concentrations, obtained for every cell of a regular grid. These indices are the sum of calculated concentrations that exceed their MOPC. Thus, the concentration indices are only calculated for the cells, in which calculated concentrations exceed MOPC. Surface concentrations, averaged at 20-30 minutes, are determined in accordance with the modified equation of atmospheric diffusion for the so-called normal unfavorable weather conditions occurring during an unbalanced (unstable) stratification of the atmosphere.

City Diffusion Model: Modified Equation of Atmospheric Diffusion. Calculation of maximal concentrations indices under normal unfavorable weather conditions may be performed using differential equation with partial derivatives of atmospheric diffusion with the boundary conditions.

City Regular Grid. The city, or part of it (district), can be represented as a rectangular part of a rectangular grid covering the city. The origin (0,0) of the coordinate system is located in the lower left-hand corner of the grid. Three parameters characterize the regular grid:

- number of grid cells along the X-axis
- number of grid cells along the Y-axis
- size of grid cells.

The steps of the grid (dimensions of the grid's cells) may be chosen arbitrary, but they equal 0.5 x 0.5 km, or 1 x 1 km as usual. In principle, the system can support calculations with 50 x 50 m grids for very small towns. The regular grid size must not exceed 35 x 35 (1225 cells in total).

Transport and Transformation of Sulfur Dioxide in the Atmosphere. There is no use in determining sulfur compounds concentrations averaged at short periods in meso-scale areas, while treating the territory as a large-step regular grid, because these concentrations exceed MPC (maximal permissible concentration) very rare. Therefore, annual concentrations of sulfur compounds, as well as dry and wet sulfur dioxide and sulfates depositions are calculated for meso-scale regions.

Regional Pollution Estimates. The state of the region atmosphere basin cannot be estimated properly on the basis of a single criterion. Therefore the Inter-CLAIR system makes use of the set of relatively independent estimations, that is different models are used to get the set of estimates.

City/regional Optimal Strategies. In accordance with the multi-criteria concept of the pollution estimation in the Inter-CLAIR system after selecting the menu item "Decisions Optimization" a set of bicriteria optimization problems is solved. Each of this optimization problems treats some costs and pollution estimates as criteria. There are two basic types of costs in the Inter-CLAIR system, which are to be minimized: capital outlays; total expenses. There are up to sixteen types of criteria in the Inter-CLAIR system, which are to be minimized for the multi-criteria problem optimization:

- emissions of pollutants
- integral economic damage
- concentration indices
- integral concentration index.

Coniferous Forests Danger. An approximate model for calculating the potential damage index of coniferous forests is being applied to the simulation of conifer forest impact in the Inter-CLAIR system. The simplest "dose-effect" model, permitting the calculation of the index of potential damage by sulfur compounds deposition to the coniferous forests of a region, illustrates the expediency of using the Inter-CLAIR system to protect the atmosphere from pollution.

Economic Damage Model. Economical damage is the indicator which quantitatively determines the industrial expenditure (costs) increase due to changes in the environment. As these changes are actually the results of the industrial processes, this economic damage represents the actual quantitative feedback from the environment to the Industry. The estimation cannot be made in general and global economic terms but in concrete ones, e.g., for the level of single enterprise, a particular city, a region.

There are two methods of economic damage calculations which can be used for this purpose:

Method A. This consists to identify and describe all negative effects which can be expressed quantitatively, e.g., 40% increase in costs for health insurance bills, decreased wheat fertility by about 30%, etc. This damage is called "direct damage". Knowing the average prices of products' losses this damage can be presented in economic terms (using local national currency). The method can be considered attractive from the theoretical point of view and can bring necessary estimations, but only when all effects are studied and incorporated in full detail. Unfortunately the last condition is not feasible if one takes into consideration that each negative effect should be estimated for all pollutants and all particular products as recipients.

Method B. This consists in accumulating all information and measured data for several years, and then to combine all the emissions into one. This approach provides the possibility to identify and compare emissions quantitatively on a ton per year basis. In other words it means that the negative impact of each pollutant should be expressed relatively to the one pollutant selected for this purpose.

More Details on the Use of Inter CLAIR

The experience gained during the preparation and calculation of numerous case studies, using Inter-CLAIR, has been included in a specialized "case studies" data base structured by model type (e.g., city or regional model) and the specific case study. The information inside the data base refers to a full description of the case study, output results (numerical and graphical), and includes also particular comments.

The set of environmental simulations, know-how and decision support tools integrated in the knowledge-based system Inter-CLAIR can be effectively applied to a number of complex ecological, environmental and innovation problems related to large industrial areas and energy production systems, as well as to other urban and regional problems, requiring the development of well balanced ecological solutions made under economical and technical constraints. The Inter-CLAIR system provides for the possibilities to estimate critical loads and to perform an environmental impact assessment, by taking into consideration all major existing technological options.

Summary (Chapter 2)

Atmospheric dispersion models and methods play an important role in assessing risk from normal operation, or from accidental releases. They are used to predict the concentrations of airborne pollutants at various distances and directions from the point of release. In the process of model selection and implementation of integrated regional risk assessment studies, the analyst needs guidance and assistance.

A large number of atmospheric dispersion codes were reviewed and discussed in the context of integrated regional risk assessment. The present chapter introduces various types of models used in this respect for air pollution assessment (e.g., long range atmosphere models, medium and short range atmosphere models). More complex codes, related to atmospheric dispersion in nuclear power plant siting, or accidental radionuclides' releases, were also reviewed and tested during the international bench mark experiment "SIESTA" (1987-1993). International literature on the subject has been reviewed and the reader has access to a comprehensive list of compiled reference.

Details are also given on specialized "Decision Support Systems" for environmental impact assessment and risk estimation (e.g., Inter-CLAIR specialized software).

References (Chapter 2)

[1] Yadigaroglu, G., Munera, H.A., Transport of pollutants. Summary review of physical dispersion models. In: Nuclear-Technology-USA, , 77(2) (1987) 125-149.
[2] Guide to Reference and Standard Atmosphere Models, American Institute of Aeronautics and Astronautics; American National Standards Institute, Inc., ANSI/AIAA, G-003-1990, August 6, 1990.
[3] Desiato, F., APOLLO, A Lagrangian Model for the Study of the Long Range Transport of Atmospheric Pollutants, Report No. RT/DISP/93/02 (1994). ENEA (Ente per le Nuove Technologie, L'Energia e l'Ambiente, Direzione Centrale Sicurezza Nucleare e Protezione Sanitaria), Roma.
[4] ApSimon, H.M.; Goddard, A.J.H.; Wrigley, J., Long range atmospheric dispersion of radioisotopes, Atm. Env. 19 (1985) 99.
[5] van Rheineck Leyssius, H.J.; van Jaarsveld, H.J.A.; de Leeuw, F.A.A.M., A lagrangian model for the real time simulation of atmospheric transport and dispersion of accidentally released materials. NATO/CCMS, 17th ITM on Air Pollution Modelling and its Application, Cambridge (1988).
[6] Davis, W.E.; Olsen, A.R.; Didier, T.B., Assessment of radionuclides air concentrations and deposition for the Chernobyl reactor accident, NATO/CCMS, 17th ITM on Air Pollution Modelling and its Application, Cambridge (1988).
[7] Buckner, M.R., Proceedings of the First SRL Model Validation Workshop, DP-1597 (1981).
[8] Eliassen, A., A Review of long-range transport modeling, J. Applied Meteorology, 19 (1980) 231.
[9] Wrighley, J.; ApSimon, H.M.; Goddard, J.H., meteorological data and the mesos model for the long-range transport of atmospheric pollutants, Proc. Symp. Long-Range Transport of Atmospheric Pollutants, Sophia, Bulgaria, Octobrer 1-5 (1979), pp. 281, World Meteorological Organization (1979).
[10] ApSimon, H.M.; Goddard, J.H; Wrighley, J., Estimating the possible transfrontier consequences of accidental releases: The MESOS model for long-range atmospheric dispersal, Proc. Seminar Radioactive Releases and their Dispersion in the Atmosphere Following a Hypothetical Reactor Accident, Risoe (Roskilde), Denmark, April 22-25 (1980), Vol. 2, pp. 819, Commission of the European Communities, Luxemburg (July 1980).
[11] Hanna, S.R., Review of atmospheric diffusion models for regulatory applications, Technical Note No. 177, World Meteorological Organization, Geneva (Switzerland), WMO-No. 581 (1981).
[12] McCarthy, G.E.; Taylor, P.M.; Aboba, B.D.; Subbaiah, M.V., Review of selected computational programs for atmospheric dispersion. In: Heat and mass transfer in frost and ice, packed beds, and environmental discharges. HTD-Volume 139, p. 107-118 (1990), American Society of Mechanical Engineers, 345 East 47 St., New York, NY (1990).
[13] Miller, C.W.; Little, C.A., A review of uncertainty estimates associated with models for assessing the impact of breader reactor radioactivity releases, Report ORNL-5832, Oak Ridge, Tenn., Oak Ridge National Laboratory, (1982).
[14] Regulatory Guide 1.109, Calculation of Annual Doses to Man from Routine Releases of Reactor Effluents for the Purpose of Evaluating Compliance with 10CFR Part 50, Appendix I, U.S. Nuclear Regulatory Commission, Office of Standards Development (Rev. 1, Oct. 1977).
[15] IAEA Procedural Guidelines for Integrated Health and Environmental Risk Assessment and Safety Management in Large Industrial Areas, International Atomic Energy Agency, Division of Nuclear Safety, Department of Nuclear Energy and Safety, P.O. Box 100, A-1400 Vienna, Austria, (1995).
[16] Egan, B.A., Turbulent diffusion in complex terrain. Lectures on air pollution and environmental analyses. Ed.: D. Haugen, Am. Meteorol. Soc., 45 Beacon street, Boston, MA, p.112-135.
[17] Brigg, G. A., Plume Rise Predictions, in Lectures on air pollution and environmental analyses, Workshop proceedings, Boston MA, Sept. 29 - Oct. 3, 1975, Am. Meteorol. Soc. , 45 Beacon street, Boston, MA, p59-111.
[18] McCarthy, G.E.; Taylor, P.M.; Aboba, B.D.; Subbaiah, M.V., Review of selected computational programs for atmospheric dispersion. In: Heat and mass transfer in frost and ice, packed beds, and environmental discharges. HTD-Volume 139, p. 107-118 (1990), American Society of Mechanical Engineers, 345 East 47 St., New York, NY.; RN: CONF-900619-- (1990).
[19] Egan, B.A., Turbulent diffusion in complex terrain. Lectures on air pollution and environmental analyses. Ed.: D. Haugen, Am. Meteorol. Soc., 45 Beacon street, Boston, MA, p.112-135.

[20] Briggs, G.A., Plume Rise Predictions, in Lectures on air pollution and environmental analyses, Workshop Proceedings, Boston MA, Sept. 29 - Oct. 3, 1975, Am. Meteorol. Soc. , 45 Beacon street, Boston, MA, pp. 59-111.

[21] Yadigaroglu, G., Munera, H.A., Transport of pollutants. Summary review of physical dispersion models. In: Nuclear-Technology-USA, May 1987, Vol. 77(2) p. 125-149.

[22] Herrnberger, V. R. D.; Doria, P., Project RADAU: Atmospheric Dispersion Model Performance Evaluation and Selection; Part 2, Appendices, PSI/TM-23-93-01/Part 2 (January 1993).

[23] Gassmann, F.; Bürki, D., Experimental investigation of atmospheric dispersion over the Swiss plain - Experiment "SIESTA". Boundary-Layer Meteorology 41 (1987) 295-307.

[24] Martens, R. et al., Bestandsaufnahme und Bewertung der derzeit genutzten Atmosphärischen Ausbreitungsmodelle. GRS-1-1300, D-5000 Köln, 1987.

[25] ENEA (Ente per le Nuove Technologie, L'Energia e l'Ambiente, Direzione Centrale Sicurezza Nucleare e Protezione Sanitaria), Roma.

[26] Lange, R., ADPIC - A three-dimensional particle-in-cell model for the dispersal of atmospheric pollutants and its comparison to regional tracer studies. J. Appl. Meteor., 17 (1978) 320.

[27] Sherman, C.A., A mass-consistent model for wind fields over complex-terrain. J. Appl. Meteor. 17 (1978) 312-319.

[28] Lange, R.A., A comparison of the Monte Carlo and the flux gradient method for atmospheric diffusion. Proceedings of the 18th NATO/CCMS Int. Technical Meeting on Air Pollution Modelling and its Application, Vancouver, Canada, 1990.

[29] Desiato, F., ARCO - A Particle Model for the Study of the Atmospheric Dispersion under Complex Terrain Conditions, Report No. RT/DISP/93/03 (1994). ENEA (Ente per le Nuove Technologie, L'Energia e l'Ambiente, Direzione Centrale Sicurezza Nucleare e Protezione Sanitaria), Roma.

[30] Taibi, S.; Castiglia, F., Some considerations on generalized Gaussian plume model. Report No. CONF-9109311, 46th ATI (Italian Thermotechnics Association) National Congress, Cassino-Gaeta, Italy. 25-27 Sept. 1991. Vol. 1. pp. 57-71, Session 1b.

[31] Chino, M., Manual of a suite of computer codes, EXPRESS (EXact PREparadness Supporting System). JAERI-M 92-082, Tokai-mura, Japan, June 1992.

[32] Flassak, T.; Moussiopulos, N., CONDOR3, Calculation of non-divergent flowfields over rough terrain, level 3.0, Version PSI_3.37 (1990). Compiled by: Kaufmann, P.; Gallus, M., CH-5232 Villigen PSI, Swizerland.

[33] Gidhagen, L., Indic AIRVIRO: Eulerian advection-diffusion grid model. Indic AB, S-600 45, Norrköping, August 1992.

[34] Janicke, L., Ausbreitungsmodell LASAT. Handbuch Version 2.17, D-77770 Ueberlingen, Juli 1992.

[35] Massmeyer, K., et al., Bereitstellung eines Programmsystems zur Realitätsnahen Simulation der Ausbreitung und Ablagerung luftgetragener Schadstoffe MOSES. (Fachband 5), GRS-A-1632/V, GRS, D-5000 Köln, Januar 1991.

[36] Mellor, G.L.; Yamada, T., A hierarchy of turbulence closure models for planetary boundary layers. Journal of Atmospheric Science, 31 (1974) 1791-1805.

[37] Perdriel, S., Note de principe du code HERMES. EdF, F-78400 Chatou, Réf: HE-33/90.04, Février 1990.

[38] Perdriel, S., Notice d'utilisation du code HERMES. EdF, F-78400 Chatou, Réf: HE-33/91.01, Janvier 1991.

[39] Rodriguez, D.J., et al., User's guide to the MATHEW / ADPIC models. UCRL-MA-103581, Livermore, California 94550, April 1992.

[40] Thykier-Nielsen, S.; Mikkelsen, T., RIMPUFF users guide. Version 30, Risoe National Laboratory, Danmark, December 1991.

[41] Störfallverordnung, StFV, 1. April 1991.

[42] Handbuch I + II zur Störfallverordnung StFV, BUWAL (Bundesamt für Umwelt, Wald und Landwirtschaft), Februar 1992.

[43] Culkowski, W.M., Patterson, M.R., ATM. Atmospheric transport & diffusion model, (NOAA Atmospheric Turbulence and Diffusion Laboratory, Oak Ridge, TN (U.S.A.), April 1976.

[44] Seigneur, C., Understand the basics of air-quality modeling. Chemical-Engineering-Progress-United-States, 88(3) (1992) 68-74.

[45] Addis, R.P.; O'Steen, B.L., 2DPUF - A sequential Gaussian puff model, Report No. WSRC-RP--90-1208 (1990). US Govt. Printing Office Dep.

[46] Wells, R.; Palma, T.; Kowalesky, K., A screening model for estimating pollutant impacts in rooftop turbulence zones. In: B.W. Gay, Jr. (ed.). EPA/AWMA international symposium on measurement of toxic and related air pollutants, 1991. pp. 177.
[47] Taibi, S., Gaussian plume model: Sensitivity analysis. In: 44th ATI (Italian Thermotechnics Association) national congress, Cosenza (Italy). 12-15 Sept 1989. Volume 2. p. 63-75, Session 3a. Report No. CONF-8909251-- (1989).
[48] Williams, M.; Yamada, T.A microcomputer-based forecasting model: Potential applications for emergency response plans and air quality studies. in: Journal of the Air and Waste Management Association (USA), 9 (1990) 1266-1274.
[49] Chitgopekar, N.P.; Reible, D.D.; Thibodeaux, L.J., Modeling short range air dispersion from area sources of non-buoyant toxics. In: Journal of the Air and Waste Management Association (USA), 8 (1990) 1121-1128.
[50] Overcamp, T.J., Diffusion models for transient releases. Journal of Applied Meteorology USA, (Dec 1990) 1307-1312.
[51] CTDMPlus code for Complex Terrain, Report No.: PB--90-243809. Report No. PB--90-504119/XAB (June 1990).
[52] Offshore and coastal dispersion (OCD) model. Version 3.0 (revised 1988) for microcomputers. 23 Aug 1990. Report No. PB--91-505230/XAB.
[53] Yamada, T.; Bunker, S., Three-dimensional forecasting models on a desk-top computer, Report No. LA-UR--90-1386, CONF-9009134--1 (1990).
[54] Winges, K.D., Fugitive Dust Model (FDM) user's guide (revised). Volume 1. User's instructions. Final report. Report No. PB--90-215203/XAB. TRC Environmental Consultants, Inc., Mountain lake Terrace, WA (USA). May 1990.
[55] Zannetti, P., Simulating short-term, short-range air quality dispersion phenomena. In: D.N. Cheremisinoff, (eds.), Encyclopedia of environmental control technology. Gulf Publishing Company. Houston, TX (USA), 1989.
[56] Petersen, B., PAL: Point, area and line-source algorithm. EPA/ORD (Environmental Protection Agency/-Office of Research and Development) air-quality simulation model (for microcomputers). Model-Simulation. 1 Aug 1989. Report No. PB--90-500844/XAB.
[57] Petersen, B., PAL: Point, area and line-source algorithm. EPA/ORD (Environmental Protection Agency/-Office of Research and Development) air-quality simulation model (for microcomputers). Model-Simulation. 1 Aug 1989. Report No. PB--90-500844/XAB.
[58] Petersen, B., PEM: Pollution Episodic Model. EPA/ORD (Environmental Protection Agency/Office of Research and Development) air-quality simulation model. Report No. PB--90-500760/XAB, 1 August 1989.
[59] Bice, B.; Yamada, T.; Bunker, S.; Williams, M.A., Desk-top computer forecast for emergency response planning. In: J.C. Allred, R.C. Eckhardt and A.S. Nichols (eds), Proceedings of the Conference on Technology-Based Confidence Building: Energy and Environment, pp. 493-497, Nov. 1989. Rep No. CONF-8907103--, LA--11728-C.
[60] Tesche, T.W.; Kapahi, R.; Honrath, R.; Dietrich,W., Improved dry deposition estimates for air toxics risk assessment. In: Proc. Annual Meeting Air Pollution Control Association - USA, 5 (1987) 1-23. (Report No. CONF-870695--).
[61] Mersch, J., CDM2: Climatological Dispersion Model (Version 2.0). EPA/ORD (Environmental Protection Agency / Office of Research and Development) air-quality simulation model (for microcomputers), 1 Aug 1989. Report No. PB--90-500406/XAB.
[62] Desiasto, F.; Inghilesi, R., AMETISTA , A Model for the Evaluation of Long-term Atmospheric pollution Caused by Stack Emissions, Report No. RT/DISP/93/07 (1994). ENEA (Ente per le Nuove Technologie, L'Energia e l'Ambiente, Direzione Centrale Sicurezza Nucleare e Protezione Sanitaria), Roma.
[63] Hanna, S.R.; Paine, R.J., Hybrid plume dispersion model (HPDM) development and evaluation, J. Appl. Meteor., 28 (1989) 206-224.
[64] Mersch, J., PTPLU air-quality simulation model. EPA/ORD (Environmental Protection Agency / Office of Research and Development), Report No. PB--90-500331/XAB, 1 Aug 1989. The model is contained on 5 1/4-inch, double density (360K) diskettes, compatible with the IBM PC microcomputer, ready for uploading to an IBM 3090 computer.

[65] Bigalke, K.; Schatzmann, M., Berechnung der Anfangsbedingungen für den Stofftransport in mesoskaligen Zirkulationsmodellen. GKSS-Forschungszentrum Geesthacht G.m.b.H., Geesthacht-Tesperhude (Germany, F.R.). Report No. GKSS--89/E/24 (1989).

[66] Dop, H.; de Haan, B.J.; Engeldal, C., The KNMI mesoscale air pollution model (1982). Scientific Report W. R. 82-6, KNMI, De Bilt (NL).

[67] Horst, T.W., A surface depletion model for deposition from a Gaussian plume. Atm. Env. 11 (1977) 41-46.

[68] Schlünzen, H., Das Mesoskalige Transport und Strömungsmodel MERAS. - Grundlagen, Validierung, Anwendung . Hamburger Geophysik. Einzelschriften, Reihe A, Heft 88 (1988), Hamburg.

[69] Päsler-Saur, J., Comparative calculation and validation studies with atmospheric dispersion models. Bericht No. 4164 (1986), KfK, Karlsruhe.

[70] Sehmel, G.A.; Hodgson, W.A., A model for predicting dry deposition of particlea and gases to environmental surfaces. PNL-SA-6721 (1978), Batelle, Richland WA.

[71] CLAIR is Trademark of GOLEM GmbH, Vienna, Austria.

[72] Gheorghe A.V.; Litwin, V.; Golovanov, S., Inter-CLAIR, Environmental decision support system for air quality risk assessment and innovation investment management in large industrial complexes and energy production systems. In: G. Beroggi et al.,(eds.), Entscheidungsinstrumente im Risikomanagement - Möglichkeiten und Grenzen von neuen Methoden und Technologien zur Unterstüzung des risikomanagements. VDF Hochschulverlag AG an der ETHZ, 1995.

CHAPTER 3

MODELS AND RISK ASSESSMENT METHODS FOR THE HYDROLOGICAL CYCLE

3.1. General Information on Model Types

State Governments and the general public have become increasingly concerned about the health risks associated with emissions of toxic pollutants discharged into the environment. Waste substances may enter the hydrological cycle either by direct release to surface and groundwater, or indirectly through contacts with the atmosphere and the soil. Examples of indirect processes are the wet and dry deposition already discussed in the precedent chapter. After being released to the hydrosphere, substances may undergo mixing and dispersion.

A given water system is composed of a number of elements with physical characteristics that can be described mathematically. The concentration of a particular substance depends physically on the advection and dispersion characteristics of the water body and the various physical, chemical, radiological, or biological transformations it undergoes. A water system can be described by a set of partial differential equations with variable coefficients. Each term of this set of equations represents a specific aspect of a particular phenomenon. The relationship between the flux of mass in a water body and the source sinks of mass can be expressed through the continuity equation:

$$\underset{(1)}{\frac{\partial C}{\partial t}} + \underset{(2)}{\frac{\partial}{\partial x}(V_x C)} + \underset{(3)}{\frac{\partial}{\partial y}(V_y C)} + \underset{(4)}{\frac{\partial}{\partial z}(V_z C)}$$

$$= \underset{(5)}{\frac{\partial}{\partial x}\left(\varepsilon_x \frac{\partial C}{\partial x}\right)} + \underset{(6)}{\frac{\partial}{\partial y}\left(\varepsilon_y \frac{\partial C}{\partial y}\right)} + \underset{(7)}{\frac{\partial}{\partial z}\left(\varepsilon_z \frac{\partial C}{\partial z}\right)} + \underset{(8)}{\sum_{i=1}^{n} S_i} \qquad (3.1)$$

where

C	= concentration of substance
t	= time
x, y, z	= Cartesian coordinates, respectively, longitudinal, lateral, and vertical directions
V_x, V_y, V_z	= velocities in the x, y, and z directions
$\varepsilon_x, \varepsilon_y, \varepsilon_z$	= dispersion coefficients in the x, y, and z directions
S_i	= sinks and/or sources

Term (1) is the dynamic term that represents the change in concentration with respect to time. Terms (2), (3), and (4) are the advection terms in the three Cartesian coordinate directions. The product V_xC describes the flux in the longitudinal direction resulting from advection or horizontal movement of the fluid containing the mass of contaminant. The terms (5), (6), and (7) represent the flux resulting from dispersion, which is assumed to be proportional to the gradient concentration in the direction of decreasing concentration. The advection and dispersion terms $\left(\varepsilon_x \partial C/\partial_x - V_xC\right)$ together represent the total flux which defines the material moving in and out of an elemental volume. Term (8) accounts for sources and sinks of a substance. In principle, Eq. (3.1) must be solved simultaneously for all phases. This includes the dissolved phases where the substance is sorbed to either sediments or biota in the water column or to sediment in the bed. Since Eq. (3.1) cannot be solved analytically, numerical methods are usually employed. However in certain cases various simplifying assumptions can lead to a form of Eq. (3.1) that can be solved analytically.

The main objective of *surface water transport models* is to evaluate the concentration of a substance in water as a function of time and position. The aquatic environment can be divided into a number of media submitted to different sub-regimes, each requiring a different kind of model:

Surface Waters

- seas, lakes and reservoirs
- estuaries
- rivers and canals
- surface runoff of rain

Subsurface Waters

- stationary waters
- flowing waters

Three *phases of dispersion* are recognizable in surface waters:

■ The first phase is the *initial mixing of the waste with the water body.* This can be simulated by near-field models. In some cases, near-field modeling is part of a more complex thermo-hydraulic model of the release process. For discharges into rivers, the initial mixing takes place in the first hundreds of meters where a significant density difference exists.

■ The second phase covers the *process between initial discharge and complete mixing.* Depending on the receiving water body, this phase may be associated either with time or distance needed to complete the process. In lakes for instance, the time for complete mixing may be months or years. For discharges into rivers, complete mixing may be quickly attained after a few thousand meters or it may be slow and require up to 10 km.

■ The last phase covers the *long-term transport of substances after complete mixing* with the receiving water body. The time scale involved depends on both the water body and the waste substance. Transport during phase 2 and 3 is represented by far-field models.

All three phases of mixing are relevant for risk analysis. Modeling of *phase 1* is important for the study of occupational risks and for specific local risk reduction measures. Analysis of risk to the general population requires modeling of *phase 2 and 3*, with emphasis on the latter.

The types of water systems that may be located near a power plant, a chemical plant, or waste repository include streams, rivers, estuaries, open coasts, impoundment, reservoirs, and lakes. It should be noted that there may be large variations in characteristics among specific water bodies within these general classes. Some models depending on their formulation, may be appropriate for describing more than one type of water system.

Review of models have been published by Onishi[1] et al., (1981), EPA[2] (1979), Grismud[3] et al. , (1976), and Orlob[4] (1977). A summary of literature reviews has been presented by Bolten[5] et al., (1983)

Three general classes of models have been designed for various purposes (Onishi[1] et al., 1981). The first class, the *water quality model,* focuses on the interaction of waste substances and life forms present in water systems. The second class, the *radionuclides model*, was designed to track releases of radioactive elements from nuclear power plants. The third class, the *sediment model*, addresses movement of sediment through water systems. Most models that calculate the only the sediment transport were designed to examine the deposition and scouring characteristics of rivers and estuaries. Models designed to deal with *sediment-only transport* are of two types. The first used primarily for suspended sediments, is deterministic and uses the advection diffusion equation. The second, mainly for bed load sediment, is stochastic and uses distribution functions. Such type of model does not always include a description of contaminant transport. Although radionuclides' transport models specifically address radioactive substances, most will be appropriate for modeling non-radioactive substances without change as well.

Water quality and radionuclides' transport models can be divided into a number of general types. Radionuclides' transport models generally contain an additional term describing the first order decay of the radioactive elements.

The model described below Eq. (3.2) is the general advection-diffusion equation with a sink/source term. It differs from Eq.(3.1) in that the longitudinal, lateral, and vertical velocities are constant.

$$\frac{\partial C}{\partial t}+V_x\frac{\partial C}{\partial x}+V_y\frac{\partial C}{\partial y}+V_z\frac{\partial C}{\partial z} = \frac{\partial}{\partial x}\left(\varepsilon_x\frac{\partial C}{\partial x}\right)+\frac{\partial}{\partial y}\left(\varepsilon_y\frac{\partial C}{\partial y}\right)+\frac{\partial}{\partial z}\left(\varepsilon_z\frac{\partial C}{\partial z}\right)+\sum_{i=1}^{n}S_i \tag{3.2}$$

where the variable are defined as in Eq. (3.1). This model type is appropriate for conservative and non-conservative substances.

A variation of this model type without a sink/source term would handle only conservative substances.

$$\frac{\partial C}{\partial t}+V_x\frac{\partial C}{\partial x}+V_y\frac{\partial C}{\partial y}+V_z\frac{\partial C}{\partial z}=\sum_{i=1}^{n}S_i \tag{3.3}$$

This model (3.3) is the advection equation and does not contain a dispersion term. It is therefore appropriate only fast-moving streams or rivers.

Again if the sink/source term is present, Eq. (3.3) can treat conservative and non-conservative substances.

$$\frac{\partial C}{\partial t}=\frac{\partial}{\partial x}\left(\varepsilon_x\frac{\partial C}{\partial x}\right)+\frac{\partial}{\partial y}\left(\varepsilon_y\frac{\partial C}{\partial y}\right)+\frac{\partial}{\partial z}\left(\varepsilon_z\frac{\partial C}{\partial z}\right)+\sum_{i=1}^{n}S_i \tag{3.4}$$

This model (3.4) is the dispersion equation, and does not contain an advective term. It is therefore appropriate for describing water bodies, such as some lakes and ponds, where there is little water movement.

The next model (3.5) does not contain coordinates, and is assuming complete mixing. It is appropriate for use only in modeling some lakes where concentration of a substance changes with time but not with location.

$$\frac{\partial C}{\partial t}=\sum_{i=1}^{n}S_i \tag{3.5}$$

Although Eq. (3.6) is of the same general form as Eq.(3.5), it uses Lagrangian routing method, in which the longitudinal coordinate system is assumed to move with the flow velocity.

$$\frac{dC}{dt}=\sum_{i=1}^{n}S_i \tag{3.6}$$

If we let $x = x_0 + V_x t$; $y_0 = y_0 + V_y t$; and $z_0 = z_0 + V_z t$, then we obtain:

$$\frac{dC}{dt}=\frac{\partial C}{\partial t}+V_x\frac{\partial C}{\partial x}+V_y\frac{\partial C}{\partial y}+V_z\frac{\partial C}{\partial z} \tag{3.7}$$

The Lagrangian technique simply involves integrating this equation. Solutions are time-dependent and one-dimensional and are based on the travel time from one location to the next. This model represented by Eq. (3.7) can be used for uniform, nontidal rivers.

Finally, the last type of models describe particle migration statistically, rather than through mathematical equations. Advection, dispersion, and sink/source are represented with Monte Carlo techniques. The latter model can be applied as an alternative to full advection-diffusion equation (Eq. 3.1).

There are a number of simpler model types that are less detailed than those described above. Some models include only a one-dimensional dispersion term and a sink/source term. Such type of model can be appropriate for some streams. Simple rivers can be described using a one-dimensional advection and dispersion term together with a sink/source term. Simplified models of some estuaries contain a one-dimensional term. Most of these simpler model types provide a steady state solution to substance transport.

In addition to the complex general simulation model, which contains mass balance and energy conservation equations, with terms representing convection of the substance by water movements, diffusion, radioactive decay, and all sedimentation processes, and other source/sink terms (including all releases of the substance to the water body), there are three groups of simpler models:

- Analytical models,
- Box models (known as compartment models), and
- Phenomenological models (used to represent specific processes)

Parameters for these models are commonly obtained by direct measurement, e.g., by tracers.

Except for the simplest of water bodies, modeling water quality is sufficiently complex that it must be computerized. Many models are available as general-purpose computer software packages that can be configured by users for specific bodies of water. A few simple screening-type models are available that can estimate maximum allowable loading of important conservative and non-conservative pollutants, but can not estimate concentrations as a function of source strength. These are described below.

Surface Water Models. Models of surface water contamination are either steady state or time-dependent. They vary in complexity, containing two or three dimensions, with or without convection, and with or without sinks. The simplest models are no more than solutions to simple equations that use mixing ratios (perhaps time dependent) and some removal constants. These are usually sufficient only for routine effluents, and can lead to gross estimation errors even in simple cases. TABLE 3.1. summarizes examples of available models used to assess exposure to radio-nuclides listed from simple to complex.

Relatively simple, straightforward models are available for estimating concentrations in rivers and streams. More complex models are needed for lakes, reservoirs, and estuaries, because they are readily stratified and large enough to support complex patterns of flow.

Subsurface Models. They are simple in concept, but complicated in execution because of the potential complexity of the subsurface structures.

Transformation Processes

A proper description of the physical characteristics of the selected water body is of the utmost importance in the choice of a surface water model. Further the physical and chemical behavior of the particular pollutant being modeled must be considered and the model must be general enough to deal with a number of different chemical classes (e.g., inorganic metals and pesticides). In water models that contain no description of transformation processes, the downstream concentration of the original chemical pollutant will be overestimated, whereas the concentration of the transformation product will be underestimated. *This provides a conservative value for dose-response models, except when the conversion product is more toxic than the original species.*

TABLE 3.1. Examples of the range of models used to evaluate changes in concentration of radio-nuclides in aquatic environments

Simple Aquatic Models			
Model Type	**Kind**	**Application**	**Assumptions**
Box model	Lakes Impoundment	Used to predict Sr^{90} in Great Lakes	Completely mixed above and below thermocline
Time dependent Box models	Lakes Rivers	Used to predict Pu in great lakes	Completely mixed with sediment interaction
		Used to describe pesticides in rivers (Cs^{137} on Clinch river)	Assumes river can be divided into a series of well-mixed compartments, includes with suspended sediments
Simple flow models	Rivers	Applied to H^3 releases into Savannah River	Calculate dilution factors and radioactive decay
		Behavior of Cs^{137}	Estimates fraction absorbed to sediments
Two-dimensional flow models	Rivers	Applied to Missouri river water quality	Mixing with vertical variation in concentration and velocity averaged

TABLE 3.1 (continued)

Complicated Flow Models			
Model Type	**Kind**	**Application**	**Assumptions**
One-dimensional models	Rivers, Estuaries	Diffusion is not considered, since the cross-sectional area of the river is considered constant	Steady, one-dimensional convection with decay and source/sink terms
		Uses KD to simulate first-order exchange between sediment and water	Simulates transport of trace contaminants in dissolved form and on particles
		Couples water transport using three sub-models. Has been used to simulate the behavior of Cs^{137}, Sr^{90} in Clinch River	Combines Wisconsin hydrologic transport model with a sediment transport model

TABLE 3.1 (continued)

Complicated Flow Models			
Model Type	**Kind**	**Application**	**Assumptions**
Two Dimensional		Applied to canyons in Los Alamos National Laboratory to simulate migration of Pu^{239}	Simulates transport of sediment and Dimensional contaminants with interactions
		Handles linear or quadratic transport and interaction of velocity and depth with sediment approximations disruption so as to be compatible with other hydrodynamic models.	Finite element model including sediment of contaminant
		Has been used to predict migration of Pu^{239} (Kepone)	Includes advection and dispersion, longitudinal and lateral wave motion to re-suspend sediments and sediment cohesion, sediment deposition and re-suspension, sediment sources, and mixing

TABLE 3.1 (continued)

Complicated Flow Models			
Model Type	**Kind**	**Application**	**Assumptions**
Three dimensional		Applicable to cases of non-steady state flow particularly in estuaries; used to estimate bacteria distribution distribution in the NY Bay.	Numerical model that computes a velocity field from vertically integrated two-dimensional equation of mass and momentum conservation which then becomes advective mechanism.
		Has been used to predict migration of Cs^{137} in the Hudson River Estuary with K_D changing with salinity	Finite difference computing unsteady distribution of flow, water temperature salinity, sediment, dissolved contaminants and particulate contaminants

Photolysis. If a substance absorbs solar light it can be photochemically transformed into another species. The photolysis rate of a particular species depends on a number of factors including the molar extinction coefficient, the intensity of light, the quantum yield, and the concentration of the chemical species. Direct photolysis of a chemical can occur through absorption of solar light at wavelengths above 290 nm. Indirect photolysis can occur through photosensitized reactions with organics in natural water systems. The presence of other materials may retard (e.g., particulate matter) or accelerate the photolysis rate.

Oxidation/Reduction Processes. Redox processes can be important in some water systems. These reactions may control the valence state of a chemical species. Metals, such as iron and sodium, are reducing agents; nonmetallic elements, such as oxygen, fluorine, and chlorine, are oxidizing agents. Oxidation is defined as a loss of electron, or a gain of in positive valence; reduction is defined as a gain of electron, or a loss in positive valence.

Hydrolysis. It is a form of complexation where the OH^- ligand replaces a leaving group, giving greater stability to a chemical species. The reaction can be catalyzed by an acid or a base, or both. Hydrolysis reactions of metallic cationic (positively charged) radionuclides are common. Ligands other than OH^-, such as SO_4^{-2}, HCO_3^-, CO_3^{-2}, and Cl^-, which are present in sea water, can also form complexes with some chemical species. Although complex formation generally increases solubility, there is some evidence that complexing may enhance adsorption to sediments.

Volatilization. Volatilization of a chemical species in water system occurs at the air-water interface. It should be mentioned here that some chemicals, notably the chlorinated hydrocarbons, are extremely volatile.

Biotransformation. The presence of microbes in water systems can cause biotransformation of chemical species.

Chemical Sorption. Chemicals in water systems can be sorbed onto sediments suspended in the water or sediments in the bed. Sediments act as a sink for sorbed material, removing it from the water column. The material adsorbed remains a dormant source of pollution, since it can be desorbed from the sediment at a later time. Ion exchange is an important process that contributes to sorption of chemicals. This phenomenon is the chemical interaction between a charged solid matrix and an oppositely charged solution species. Cationic (positively charged) species are good candidates for sorption, since the surface of most geologic materials and soils have an inherent negative charge.

There is a wide variation in partition coefficients for different substances. In water models that contain a description of sediment sorption, partition coefficients should be chosen judiciously.

> ***A Word of Caution***
> *Unlike atmospheric dispersion models, many water quality models are not readily adjustable to conditions different from those for which they were designed. They tend to be highly site-specific. Thus, accommodating site-specific conditions may require gross revisions of existing models or use of models specifically designed to be general-purpose and easily configured by users.*
>
> *It should be stressed here that model verification is an extremely important consideration. One method involves the use of tracer dye to measure the fraction of sediments sorbed. Another method is laboratory testing of the rate constants for the different transformation processes for organic chemicals. A third method of verification involves doing an actual field experiment.*
>
> *Frequently, before a model is used widely, it is tested against laboratory measurements or in an existing water body with one or more contaminants. The model results are then compared with measurements, and the model parameters may be adjusted sot hat better agreement is obtained. Since the fate of pollutants in water systems is so complex, this empirical approach may be the only method for addressing water pollution problems.*

3.1.1. RIVER MODELS

Frequently, there is a continuous routine release of substances to a river. After some time, such discharges lead to uniform concentrations at any river cross section located sufficiently downstream from the point of discharge. Uniform mixing may be attained in a shorter distance when the discharge method provides for some initial mixing, when, for instance, the waste is released as a vertical line source.

Continuous and long-lasting discharges into a river may also result from accidents at nuclear or chemical plants. In such cases, the most likely path from the plant site to the river is via groundwater. There is therefore an exchange area between surface and groundwaters, usually modeled as a relatively ill-defined area source in the bottom of the river.

> ***Model for Assessing Microbial Pollution in Rivers***
> *Modeling the microbial pollution in streams is done by using an indicator organism such as coliform. A simple model used is:*
>
> $$N = N_0 \cdot e^{-k \cdot t}$$
>
> *where*
>
> N = *represents the number of coliforms per 100 ml;*
> N_0 = *initial number of coliforms per 100 ml;*
> k = *off rate constant per day (k = 1.0 ... 1.8, in medium sized ones at 10 °C);*
> t = *times [days]*
>
> *The correction equation for a given temperature is:*
>
> $$k = k_{20}\, 1.075^{(T-20)}$$

Steady Discharge Into a Plug Flow Channel

Steady discharge into a plug flow channel with constant linear velocity u leads to the analytical equation:

$$C = \frac{qC_0}{Q} e^{\left(-\frac{Kx}{u}\right)}$$

where

q	=	*discharge volumetric flow rate*
C_0	=	*concentration of waste in the discharge (mass of pollutants or activity of the radioactive material per unit volume*
Q	=	*river volumetric flow rate at the discharge point*
x	=	*distance downstream from the release point.*

The removal constant K (in inverse time units) should include all mechanism relevant for the substance under consideration. For example, sedimentation and radioactive decay contributions λ_s and λ. If chemical processes leading to a decrease of the pollutant are also present, then an effective decay constant λ_e may be substituted for λ.

Several European countries have adopted the above mentioned model[6] to calculate average annual exposures produced by radioactive effluents. The *Regulatory Agency of the Federal Republic of Germany* allows the removal constant K to include radioactive decay only.[7] This assumption is conservative since sedimentation may remove a considerable fraction of the discharge waste.

Rivers are modeled as linked segments between nodes where there are important changes, such as a large discharge, a large intake, entry of a tributary, or a large change in the physical characteristics of the river. *Within segments, all conditions are assumed constant except for the flow-controlled time of transit to the next downstream node.* Non-conservative substances, such as decomposing organics and the associated oxygen uptake, pathogens, radionuclides with short half-lives, and substances with high deposition, biological accumulation, or chemical reaction rates, are estimated as a function of time while in each segment. Conservative substances accumulate between sources.

Some *river pollution problems*, for which maximum concentration is of special concern (such as heat), are *modeled as plumes* for short and possibly long distances downstream of the discharge point.

Organic and nutrient loading of rivers are particularly important, and a broad range of helpful equations and models is available to assist in determining the *self-purification* capacity of a river and the *maximum organic loading* that can be accommodated while maintaining dissolved oxygen levels at specified minimum levels.

3.1.2. LAKE MODELS

Classification of Lakes

Lakes are generally classified as *oligotrophic* (low nutrients, always oxygenated) and *eutrophic* (high nutrients, can become anoxic). Oligotrophic lakes tend to be nutrient limited and therefore do not support abundant growth of plants. *Within limits, these lakes can absorb exogenous nutrients and oxidize organic material without damage.*

> ***Definition***
> *Lake pollution is caused by the presence of a chemical substance in the lake at a concentration greater than the natural one. When this substance is an algae nutrient, lake pollution is termed eutrophication, while the general term "chemical pollution" is adopted for substances which do not enter or enter in very small quantities the metabolism of aquatic plants (e.g., heavy metals).*

Eutrophic Lakes. Eutrophic lakes have *large amounts of nutrients* which support abundant growth of algae and other aquatic organisms. Dead plants and animals sink to the lower levels of these lakes, where decomposition by microorganisms depletes or eliminates oxygen, with associated killing of oxygen-dependent species. *Eutrophic lakes cannot absorb large quantities of exogenous nutrients and organic materials without damage.*

Oligotrophic Lakes. Oligotrophic lakes are usually phosphate-controlled. Vollenweider has developed a simple equation that can be used to estimate the loading of nutrient phosphates that can be added to an oligotrophic lake without reaching a critical level. Eutrophication can be expected when phosphorus loading reaches two to three times the critical level.

> ***A Word of Caution***
> *The regime of water flows in lakes can be very complicated with some parts of the water really stagnating while other parts are flowing like rivers without mixing with the surrounding waters. For this reason, simple calculations assuming complete mixing and using material balance equations may be completely misleading. In addition the water in the lake is strongly stratified. In large and deep lakes this stratification is completely disturbed according to the season. Extreme prudence has to be exerted if pollutants or heat are considered to be discharged in to the lakes.*

Power plants often use small ponds for cooling purpose, either isolated or connected to rivers. In case of contamination, complete mixing is quickly attained. The time-dependent waste concentration is obtained from a simple mass balance. For thermally stratified lakes, the simplest approach is to assume complete mixing in each compartment. To model a small lake, two compartments are sufficient.

Great lakes, seas, and oceans share the common feature that dispersion of waste is constrained by one side only: the coastal line. A continuous release to a large lake produces a plume that moves parallel to the shore for several days, sometimes as far as 25 km from the point of discharge. The process may continue until a shift in the wind reverses the water current into the opposite direction. The cycle is then repeated. Stagnation during the process may be of a few hours' duration. Dispersion of waste within the parallel-to-shoreline plume is rather small. Reversal processes on the other side are accompanied by massive exchanges with offshore water. Such a mechanism efficiently removes pollutants from the coast. In some occasions, the reversal may be accompanied by fumigation.

Modeling Lake Pollution

The state of a lake is determined by the distribution of various abiotic chemical substances, which are dissolved in water or aggregated to suspended matter, as well as by the evolution of biologic populations living in the water body. Although various chemical elements (C, O, H, N, P, Si, Mg, Ca, and oligo-elements) are utilized by aquatic plants, phosphorus is considered as the most important ecologically. For this reason mathematical *modeling of lake eutrophication is primarily concerned with the phosphorus cycle and mass balance in the water body.* The basis of all macroscopic phenomena which take place in a lake and are relevant to the eutrophication process, is the *photosynthetic reaction*, which under the influence of solar radiation energy, transforms inorganic phosphorus, dissolved in water as the orthophosphate ion $(PO_4)^{-3}$, into a component of algal cell protoplasm.

Model for Assessing Critical Phosphorus Loadings in a Lake

Phosphorus concentration is an important parameter which characterizes the oligotrophic lakes. Critical phosphorus load above which eutrophication conditions may begin to develop is given by:

$$L_c = 10 \cdot q_s + \left[1 + \sqrt{Z / q_s}\right]$$

where

L_c = *critical phosphorus load [mg Phosphorus / m^2.year]*
q_s = *Q/A - overflow rate [m/year]*
Z = *lake mean depth [m]*
Q = *annual inflow rate [m^3/year]*
A = *lake area [m^2]*

Eutrophic conditions may be expected when the actual phosphorus load L equals 2 to 3 times L_c.

The maximum annual discharge of phosphorus B [t/year] into the lake, above which eutrophic conditions may begin to develop is:

$$B = 10^{-9} L_c A$$

Value of B has to be used along with inventory results yielding the total anticipated phosphorus input load into he lake.

Chemical substances which are adsorbed on individuals of lake biota can have toxic effects on it, and interfere with the eutrophication process. Part of the pollutants are adsorbed on settling materials, and eventually to lake sediments, from which they are remobilized only on very particular conditions. The dissolved toxic substances undergo advective and diffusive (turbulent) transport, and they are flushed by outflowing waters. In a general way a pollutant distribution model can be considered as a mathematical tool to link the pollutant concentration in lake to that in inflowing waters. Once the toxic threshold of the pollutant with respect to a particular biologic population is established, the model enables the evaluation of the highest pollutant concentration in inflowing waters which is still compatible with the life of this population in the lake waters.

The mathematical models used for ecological lake modeling are concerned with:

- Modeling the *lake as a mixed reactor:* the model consists of a single differential equation for the concentration of total phosphorus transported into the lake by inflowing water, homogeneously distributed throughout the lake water body, and lost from this through outflow and sedimentation.
- Evaluation of *water mean residence time* in a lake
- Modeling *lake sediment interaction*
- Modeling *thermal stratification* and *vertical mixing*
- Modeling *alga growth* and *biochemical interactions.*
- Modeling *eutrophication* and *eutrophic lake restoration*
- Modeling *heavy metal distribution in the lake-sediment system*, and *heavy metal distribution in the epilimnion-hypolimnion system.*

TABLE 3.2. Potentially important characteristics affecting subsurface movement of pollutants

Boundary conditions	• Distribution of hydraulic head • Recharge and discharge points • Locations and types of boundaries • Hydraulic conductivity
Material constants	• Porosity • Transmissivity • Extent of hydrogeologic units • Adsorption-desorption
Attenuation mechanisms	• Ion exchange • Chemical complexing • Nuclear decay • Ion filtration • Generation of gases • Precipitation-dissolution • Biodegradation • Chemical degradation
Molecular diffusion and hydrodynamic dispersion	• Transverse • Longitudinal • Vertical
Pollutant concentration	• Initial and background • Boundary conditions

The experience in modeling lakes as ecosystems and the mathematical models used have been described by Rossi[8] (1991), on behalf of the Commission of the European Communities, and presented in book form.

3.1.3. SUBSURFACE MODELS.

Contamination of subsurface aquifers is modeled in two phases:

- vertical transport through the unsaturated zone; and
- plume-like spreading and transport through the aquifer.

Vertical transport. Vertical transport of pollutants through the *unsaturated zone* above an aquifer is a function of the physical and chemical characteristics of the pollutant and the percolation characteristics of the soil. Many organic pollutants are relatively nonpolar and hydrophobic, so they tend to sorb into soils and migrate more slowly than polar pollutants. Inorganic chemicals can precipitate out. Some low-density organics can even float. *Soils differ greatly in their physical and chemical characteristics and their interactions with specific pollutants.*

> ***A Word of Caution***
> *These problems with data quality, plus the long time spans involved in groundwater movements, have hindered verification of models for groundwater transport. Most are not fully verified. The reliability of model results therefore depends heavily on site-specific conditions and analysts' ability to account for them adequately in the coefficients supplied to the models. Much professional judgment is required.*

Plume-like Spreading. Similarly, once in the aquifer, pollutants form *plumes* by diffusion and transport in gravity-driven water flow. Again, the rates of movement are controlled by the physical and chemical characteristics of the pollutant and the geo-hydrology of the aquifer. *Modeling movement in sand is simple; modeling movement in fractured rock or solution cavities is orders of magnitude more complex.*

Sub-Surface Movement of Pollutants. Potentially important characteristics affecting subsurface *movement of pollutants* are shown in Table 3.2. These characteristics are incorporated into models by combining a groundwater flow equation and a chemical mass transport equation. *There are separate models for unsaturated and saturated zones, but they are often linked in comprehensive computer codes.*

Groundwater flow modeling requires powerful computer models and the availability of a great number of data. Hydro-geological data may be obtained from cores taken in boring and from classical hydraulic tests (pumping, water injection, water table level measurements etc.). These tests should possibly be complemented with more specialized tests such as water sample chemical analyses, temperature measurements, tracing substances flow measurements, isotopic sampling e.g., for O^{18} deficit measurement, H^3 content etc.

It is common that the geophysical and hydrological characteristics that must be modeled vary by large amounts over short distances. And because sampling requires expensive drilling programs, data are often sparse.

Effluents from Refineries. Oil refineries placed on-shore sometimes on large industrial areas are the cause of environmental impacts due to normal or accidental conditions. The impurities in effluent water from refineries are of the following sort:

- in solution (e.g., soluble salts and organic compounds, H_2S, Ammonia)
- insoluble material (e.g., higher-molecular-weight oil fractions and suspended solids).

Certain elements of the mechanism of the *fate of oil in the marine environment* are:

- evaporation
- dissolution
- adsorption
- entry into sediments
- hydrocarbons in marine lists.

The process of data gathering and the application of associated models for this problem requires specialized information and knowledge.

There are basically two types of models used for simulating groundwater flows:

Direct Models. According to the scope of the investigation and the data available, direct models are used. In these models the physical data of the aquifer and water boundary conditions are entered while flows are calculated.

Inverse Models. Conversely, when the aquifer data is uncertain, an inverse model can be used in which the permeability (or transmissivity) of the rock structure is calculated from flow data actually measured or assumed for simulation purpose.

3.1.4. MODELS FOR ANALYSIS OF ANTHROPOGENIC COMPOUNDS IN AQUATIC SYSTEMS

A few representative examples of a large number of ecotoxicological models are mentioned below:

- *Environmental Software* (Ambrose and Barnwell[9], 1989)
- *Multi-Media Screening Model* (Cohen[10] et al., 1990)
- *Evaluation of Pesticide Groundwater Pollution* (Jury[11] et al., 1987)
- *RIVERSIM* (Reichert and Wanner[12], 1990).
- *ECOTOX* (Jorgensen[13], 1990)
- *EXAMS* (Burn[14], 1985);

▪ *EXAMS.* This code was developed to determine the dynamic behavior of organic substances based on simple, measurable physico-parameters. [15] (PC-version available).

■ *E4CHEM.* It is a package comprising various mathematical models for calculating the concentration of chemical substances in various environmental systems and the consequent risk to the system (Matthies and Trapp[16], 1988). The subprogram EXWAT contains a compartment model for the transport and transformation of substances in surface waters. This model, however, only allows equilibrium calculations. Other models are foreseen for the treatment of air (EXAIR), soil (EXSOL), organisms (ETTOX, ETSYS), and the calculation of emission rates based on user data (RLTEC).

The next two programs, MASAS[17] and CHEMSEE[18], both can be used to set up mathematical models which simulate the temporal and spatial behavior of trace substances in lakes. For that purpose the real world system must be translated into a model system. These user-friendly program were developed for use with Macintosh personal computer and written in Modula-2 (*MacMETH development system*[19]). They make use of a software library, *Dialog Machine.*[20] Other versions for MS-DOS operating system are in development:

■ *MASAS.* It is a program specifically designed to investigate the behavior of a given anthropogenic organic trace substance. It comprises 21 interconnected modules. MASAS computes the concentration of the substances of interest in the water column (and optionally, in the sediments) as a dynamic model variable. The substances are stored on file, and are characterized by physico-chemical parameters(vapor pressure, water solubility, Henry coefficient, etc.) and reactivities (hydrolysis rates etc.) Storage of the reactivities of individual species of a substance (dissolved neutral, particulate neutral, dissolved anion/cation) is also possible. The lakes are characterized in terms of morphometric (volume, isobath areas, etc.), hydraulic (rate of throughflow, depth of epilimnion, etc.), and physical and chemical parameters (temperature, pH, particle concentration etc.). MASAS support an iterative form of model development. Based on initially simple models, models of increasing complexity can be constructed, depending on the type and extent of lake and substances data available. Various types of models (one-box, two-box, combined-box, n-box) with different degrees of spatial resolution are available to describe the vertical variation in the concentration of the substance being studied. Processes can be described on different hierarchical levels. Information fields show the user which parameters are missing, and which routines are available to approximate the missing data.

■ *CHEMSEE.* It is a flexible "model construction kit" for chemical processes in lakes comprising 5 modules. Any number of variables can be defined by the user to represent the concentrations of substances and particles in the water column, and of a substance in the sediments.

Underlying both programs is the mathematical description of the behavior of a substance in a lake in terms of a one-dimensional vertical model. Such a model, in which horizontal concentration differences are neglected, is suitable for the description of a reactive substance in deep lakes. Physical processes occurring in a lake are not influenced by the presence of trace substances and are therefore represented in the model by corresponding input parameters (e.g., epilimnion depth, temperature, etc.). Transport and transformation processes (loading, adsorption, sedimentation, air-water and sediment-water exchange, chemical reactions) can be defined interactively and assigned to the various model variables. The results of simulation runs are output both graphically and on file, and the interactively defined models can be saved for future use.

3.1.5. FUGACITY MODELS

Since the Second World War there has been growing concern about contamination of the environment by man-made chemicals. Some chemicals such as pesticides may be specifically designed to alter the assembly of biota present in natural or agricultural ecosystems. They may be organic, inorganic, metallic, or radioactive in nature. There are two procedures for calculating the behavior of chemicals in the environment:

- Conventional procedures which are based on manipulations of concentrations.
- Procedures using the concepts of activity and fugacity to characterize equilibrium between environmental phases such as air, water, and soil. Organic chemicals are more susceptible to generalization, using this approach, than metals or inorganic chemicals.

The fugacity approach is very convenient in environmental chemical equilibrium or partitioning calculations. It is also a convenient quantity for describing mathematically the rates at which chemicals diffuse, or are transported between phases (e.g., volatilization of pesticides from soil to air). The transfer rate can either be expressed as being proportional to, or driven by the fugacity difference which exists between the source and destination phases. It is relatively easy to transform chemical reactions, advective flow, and non-diffusive transport rate equations into fugacity expressions, and build up sets of fugacity equations describing the complex behavior of chemicals in multi-phase, non-equilibrium environments. Mathematical procedures, specific literature citations, models and computer programs *applying the concept of fugacity*[21, 22, 23, 24] can be found in the excellent book by D. Mackay[25] (1991).

Further literature of interest can be found in *Chemical in the Environment* by Neely[26] (1980), *Environmental Exposure from Chemicals* by Neely and Blau[27] (1985), *Environmental Risk Analysis for Chemicals* by Conway[28] (1982), and in the various issues of the *Encyclopedia of Environmental Chemistry* edited by Hutzinger[29]. Further the following handbooks may be of great utility: *Handbook of Environmental Data on Organic Chemicals*[30] (1983).

Among the literature describing fate/fugacity models, we find: Southwood[31] et al., (1989); the Bioaccumulation model by Thomann[32] (1989); a multimedia compartment model by Schramm et al.,[33] (1987), which can be used to estimate the fate of lipophilic compounds in plants; a pharmakokinetic model by Paterson and Mackay[34] (1987); and a fate model describing exposure and human uptake of persistent organic chemicals by Paterson and Mackay[35] (1989).

Screening-level models are quite useful and a model for estimating *aerobic biodegradability* has been developed based on a survey of expert knowledge (Boethling and Sabijic[36] 1989).

3.2. Risk Assessment for Groundwater Pollution Control

3.2.1. INTRODUCTION

The *protection of groundwater resources* has emerged in recent years as a top priority for natural resource management around the world. The present state of social and economic development is characterized by a sharp increase in groundwater exploitation. The problem of determining the limits of groundwater exploitation is made more complex by the increasing rate of groundwater development. Two distinctive features of groundwater are:

- *its renewal*
- *the potential change in quality caused by population increase and activities.*

> *Pollution is understood as deterioration in water quality caused by groundwater exploitation and/or man's activity.*

The rate of groundwater development should satisfy a given demand of defined water quality for either an unlimited period of time (drinking water) or a planed period of time (as in mining activities). The rational limits of the rate of groundwater exploitation should be such that:

- protection against depletion is provided
- protection from pollution is provided
- negative ecological effects are reduced to a minimum
- economic efficiency of exploitation is attained.

Groundwater systems are often connected with surface waters, and the possibility for their exploitation in many cases depends on the conjunctive management of both surface and groundwater. In a model of an aquifer, all planed and existing effects on the aquifer (mine drainage, infiltration from irrigation fields, seepage from reservoirs, canals, etc.) must be included.

The exploitable groundwater resources may be supplied from:

- aquifer storage
- natural resources (recharge)
- captured resources (leakage from other aquifers and capture from surface streams and lakes as well as artificial recharge).

Human activities which influence groundwater are:

■ Activities which recharge the system and cause a rise in the water level (for instance irrigation).

■ Activities which increase the discharge from the system, and lower water levels (i.e., draining of agricultural land, draining of swamps and bogs, draining of soil and rocks for the purpose of excavating or tunneling, and the draining of open pits and underground mines.

- Activities which produce changes in groundwater quality (wastes and residuals).

The quality problems fall into four broad categories: waste management, non-point source pollution, general water quality, and use of aquifers as storage reservoirs.

Waste or residuals can be disposed to the atmosphere, streams, and other surface-water bodies or into or on the ground (problem of radioactive wastes, toxic industrial wastes). Disposal either as solid or liquid into the solid earth has associated hazards of contamination and transport by groundwater.

The Most obvious role that models play in management is as predictive tools. If, in addition to a simulation routine, a model contains an optimization procedure, it enhances the analytical capabilities used in the preparation of a decision. Such a sophisticated tool is often called a *management model.* For many field problems the initially available data (generic and field data) are insufficient for a detailed quantitative prediction. Using estimates in conjunction with a model has often resulted in a very good first impression of the system's behavior. Thus models are useful in identifying existing trends and changes in trends which would result from alternative actions even if they cannot forecast accurately the actual detailed outcomes. Once the risk of groundwater contamination is properly assessed, including its extreme and catastrophic ramifications, then its management can be appropriately pursued on the basis of detection and monitoring, prevention, abatement and containment, correction and restoration.

3.2.2. RISK ASSESSMENT METHOD

Fossil fuel power plants and other processing facilities can release toxic materials into the air, soil, surface water, and groundwater. Once emitted, these pollutants are dispersed and transported throughout the region, while undergoing degradation and decay, adsorption, transformation, and biological uptake processes. Pollutants may also be transferred between media such as when particulate material emitted from a power plant stack is deposited on land. To calculate the ultimate chronic health risks associated with such emissions, the method must be able to follow them throughout the environment and to calculate the long-term concentration in all media of the region surrounding the emission source. The pollutant discharges into the different media are dispersed, transformed and transferred between media. Pollutant concentrations in the environment are passed through food webs and other exposure pathways to generate exposure rates by type of intake for the regional population. These exposure rates are combined with a regional population model or other information and with dose-response models for the pollutant to calculate the overall risk associated with the original pollutant emission levels. Complex phenomena are involved in pollutant transport and transformation processes. These phenomena make it difficult to calculate the necessary pollutant concentrations.

Assessing the risk associated with groundwater contamination is a major task that suffers in many instances from a critical lack of an appropriate data base. Existing data bases do not provide sufficient grounds to quantify the type and shape of the probability density functions of the random variables associated with the modeling of groundwater contamination. Management decisions associated with extreme and catastrophic impacts cannot be based on risk assessment approaches that use averaging, commensuration, and aggregation of "high-frequency low-impacts events" with "low-frequency high-impact events."

■ Generally the first step in conducting a risk assessment for groundwater management involves a review of all available site environmental monitoring data in order to select potential chemicals of concern on which the assessment will focus. The next step, hazard identification, involves identifying chemical-specific human health and ecological effects criteria. In the absence of human data to describe low-dose effects, the frequently used approaches for dose-response characterization are separated for threshold (non-carcinogenic) and threshold (carcinogenic) effects. These approaches generate health effects criteria to be used in the calculation of risk.

TABLE 3.3. Assessment of the quantitative effects from continuous sources

Assessment of the Quantitative Effects from Continuous Sources
■ Source of Continuous Pollution ■ Dispersion ■ Residual Toxic
■ Effects: ⇒ Population Sensitivity ⇒ Environment Sensitivity

↓

Resulting Impact

■ A method has been presented by Lange[37] (1989), to assess the risk due to several activities in the inflow area of a receptor, taking into account the most inadmissible chemical substances (pollutants). The source-path-receptor concept is used. Each activity may lead to a range of accidents and therefore to a range of risk values. Instead of using the single concentration value derived from analyses of the most credible accident at the site of the polluting activity (and its probability of occurrence), the method encompasses the complete range of reasonable accidents and their resulting concentrations per polluting substances. This leads to curves of concentrations and their probability of occurrence, from which curves of the effect on the receptor and its probability of occurrence can be derived. These so-called P-E (Probability-Effects) curves define the risk of the described polluting substance to the receptor as a result of all reasonable accidents at the site of the activity. By manipulating such curves, it is possible to carry out a summation of risks due to different pollutants or due to different polluting activities under certain conditions. For assessing the relation between P_i and C_i different outflow processes (source) can be valuable. The ratio between a certain concentration (C_i) and the maximum concentration ($C_{i\text{-}max}$) allowed for the receptor is assumed to represent a value for this (in)admissibility or effect.

■ Following hazard identification and dose-response modeling, potential pathways by which human populations may be exposed under current or potential future land-use conditions are identified. To evaluate an exposure at the exposure point, the concentration of chemicals of concern must be evaluated (measured concentrations, fate and transport modeling results). The concentration found must then be compared to applicable or relevant requirements and standards.

Bolten[5] et al., (1983) have developed a methodological framework for assessing the risks from toxic pollutants emitted by a hypothetical coal-fired plant and applied it in a case study of arsenic emissions from such plant. The models in the framework can determine the chronic human health risks associated with different costs of controlling these emissions.

In a second report Bolten[38] et al., (1985) describe alternative models for risk assessment of toxic emissions. The general framework uses a composite multimedia modeling approach for environmental calculations. This approach applies independent models for each medium sequentially to determine the distribution of a pollutant in the air, soil, groundwater, and surface water. The inputs and outputs of the different models are coupled to provide the interactions between environmental media

The selection of models encountered in the literature is a difficult task. For surface water transport, hundred of models could be used. These models differ in many ways and can be characterized by these properties (among others):

- Whether they are steady-state or time-dependent
- The types of water bodies to which they apply
- The number of kinds of pollutants included
- How transformation and decay processes are treated
- Whether they emphasize sediment processes, water quality, or radionuclide transport.

Most models have been developed for a particular purpose and may not be suitable for toxic chemical risk calculations. The models included in the framework proposed by Bolten[38] et al. (1985), *computer program RATE - Risk Assessment of Toxic Emissions,* calculate pollutant concentrations and transfer rates for the air, soil, and surface water bodies. The models can run with limited data and are designed to calculate upper and lower bounds to exposure rates and concentrations with extensive sensitivity analysis. The results are of the type "level I analysis". Constant mean exposure rates and stationary, static populations are assumed.

Calculation results have shown that assuming a static population may overestimate the health risk, if birth, death, and migration rates are large. Similarly, although some dose-response models are intensive to variations in exposure rates, others will provide significantly different results when used with time-dependent dose information. The health risk may be either over- or underestimated, depending on the nature of the pollutant and its effects.

By performing this screening analysis (level I), the user can isolate the important aspects of a problem and determine where further detailed analysis (level II) and resources are necessary.

3.2.3. POTENTIAL EXPOSURE PATHWAYS

Human Exposure to Groundwater

To evaluate *exposure to groundwater*, standard intake assumptions are generally employed, e.g.,:

- an average adult ingests 2 liters of water a day over 70 year lifetime;
- *average body weight is assumed to be 70 kg.*

Additionally, *inhalation exposure* to contaminants in groundwater may occur through use of water in day-to-day activities such as cooking, bathing, washing of dishes and clothes, or showering. *Dermal exposure* are also possible. Exposure through showering may be quantified using the *model of Foster and Chrostowski*[39] (1987). For many volatile organics the additional risks through inhalation of contaminants while showering may be similar to the risks associated with ingestion. Dermal exposures are generally small compared to ingestion or inhalation. Failure to assess these pathways could lead to groundwater risk management decisions not protective of public health.

In many locations groundwater discharges into surface water bodies create additional potential pathways of exposure. This is a particular concern for small water bodies (with aquatic life, but with to high concentrations of pollutants). Contaminants that have high octanol-water partition coefficients have a potential to bioaccumulate and generally are of particular concern.

> ***Toxicity of Contaminants with High Octanol-Water Partition Coefficients***
> *These chemicals may not only be toxic to aquatic life, but may potentially cause risks to other organisms higher in the food chain or to human that ingest fish from these surface bodies on a regular basis.*

It must be kept in mind that change in local pumping conditions due to the installation of a new municipal well may have an influence on contaminant migration and thus groundwater which is not currently used as a drinking water may become a supply source in the future.

In evaluating potential exposure to contaminants present in groundwater, it is important to define the aquifer(s) to be evaluated. In many instances, particularly in areas with fractured bedrock, domestic wells are open to more than one water bearing zone and estimation of the potential future pollutant concentration in these wells would be dependent upon the well construction. This type of information is not always available. USEPA's groundwater strategy (USEPA[40], 1986) suggests that the risks associated with potential water supplies should be evaluated as if the water supplies were actually in use. Thus aquifers with natural mineral contents low enough for potability and with high enough potential yields are all considered potable whether in current use or not.

3.2.4. QUANTIFYING EXPOSURE AND RISK FROM GROUNDWATER

A good risk assessment methodology, with case examples, has been commented by Anderson[41] et al., (1990) and is summarized below. Site specific information is essential in evaluating the potential risk associated with exposure to contaminants present in groundwater:

■ *Screening Approach for General Hydrological Parameters*. These can often be obtained from available publications (Walton[42], 1985) but caution must be exercised. When evaluating hazardous waste sites, there is frequently little or no information available regarding the history of chemical disposal in potential source area.

■ *Estimation of Chemical Loading to Groundwater from the Unsaturated Zone*. A variety of approaches can be applied, ranging from a simplified steady-state soil-water partitioning model to time-dependent models that consider linear adsorption/desorption without accounting for dispersion or more complex compartment numerical models that incorporate time-varying transport, advection, and dispersion such as *the "Pesticide Root Zone Model"* (PRZ: USEPA[43], 1984). A word of caution: applying these models to the same site may result in soil pore water concentrations that may vary as much as several orders of magnitude.

■ *Estimation of Contaminant in the Groundwater*. The output from the unsaturated zone model may then be coupled with groundwater models to predict concentrations of chemicals of potential concern at potential exposure points. Again these models may range from simplified mass-balance mixing models through analytical solutions of transport equations to complex three-dimensional numerical models. *In order to obtain good result it is necessary to supply the model with site specific data, which shall replace the default values provided by the model.* In many instances, a steady-state assumption is not applicable.

Once the concentration of chemicals of concern in the groundwater have been calculated several types of analysis can be performed:

■ *Non-Ingestion Exposure to Volatile Organic in Groundwater*. Make an assessment of the analytical data for residential groundwater. Carcinogen have to be considered as more important than other pollutants. Then calculate chemical *concentrations in the air* that people might be exposed to (by taking a shower, etc.) from using water containing these volatile organics.

■ *Estimate Also the Dermal Exposure*. This is proportional to the lipophilicity of the chemicals. Dose can be derived on the basis of benzene doss and the ratio of the octanol-water partition coefficients of benzene to those of the chemicals in question (according to the method of Foster and Chrostowski[39], 1987). Finally estimate the risk using the cancer potency factors published in the *Superfund Public Health Evaluation Manual* (EPA[44], 1986).

■ *Potential Risks Associated with Potential Biotransformation of Chemicals.* As the concentration of a particular compound present in groundwater may decrease through biodegradation, the total risk resulting from exposure to the groundwater may not necessarily decrease. Some compounds (i.e., vinyl chloride) are transformed into more toxic compounds through biotransformation.

Exposure models that are appropriate for utility application should provide the link between the interactive media models, on the one hand, and the pharmacokinetic and dose-response models on the other hand. Such a model would be adequate if it could determine average toxicant concentrations for a particular time frame and population group for each exposure pathway.

Chemical and physical transformation of the toxic chemicals originally released from a stack can take place in and among all media (air, land, surface water, groundwater). These transformations will ideally be considered in the individual media transport models, or when different models are coupled together. An exposure model should contain two major elements:

- the exposure pathway which describes the various methods of human exposure, and
- the distribution pathway, which addresses production and usage characteristics of an appropriate population group.

3.2.5. AQUIFER RESTORATION TECHNIQUES

Due to the quite often significant impact of petrochemicals manufacturing and storage facilities on the quality of the surrounding ground and groundwater quality, there is a need for clean-up and rehabilitation measures to prevent further spreading of leaking pollutants. A variety of techniques exists for containing and treating contaminated groundwaters.

Currently contaminant-source control and groundwater treatment objectives are difficult to achieve. Thus, effective aquifer restoration and risk reduction will likely involve both the use of improved, innovative treatment techniques and definition of site-specific, risk-based cleanup guidelines. Techniques currently available for controlling, extracting and treating contaminated groundwaters include physical containment techniques and physical, chemical, and biological treatments. Among the strategies that can be followed to achieve these ends, we find various technologies such as pumping, stripping, isolation, biotechniques, excavation and immobilization, associated with different financial costs. The identification of the optimum strategy is the key to a successful pollution abatement policy.

A reference that presents a detailed discussion of the various technologies was published by *US Environmental Protection Agency* (USEPA[45], 1985). Additionally, a model has been developed by the *American Petroleum Institute (API)* for estimating the costs and duration of treating groundwater contaminated with petroleum hydrocarbons (API[46], 1986).

The past few years have demonstrated that the use of *barrier-technology* with respect to remedial actions is a necessity (containment strategies). This simplified approach considers the migration of contaminants in terms of emissions across a specified boundary by means of a groundwater transport instead of using concentration developments. In such a case environmental risk assessment can also be done using stochastic modeling (Van Ree[47] et al., 1989). An environmental design shall be characterized by its emission specifications including the probability density function of the actual emission. *Monte Carlo simulation* provides the final step in the derivation of the (cumulative) probability function.

Among the *physical containment techniques*, we have:

■ *Impermeable Barriers, Used to Contain, Capture, or Redirect Groundwater Flow.* slurry walls, grout curtains, sheet piling, and hydraulic barriers. The Principle of this latter method is to modify hydraulic gradients about contaminated waters through depression and recharge of the water table (pumping / recharge strategies).

■ *Surface Water and Leachate Control.* The most commonly used controls are diversion and collection systems, grading, capping, and revegetation.

■ *Groundwater Treatment Technologies.* These include several physical, chemical, and biological processes.

The determination of safety in groundwater cleanups is generally based on a case-by-case comparison of levels present at the site to existing health-based standards or pollution levels developed by risk assessment. The method includes the risk assessment *per se*, consisting of:

- hazard assessment
- dose-response assessment
- exposure assessment using worst-case assumption
- risk characterization indicating a quantitative risk for a given exposure.

It must be noted that a *quantitative risk value* is not a realistic prediction of the effect of exposure, rather a health-protective upper bound on the risk (from a lifetime, usually 70 years, of human exposure to a chemical). This value is compared to the acceptable risk range (10^{-4} - 10^{-7}). The true risk could be very small or even as high as the upper bound estimate provided in the risk assessment. For a given exposure concentration, *EPA's risk assessment methodology* also results in estimated risks that are an order of magnitude higher than the risks estimated using the *risk assessment methodology of the Food and Drug Administration and the Center for Disease Control.*

The *quantification of soil pollution problems* can be made in terms of risk analysis by applying the *"source-path-target methodology"* developed by Loxham and Van Ree[48] (1987), where the problem is broken down by characterizing an emission source by its strength and nature, identifying local and far-field targets by their associated dose-effect relationships, or possibly their maximum allowable concentration limits, and quantifying the dynamics on the pathway between them.

The *system analysis model REFCON* can be used for the near-field (Loxham[49] et al., 1989). This is a multiple soil and groundwater zone model which considers a manufacturing or storage facility situated above or in the vicinity of a target aquifer which is protected by one or more geological barrier formations. The code aims to predict the time-dependent fluxes and concentrations in the surface water, treated drainage, and groundwater systems outflows as function of pollution management strategies applying to the related facility. It is based on so-called transfer function approach (Green's function associated with each contaminant). The actual emission flux and concentrations into the far-field are obtained by convoluting this transfer function with the results of actual releases or release scenarios from the facility in the near-field. The transfer function must be established by experiment , observation, assumption, or sub-process modeling. The model can be used alone or in combination with more detailed migration models.

Example of Event Tree Analysis for Groundwater Pollution
An example of an event tree for risk assessment regarding groundwater pollution has been given by Shy and Rios[50] (1989).

Example of Event Tree for Assessing Groundwater Pollution Risk

A) Making of urbanization policy
B) Design of facilities
C) Fabrication and installation of the facilities in question
D) Training and education of operating and maintenance personal
E) Maintenance of the facility
F) Operation of the facility
G) Acts of God (weather, storm, etc.)
H) Malicious events (man-made by inadvertence)
I) Hazardous material(s) spilled, its path and whether it is contained or not
J) Cleanup and solubility of the spill
K) Pollutant interaction with the aquifer including removal, retardation, transformation, and dispersion
L) Event (L) is considering the contaminated water/plume reaching the aquifer/well
M) A set of events (M) describes remedial actions, like contaminant removal
N) The water may be prepared and treated for use in potable systems (N),
O) The water is being pumped into the distribution system (O).

As a result of this analysis the distribution system may or may not be contaminated.

Figure 3.1. Event tree for risk assessment of groundwater pollution

In this example *(Figure 3.1)* the specification of the type and amount of future urbanization development (land use policy) inside the recharge zone is the critical initial event. Each node (event) has two possible outcomes: one with favorable or desired consequences and the other with negative or generally unwanted consequences. However, the exact consequence of the negative branch may depend on the circumstances responsible for the negative outcome at the node. Positive results of an event are those that are likely to preclude water contamination. The events of the chain under consideration (listed below) are represented by a letter (A-O): The coupling of event tree analysis with fault tree analysis provides powerful, flexible procedures for detailed risk estimation (Vesely[51] et al., 1980).

3.2.6. EXPERT SYSTEMS FOR GROUNDWATER CONTAMINATION

Expert systems are presently being developed for assessing the risks of water pollution. Such systems are based on if-then-else rules, in a way similar to the *PROSPECTOR* expert system[52] for mineral exploration. In general such an expert system has to deal with uncertainty, and criteria must be defined to transform qualitative and/or quantitative information into a calculable unit.[53]

Expert system may be most useful in cases where the system's dynamics are highly non-linear. Expert systems may have a role to play in the area of automatic programming, i.e., for constructing software for solving nontrivial problems in the quantitative interpretation of geophysical logs.

■ *DEMOTOX.* This system is designed to help evaluate the potential for groundwater contamination caused by a variety of organic contaminants using a mobility and degradation index (Ludwigsen et al.[54], 1986). The index is the ratio between the time required for a contaminant front to travel through a soil zone and the half-life for biodegradation (small values indicating a greater contamination potential). Another expert system is used for the determination of groundwater flow directions and the permeability of units at a site. The knowledge base for this prototype system is a series of if-then-else production rules and an external function that establishes groundwater direction as the solution to a three-point problem (Law et al.[55], 1986).

■ *Expert ROCKEY.* One of the most comprehensive computer system related to the problem of groundwater contamination is *Expert ROCKEY* (McClymont and Schwartz[56], 1987). The system contains a contaminant transport model, two expert systems (EXPAR and EXINS), and a plotting package. The transport model describes the transient, three-dimensional spread of a dissolved contaminant in a unidirectional flow system. This model accounts for advection, dispersion, sorption, first-order decay, and time-varying loading of the source. The EXPAR system is used for preparing a set of input data for the model. The EXINS system is a demonstration prototype to help plan a monitoring strategy for a first-stage field investigation.

Most resources planning and management tasks take place in an environment in which the basic input information, the goals, the constraints, and the consequences of possible actions are not known precisely (mostly due to insufficient data and imperfect knowledge). The *fuzzy sets theory* creates a conceptual framework to deal with imprecision and indeterminacy in non-random environments and complex problem situations, where either some variables or parameters or the relation between them is inherently ill-defined. Literature on fuzzy sets theory applied to water resources can be found in Hipel[57] (1982), and Kindler[58] (1989).

3.3. Groundwater Models

3.3.1. INTRODUCTION

There is a concern on how groundwater models should be used to address legal and regulatory concerns. Hydrologists are being caught in the middle between some major advances in science and increasing pressure from legal and regulatory bodies to use models to provide answers to specific questions. A study[59] was started jointly by the Electric Power Research Institute, the US Regulatory Commission, the Environmental Protection Agency, the US Army, and the National Science Foundation, to address two questions:

> *"To what extent can the current generation of groundwater models accurately predict complex hydrologic and chemical phenomena"?, and "Given the accuracy of these models, is it reasonable to assign liability for specific groundwater contamination incidents to individual parties or make regulatory decisions based on long-term predictions"?.*

Conclusions of the above mentioned study concerning the use of groundwater models show that few flow and transport problems can be modeled with confidence. *The most satisfactory results to date have come with models involving the flow of water or the transport of a single non reactive contaminant in a saturated porous medium.*

A model, in general, is a tool designed to represent a simplified version of reality. Using groundwater models, it is possible to predict the effects of certain actions and to test various management schemes. The validity of the predictions, of course, will depend on how well the mathematical model approximate the field conditions. A problem results of the necessary simplifying assumptions that must always be made in order to construct a model of the region representing well enough the field situations. It is essential that the user is aware of the nature of limitations of any model in simulating the physics of the groundwater system. With growing number of numerical models, it is becoming increasingly difficult for a groundwater modeler to resist the temptation of using sophisticated models. It is a good practice to start from simpler models and progressively refine the solution. Very often, groundwater systems are amenable to apply simpler models.

Sloan and Moore[60] (1984) have conducted a study to illustrate the adequacy of simple physically-based models versus complex models. The dimensionality of the problem (depending on the geological aspects of the field) is often a major criterion in selection of groundwater models. Geometry of the system and accuracy of the solution needed are the preponderant factors. In the initial stages of the study, a fairly accurate solution of the problem can, frequently, be obtained by simple one-dimensional approach. *The use of more complex models is often limited by the availability of good data.* The models are, usually, more appropriate for repeated simulations, acting more as sensors of parameter variations than as absolute predictors. It is useful to perform sensitivity analysis by conducting repeated simulations of the same problem with perturbation of various parameters. Selection and use of any groundwater model must include an evaluation of magnitude of the problem, of parameter and data availability, of site conditions (including homogeneity and isotropy), of initial and boundary conditions, and a sensitivity analysis of the results.

3.3.2. REVIEW AND CLASSIFICATION OF MODELS FOR GROUNDWATER

In risk assessment, groundwater transport analysis is normally performed in two stages - *flow, and contaminant transport* - using different models. *Flow models predict how groundwater moves under the influence of given recharge, infiltration, and withdrawal rates.* Given these results, *contaminant transport models determine pollutant transport in the groundwater.* They also include degradation, transformation, and removal of pollutants in the soil.

Groundwater flow characteristics differ in the partially saturated and saturated zones. The partially saturated zone normally extends down from the leachate source to the groundwater table; the saturated zone includes the volume below the water table. Different types of models using different types of assumptions are normally required to describe the flow and contaminants transport in these two zones.

As discussed in Whelan and Brown (1984)[61], *level I groundwater models* (used for screening purpose) assume simple geometries and uniform, homogeneous properties. Most simplified models for the partially saturated zone, for example, assume that all water and pollutants move vertically in a uniform soil column. *Level I models* for the saturated zone generally assume that regional groundwater movement occurs only in one direction and treat the groundwater system as a single-layered aquifer with uniform properties. Both types of models are frequently analytical or semi-analytical, unlike the detailed groundwater models that simulate groundwater flow with numerical solution algorithms.

Hundreds of different groundwater models have been described in the literature. Many of these models have been reviewed by various authors including Bachmat et al.[62] (1978), Anderson[63] (1979), Knox and Canter[64] (1980), Lappala[65] (1980), Onisji[1] et al., (1981), SAI[66] (1981), Koines[67] (1982), Oster[68] (1982), Nelson[69] (1982), and Kincaid[70] et al., (1983). A model survey of van der Heijde[71] et al., (1985), reviews a total of 399 numerical models of which 84 are mass transport models (*Figure 3.2*). Out of these 399 models, 337 (84%) are designed for predictive purposes, 33 (8%) for management purposes, and 29 (7%) for identification. A description of the different types of models may be found in Kinzelbach[72] (1986). Another survey was prepared by van der Heijde[73] et al., (1988).

Most of the models are prediction models, which may be subdivided into four categories: flow, subsidence, mass transport, and heat transport models:

Flow Models. They utilize information on aquifer parameters, boundary conditions and determine quantitative aspects of groundwater flow such as direction and rate of flow, changes in water level, stream-aquifer interaction and interference effects of wells. Flow models have also been developed for the unsaturated zone and for coupled saturated-unsaturated surface systems.

Number of Surveyed Models (399)

- ■ **Prediction** *(337)*
  - - **Flow** (203)
    - - Single phase
      - - Saturated (150)
        - - Lumped (7)
        - - Hydraulic (102)
          - - Single aquifer (74)
          - - Multiple aquifer (28)
        - - Hydrodynamic (41)
        - - 2-dimensional (25)
        - - 3-dimensional (16)
    - - Multi phase (15)
  - - **Mass transport** (84)
    - - Lumped (7)
    - - Distributed (77)
      - - Conservative (30)
      - - Non-conservative (47)
  - - **Heat transport** (22)
  - - **Deformation** (12)
  - - **Multipurpose** (16)
- ■ **Management** (33)
    - -Groundwater (17)
    - - Quantity (13)
    - - Quality & Quantity (4)
    - - Groundwater and Surface water (16)
    - - Quantity (14)
    - - Quality & Quantity (2)
- ■ **Identification** (29)
  - - Direct (16)
  - - Indirect (13)

Figure 3.2. Models surveyed by van der Heijde[1] et al., 1985.

Subsidence or Deformation Models. They describe the phenomenon of land subsidence which is caused by water withdrawals of groundwater. Subsidence models are needed to predict deformation-related impacts of various pumping schemes in affected areas.

Mass Transport Models. They deal primarily with questions of groundwater quality. They are used to predict the movement and concentration in the aquifer of various pollutants including radionuclides, leached solids from landfills and irrigated areas, and salt water intrusion in coastal areas. Transport models that describe the movement of pollutants without reactions are called *conservative models, respectively,* models that take into account reactions are termed *non-conservative. There are also transport models for imiscible oil/water fluids.*

Early modeling efforts in the late 1960's and early 1970's have been concerned with non-reactive tracers or contaminants. Now emphasis has shifted to incorporating homogeneous and heterogeneous reactions into the transport models (Cederberg[74] et al., 1985; Miller and Benson[75], 1983).

The description of the *transport of fine suspended matter in surface waters* is becoming increasingly important for morphological studies, as well as for the assessment of the impact of outfall discharges, dredging operations etc. on the water quality. The dispersion of such particles takes place over long-distances and depends mainly on their characteristics, a/o. the settling velocity, and the hydrodynamic features of the water body, i.e. the flow field and diffusion coefficients. Previous research has mainly dealt with the transport of suspended load in 1-D channels. Numerical models were proposed by van Rijn[76] (1986), and Celik[77] et al., (1988), which were successfully calibrated and tested against field data. These models cannot handle a horizontally variable flow field. The quasi 3-D model of Christodoulou[78] (1993), was developed for simulating the dispersion of particulate matter, and is based on a separation of variables technique, whereby the vertical distribution of sediment is solved separately for conditions of local equilibrium, following the well-established Rouse profile. A decay term depending on the near-bed concentration and the absorption/reflection index of the bottom boundary is used to obtain the horizontal depth-averaged concentration distribution.

Transport of Sediments. The mathematical models available for modeling transport of sediments are generally restricted to transport under equilibrium conditions, but in morphological processes, the sediment transport often takes place under non-equilibrium conditions. Ouillon[79] et al. (1993) have developed a 3-D non cohesive suspended sediment transport model for free surface flows. The evolution of each granulometric class is described by its own mass balance.

Heat Transport Models. These couple the flow of heat with water or steam for problems where thermal effects are important (hot springs, heat storage, and geothermal reservoirs.)

Over the past 20 years, a variety of mathematical models have been developed for the purpose of predicting vertical turbulent heat and mass transport in lakes and reservoirs. A review of such a work is given by Harleman[80] (1982). Another review has been prepared by Rodi[81] (1980).

Modeling of Vertical Transport in Lakes. In water bodies of sufficient depth, an important characteristic which affects such transports is the presence of thermal, or more generally *density stratification.* Any predictive model of vertical transport in a lake must consider the effects of temperature, and thus hydro-thermal models form the basis for mass transport models used in water quality studies. Prediction models can be either analytical, stochastic, or discretized (either analog or numerical). Recent research has centered on the "integral" approach, wherein a well-mixed layer of a depth which varies with time is assumed to exist at the lake surface. Beneath this "mixed layer", a stratified region exists. Such models are based on mechanical energy budgets for the mixed layer, and have been shown to simulate conditions more accurately than earlier diffusion models. A new area-averaged, time-variable model of vertical variations in water motion, temperature, and turbulent energy in a lake has been presented by Owens[82] et al., (1993). In the *PROBE code*[83], vertical mixing and temperature distribution is simulated by one-dimensional, horizontally-integrated, K-ε turbulence model.

> ***Note on Rate Constants and Kinetics' Formulation in Surface Water Quality Modeling***
> *Useful data are available from a US-EPA Report No. EPA/600/3-85/040 (1985) of the Environmental Research Laboratory.*

3.3.3. GROUNDWATER FLOW MODELS

A sophisticated literature exists including several textbooks devoted exclusively to groundwater flow problems (Huyakorn and Pinder[84], 1983; Remson[85] et al., 1971; Wang and Anderson[86], 1982). Haverkamp[87] et al., (1977) have made a comparison of numerical simulation models for one-dimensional infiltration. The numerical modeling of isothermal and non-isothermal flow in unsaturated fractured rock has been reviewed by Pruess and Wang[88] (1987).

Flow Models
As a system becomes more complicated through partial saturation, the presence of several mobile fluids, fracturing, or the existence of reacting contaminants, many more questions arise about the adequacy or validity of the underlying process models. The processes that control *saturated groundwater flow* are reasonably well understood, and standard models of these processes are generally believed to be able to give reliable predictions if provided with adequate amounts of data. Nevertheless, the impact of field scale heterogeneity are still widely debated, and there are few clear guidelines on how model inputs should be estimated from limited data bases or on how hydrologic monitoring programs should be designed.

A number of well-documented, robust, and flexible computer codes are readily available (Bachmat[89] et al., 1980; Bachmat[90] et al., 1978). Information is also available from the International Groundwater Modeling Center (IGWMC)[91], or from Scientific Software (Washington DC).

Infiltration Models
Among the water infiltration models, one finds the *SCS curve number method*[92] a modified *Holtman's equation*[93] the *Green-Ampt model*[94], and *Richard's equation.*[95] The Richard's equation numerical model and the Green-Ampt model give the best estimates of infiltration through different soil profiles. The *Holtman model* gives poor estimates, and the SCS curve number model gives good estimates of the accumulated infiltration volume.

In situations of chemical spills or leaking underground storage tanks, it is necessary to know the space and time distribution of the chemical concentrations under the subsurface of the soil-water environment to design appropriate remediation or treatment schemes. It is known that a concentrated organic phase might well persist for some distance from the site. The initial transport after a spill is essentially one-dimensional (vertical) in the unsaturated zone. Upon reaching the underlying water table, however, the organic flow is three-dimensional in response to the spreading and advection by the groundwater flow. If the organic phase is lighter than water, the contaminant will remain floating on the water table, otherwise it will sink. A simple adapted Green-Ampt type model can be used in such a situation.[96] *Literature and models* on this subject can also be found in Pinder and Abriola[97] (1986), Faust[98] (1985), Hochmuth and Sunada[99] (1985), Osborne and Sykes[100] (1986). All these models assume no mass transfer between phases. Baehr and Corapcioglu[101] (1984) developed a model that incorporates mass transfer between phases, adsorption / desorption phenomena and biological degradation, but applied a much simpler form of the model to predict gasoline movements into soils.

One of the major components of the hydrologic balance is infiltration. Estimates of surface runoff will depend on good estimates of infiltration into the soil. Modeling unsaturated flow in reclaimed lands is a challenge. *Unsaturated flow* is less well understood, and there have been very few field studies that extend over the scale of interest in most contamination applications, and most of these have focused on one-dimensional transport in the vertical direction. Some investigators believe that unsaturated flow can move horizontally over significant distances. The numerical demand for 3-D simulation is large and accurate three-dimensional unsaturated flow modeling capabilities (Cray computers) are not available to most consultants. Yet many important contamination problems, such as leaking underground storage tanks, infiltrating pesticides, and leaching mining wastes, affect the unsaturated zone. *Flow through fractured media* may be either saturated or unsaturated. Both types of fracture flow are difficult to predict due to a lack of geological information.

Various groundwater flow model structures with varying degrees of sophistication have been proposed, such as:

- Single horizontal, layer model (Pinder and Bredehoeft[102], 1968)
- Vertical section, asymetric model (Javandel and Witherspoon[103], 1969)
- Quasi-three-dimensional model (Bredehoeft and Pinder[104], 1970); (Chorley and Frind[105], 1978); (Premchitt[106], 1981)
- Three-dimensional model (Freeze[107], 1971); (Gupta and Tanji[108], 1976); Frind and Verge[109], 1978).

It is realized that the first three models were devised for generalized and simplified situations which are rarely found in real groundwater basins. The water bearing strata are generally simplified to be aquitards and extending continuously over the domain. This assumption may, in some circumstances, lead to significant errors. However full three-dimensional models are too complicated for simple simulation problems and do not justify the extra effort, when data are scarce.

The model developed by Boyle Engineering Company (Yoon[110] et al., 1985) is a finite element groundwater flow model, which contains a sophisticated soil moisture algorithm to simulate interactions with the surface water system. It allows each aquifer within the system to have different spatial extent, and it simulates stream aquifer interactions.

The process for defining groundwater flow paths in three dimensions is mathematically straightforward but computationally tedious. The solution is obtained numerically by means of finite-difference approximations (Trescott[111], 1975).

Sources of uncertainty in groundwater flow models are of five different kinds:

▪ Uncertain values of *point input variables* such as unrecorded pumping rates, uncontrolled emission of hazardous substances from waste disposals into groundwater, etc. Uncertain values of *spatially distributed input variables* such as recharge of groundwater by infiltration, emission of biogenic substances induced by agriculture, uptake of water by plants, sorption/desorption processes in porous medium, etc.

▪ Uncertain values associated with the definition of boundary conditions for both groundwater flow and transport problems, e.g., contacts with contaminated surface waters.

▪ Uncertain values of initial conditions assumed by decision-makers when analyzing a given scenario.

▪ Uncertainty in the values of distributed hydraulic and mass transport parameters

▪ Uncertain three-dimensional geometry of the groundwater system (presence of fractures, variable porosity).

On the basis of field observations one can identify a hierarchy of groundflow water systems that are essentially determined by the water table conditions (Toth[112], 1963). Recent development about the analysis of groundwater flow systems have been led by Engelen[113] (1986). Stagnation zones play an important role. The hierarchical arrangement of the system determines the vulnerability risks and restoration potential of the groundwater systems. Knowledge about these systems is essential for environmental and water management studies. Zijl and Brouwer[114] (1989) have developed a new code, *FLOSA* (Flow System Analysis Finite Difference), designed to cope with complex situations and avoiding the lack of accuracy of the piezometric-based models in such situations. The code is partially based on the vector field approach, but takes the water table, or piezometric levels, at the upper modeling plane, as the input variables. These data may be obtained by measurements in observation wells and/or by simulation with a conventional piezometric model. The resulting output consists of the tree-dimensional flow system and the recharge and discharge patterns and rates pertaining to them.

Reconstructor Tool. Imprecise estimates of initial positions of piezometric heads badly affect the performances of any dynamic flow model and transport model. *The use of additional information processed by a "reconstructor" improves the estimates of piezometric head and hence decisions based upon these estimates.* The reconstructor is an abstract dynamic system that tracks trajectories of the state variables (piezometric heads, concentrations of a pollutant at some points) of some real dynamic system which cannot be directly measured and are thus reconstructed from the measured ones.[115] When applying the classical Galerkin finite element approach to a groundwater flow equation, one obtains a set of ordinary differential equations which, together with the appropriate initial conditions, can be identified as the state equation of a dynamic system. In the case of groundwater monitoring, the net work equation must be supplemented with the observer equation, $\underline{y} = C\underline{x}$, in which C represents spatial distribution of measuring points. The state vector $\underline{x}$ can be subdivided into its observed ($\underline{x}_1$) and unobserved ($\underline{x}_2$) subvectors. It is then the task of the reconstructor to estimate the unknown state subvector $\underline{x}_2$ (unobserved values of piezometric head in N nodes of a finite element grid) from the known input vector (source term vector, $\underline{u}$, representing recharge/discharge processes as well as boundary conditions) and the measured part $\underline{x}_1$ of $\underline{x}$. Using this methodology can reduce the initial state error at the reconstruction point.

The transport of pollutants released into the groundwater requires reliable predictions to assess potential hazards to the public. Analytical methods have limited capability to accurately predict complex three-dimensional dispersion where geometry and soil conditions vary over a wide range. Numerical models, which are more flexible in solving complex transport processes, calculate more realistic solutions but tend to produce excessive numerical errors. Such errors lead to overestimation of the hazard, or under-prediction in areas where the consequences could be significant.

Methods of Handling Parameter Uncertainties

Various techniques are available to handle the effect of parameter uncertainty in groundwater flow. These techniques can be divided into the two broad categories: *full distribution analyses*, and *first and second moment analyses* (Dettinger and Wilson[116], 1981).

First and Second Moment Analyses' Technique. First and second moment analyses are based on the underlying assumption that all important information about the random variable or parameter is contained in only two values, the mean and the variance-covariance. The two most common approaches are the *perturbation method* and the *Taylor series expansion.*

Perturbation Methods involve the formulation and subsequent analysis of an equation that describes the perturbation. This first approach can lead to complicated results when a number of parameters are uncertain.

The Taylor Series Method is based on the expected values of a truncated Taylor series expansion. Generally, only the first-order approximation is utilized for the analysis of the first and second moments; however, second order approximations have been used to determined better estimates.

3.3.4. GROUNDWATER TRANSPORT MODELS

The mathematical methods used for solute transport problems include: standard finite difference methods, standard finite element methods, method of characteristics, and random walk method.

Finite Difference Method. the aquifer is discretized in several cells for which a pollutant mass balance over a time interval is taken. Then a system of linear equations is established. The method allows to choose the spatial and temporal weighting of concentrations used in the description of advective and dispersive fluxes across the cell boundaries. Difference models differ only in the choice of these weights. As an example of an explicit finite difference model, one finds the commercial code *SWIFT* (Dillon, et al.[117], 1978).

Finite Element Method. This method arrives at linear equation systems in two steps. First an ordinary differential equation in time is obtained by the "pivot method" for every node of the grid. The trial solution has to fulfill the transport equation with a minimal residual error (Gallerkin condition). This system is then converted into a system of linear algebraic equations by differentiating with respect to time. Usually the triangular elements is used for linear interpolation. As an example of a finite element model on the basis of triangular elements and linear interpolation with forward differentiating in time, one finds the commercial code *SEFTRAN* (Huyakorn[118] et al., 1984).

Method of Characteristics. The transport process is separated into advection along characteristic lines and pure dispersion in a coordinate system at rest relative to the advective movement (realized in praxis by moving coordinate grids). Another, less tedious way consists of particle tracking. The dispersive redistribution of concentration between particles is performed on a fixed grid, while the convective transport is calculated by moving particles which are indicators of local concentrations. As an example of model using the method of characteristics, on finds the commercial *USGS code* (Konikow and Bredehoeft[119], 1978).

Random Walk Method. It is also a particle tracking method, where each particle represents a fixed mass of pollutant. Both the advective and the dispersive transport are represented by particle movements. Advection takes place along the direction of flow field, and the dispersion calculation consists of adding a random movement, the statistical properties of which are related to the size of the dispersion coefficients. The concentration is finally obtained by counting the particles (mass) in each cell *(Model of Prickett[258] et al., 1981)*. A complete velocity term according to the *Ito-Fokker-Plank theory* may also be added.

Answering the question which model is appropriate as a groundwater transport model depends essentially on the physical aspects of the problem including criteria such as the number of spatial dimensions required, the presence of density effects, etc. The choice of the numerical method used for solving the transport equation is also important. The transport equation in two dimensions incorporates diffusion-dispersion, adsorptive retardation, and a first order degradation reaction. The purely dispersive transport equation, which is obtained by neglecting the convective term is of hyperbolic type.

Models for groundwater transport can be found in the handbook of Javandel et al.,[120] (1984). The development of mathematical simulation models for the calculation of *pollutant transport in the unsaturated zone* began in the early 1980's. Frisel and Van Veen (1981) published a survey and description of twenty-five such models. Recently deterministic models are prevailing for investigating the *effects of pollution on groundwater quality*.

A comprehensive *review of solute transport modeling* is presented by Naymik[121] (1987). Simple *reactive transport models* are in wide use. They consider sorption, biodegradation and other chemical effects. Most of these reactive solute models are based on small-scale laboratory studies, which may not accurately reproduce conditions found in the actual environment. Field-scale experimental investigations of multiphase transport are very limited, and existing laboratory-scale results indicate that this type of transport is influenced by a number of interacting factors, including viscosity, density contrasts, capillarity, and phase transitions. Although models of *multiphase transport* are available, the input parameters that they require are difficult to estimate in practice. The status of *contaminant transport modeling* depends greatly on the chemical species and phase of interest. The processes that influence the transport of dilute, non-reactive aqueous phase solutes are well understood, at least in saturated media, but the effect of spatial and temporal variability are not well understood. Mass transport is controlled by a variety of physical processes (advection, diffusion, and dispersion; radioactive decay and sorption).

Multidimensional models of these processes have been used successfully in practice. Work is still required to account for other more complicated chemical processes (e.g., oxidation/reduction, precipitation, hydrolysis, and complexation) and biological processes (e.g., bacterial degradation) in mass transport models. The main source of uncertainty in prediction lies in establishing values of controlling parameters like velocity, effective diffusion coefficient, and dispersivity, which vary spatially and can be difficult to measure or to estimate. *In general it is more difficult to calibrate a solute transport model of an aquifer than it is to calibrate a groundwater flow model.*

Models used to study contamination-related problems, including the design of remedial actions, are based, according to Anderson[122] (1984) on solution of the classic dispersion-convection equation. However this equation fails to describe contaminant movement near the pollutant source or over a short time period. Application of the equation has resulted in estimates of coefficient which are both time- and scale-dependent. Two different approaches have been adopted: they include a detailed description of the velocity, either deterministically (by considering heterogeneities e.g., stratification) or stochastically (using a random velocity field in the mass transport model).

Multigrid Technique. Reliable groundwater transport predictions require accurate spatial and temporal characterization of a hydrogeologic system. When the amount of spatial sampling that can be obtained to resolve the flow parameter variability associated with heterogeneities is scarce, the use of a multigrid technique to simulate two-dimensional flow velocity fields with high resolution can be applied to study such effects of limited sampling of heterogeneity on transport predictions.

Heterogeneity is typically characterized by various length scales which represent the distance over which a system parameter is correlated. The finite difference code *MGRID* a multigrid solution technique.[123] The code is able to solve groundwater problems with parameter distributions that have a much wider range of spatial frequency components than could be realistically measured with field tests. Multigrid is an iterative equation-solving technique that moves between sets of successively coarser grids which effectively separate the spatial frequency components of the solution to maintain a high rate of convergence. High spatial frequency components of the solution are converged on the fine grid, while lower spatial frequency components of the solution are computed efficiently on coarser grids. The method is well suited for solving very large systems of equations (e.g., 250000 unknowns).

Moment Method. A moment method for calculating groundwater flow uncertainty during the analysis of a landfill has been proposed by Sykes[124] et al., (1989).

Remarks
- For remedial investigation and landfill studies, the more practical choice for the uncertainty analysis is the first and second moment method based on Taylor series expansion.

- The uncertainty calculation provided by the moment method is based on the adjoint operator technique and may be validated using both Monte Carlo and Latin hypercube methods.

- The groundwater flow and solute migration model GWPGM3 (Sykes[125], 1987) can be used for parameter calibration, simulation and uncertainty calculation concerning remedial investigations and landfill analyses.

Groundwater transport of contaminants from a seepage basin may be calculated using a three-dimensional, *second-order moment technique* (Pepper and Stephenson[126], 1989). The numerical method calculates the zero, first and second moment distributions of concentration within a cell volume (Egan and Mahoney[127], 1972). By summing up moments over the entire solution domain, and using a quasi-Lagrangian advection scheme, concentrations are thus transported without numerical dispersion errors. Because the method maintains sub-grid scale resolution, steep concentration gradients can be accurately calculated with a relatively coarse grid. The use of fractional step allowed to reduce programming complexity, and storage (Yanenko[128], 1971).

The *GWGPM3 groundwater flow and solute migration model* has been developed by Sykes[129] (1987). *WOLFNET*[130] is a stand alone interactive calculation program of *groundwater flow networks.* It assumes pure Darcy flow although rivers and unsaturated regions can be represented. For a given data set the flow system can be solved rapidly and passed to a transport model.

The *WASTEN model* (Selim and Iskandar[131], 1980) *is used for describing the transport of agrochemicals in the unsaturated zone.* In a study by Benes and Vrba[132] (1989), it was found that such models are not suitable for prognostic calculations of the long-term transport of pollutant that pass through the unsaturated zone.

The *Vertical Horizontal Spread (VHS) Model* (USEPA[133], 1985; and 1986) is *a steady-state groundwater model,* in which the only attenuation mechanism considered is vertical and horizontal spreading; it neglects longitudinal dispersion, and chemical degradation kinetics. This model is quite often used by governmental Agencies in the USA.

Remarks

- Although models for important reactions like oxidation/reduction, precipitation, and biodegradation exist, they are difficult to characterize in terms of kinetic parameters, and largely invalidated in practical applications. Thus the transport of multiple reacting constituents such as organic compounds and trace metals cannot be modeled with confidence.

- Transport modeling in fractured systems remains a highly speculative exercise. Contaminant plumes have complex three-dimensional structures, which can be difficult to predict when soil properties are very heterogeneous and data available at a newly discovered contaminated site are scarce.

3.3.5. NUMERICAL GEOCHEMICAL MODELS

Special codes exist, which are very useful in analyzing groundwater movement in large multi-layered groundwater systems with complex stratigraphy or for studying the migration of radio-nuclides in geologic formations. Intensive risk assessment studies are performed for assessing the risk of nuclides' migration within nuclear waste repositories. Although the enumeration is not complete, some of the codes are briefly described below:

Mass Transport (radionuclides) in Geological Repository

In order to evaluate the long-term performance of a geological repository predictions must be made of the likely extent and rate of radionuclide transport following contact with groundwater. The simple groundwater transport models employed in probabilistic risk assessment codes cannot provide the level of information needed for nuclear risk assessment. Models describing the chemical speciation of radioelements within the repository environment itself and along pathways through the geosphere are needed. Such coupled models have been developed, among which the more advanced models combine a rigorous thermodynamic treatment of equilibrium chemistry with multi-component solute transport. The *method of coupling* may be characterized as *iterative* (Grove and Wood[134], 1979; Jennings[135] et al., 1982; Cederberg[136], 1985; Read[137] et al., 1985), or *direct* (Miller and Benson[138], 1983; Carnahan[139], 1987) and both methods have their advantages. The former employs an equilibrium chemistry code to calculate the speciation and saturation state of components in a geochemical system and iterates between this and a non-reactive transport model to move the solution in space and time. Irrespective of the codes chosen, however, all iterative "cell-type" models suffer from inherent lack of interaction between the transport and chemical components (Noy[140], 1983).

CHEMTARD Code. This code incorporates directly coupled geochemical transport equations, and has been developed to simulate the migration of groundwater contaminants from a disposal vault to the biosphere (Read and Liew[141], 1989). CHEMTARD has evolved from *CHEMTRN*, a well documented and widely-used simulator of chemical transport, developed at Berkeley Laboratory. CHEMTARD is capable of simulating the one-dimensional transport of aqueous chemical species by advection/or diffusion while accounting for phase transfer by reversible precipitation-dissolution, ion-exchange or surface adsorption. It incorporates radioactive decay, oxidation-reduction (redox reactions).

Two-dimensional coupled geochemical and geohydrological contaminant migration codes can be used to assess the potential impact of pollutants on groundwater quality. However, these codes are very heavy on computing resources. It must be remembered that depending upon the emphasis in the problem between the flow-field or the geochemical complexities, one- or even zero-dimensional models can be usefully applied.

FE3DGW Code. The FE3DGW (Finite Element 3-Dimensional Groundwater) code has been developed for analyzing flow through large, confined, multi-layered groundwater systems with varying number of layers, varying thickness, and constant or time-dependent source/sink terms (well, river, and lake infiltration, and pumping stresses).[142] The geologic input data for a complex, multi-layered groundwater system is reduced to hydrologic interpretation in the form of well logs at each surface node and subdivision of the entire region into two-dimensional elements.

The code was developed and assessed in the framework of the Civilian Radioactive Waste Management (CRWM) Program of the US Department of Energy (DOE), with emphasis on disposal of high-level (radioactive) nuclear wastes. The code can be used to support site characterization, evaluate groundwater flow rates, and estimate travel path and time in regional and local groundwater systems for a variety of boundary conditions. To address the uncertainty in the data base, Monte-Carlo analyses can be conducted. Several subprograms facilitate interaction between disciplines. The code solves the partial differential equation for hydraulic head in large, multi-layered systems by the finite element method. The hydraulic conductivity can be prescribed directly as input or calculated from intrinsic permeability, density, and viscosity relationships. The code uses three-dimensional Cartesian coordinates (X, Y, Z). The code accounts for heterogeneity in the aquifer permeability and porosity. Anisotropy is easily accommodated. The current code version does not account for density variation and calculates only the hydraulic heads. Its upgraded version CFEST[143] can be used for coupled flow, energy and solute transport. The current version of the code does not directly solve for the unconfined top layer. Unconfined steady-state solutions are obtained by iterative execution and upgrading the top elevation of the saturated zone. The code has been written in FORTRAN and is running on a VAX or CRAY machine. Other versions of the code exist for IBM compatible PC's.

FASTCHEM Code. It is a two-step coupled geohydrochemical code for modeling subsurface transport of solutes through porous media. The coupled code *EICM* is part of a modeling package (FASTCHEM) used to investigate solute migration from coal fly-ash and flue gas desulfurization sludge disposal sites. The coupled code EICM was designed to calculate the temporal and spatial distribution of aqueous solutes in a one-dimensional stream-tube. The EICM transport module uses the theory of Markow processes to distribute the solute in the stream-tube during the time step.

There are several numerical models which can be used to describe the dissolution and precipitation of minerals in aqueous systems in a variety of geological sitings.

Geochemical Models. These include *SOLMNEQ* (Kharata[144] et al., 1973); *WATEQF* (Plummer et al.[145], 1976); *MINEQL* (Westall[146] et al., 1976); *EQ3/EQ6* (Wolery[147], 1979), and *PHREEQE* (Parkhurst[148] et al., 1980).

The basic function of these models is to describe the stability of mineral assemblages and the distribution of dissolved species in aqueous systems. These models are usually confined to systems characterized by low temperature, low salinity and chemical equilibrium. In general there are two different approaches for modeling chemical reactions. The non-equilibrium or kinetic approach and the equilibrium approach. The latter being valid for deep sedimentary basins.

Swift II/III. This is a double-porosity flow and transport model *(Nuclear Waste Isolation Model).*[149] This simulation model accounts for advective-dispersive transport in the fractures and diffusive transport in the matrix, and can be used in conjunction with conservative tracers experiments. Swift III is a three-dimensional groundwater flow model which uses the finite difference method (*Geotrans*[150], 1986).

3.3.6. EVALUATION AND TESTING OF GROUNDWATER MODELS

The fact that many of the models used in practice have not been validated to a significant extent provides an important source of uncertainty in the predictions that come from the models. At present, given existing modeling capabilities (uncertainties in the natural variability of model parameters due to field measurements or geological particularities), it would be unwise to rely solely on any single source of information when deciding how to formulate regulations, carry out a cleanup, or protect public health.

To evaluate groundwater models in a systematic and consistent manner, the *International Groundwater Modeling Center (IGWMC)* has developed a model review and a three-level code testing procedure (van der Heijde[151] et al., 1984). *Level 1 testing* uses analytical solutions to verify the numerical technique and to illustrate certain behavioral aspects of the numerical solutions. *Level 2 testing* establishes a complete performance characterization with respect to the modeling design criteria, using hypothetical problems, controlled experiments, and code inter comparison. The aim is to verify the model's response to aquifer heterogeneity and anisotropy and its response to irregular boundaries. History-matching and post-audits are done during *level 3 testing*.

Several recent and ongoing studies have focused on *testing and bench marking numerical models* designed to simulate groundwater processes (Stanford Geothermal Program[152], 1980; Hamilton[153], 1982; Kinzelbach[154], 1987). In the nuclear field, a lot of development work is going on, especially in view of assessing the repository sites for nuclear wastes. In this context, two of the most prominent *code testing studies* are *INTRACOIN* [155] (International Nuclide Transport Code Inter-comparison) and *HYDROCOIN* [156] (Hydrologic Code Inter-comparison).

Reliable groundwater transport prediction requires accurate spatial and temporal characterization of a hydrogeologic system. However cost constraints and the desire to maintain site integrity by minimizing drilling can restrict the amount of spatial sampling that can be obtained to resolve the flow parameter variability associated with heterogeneities. The potential pitfalls in using groundwater models has been summarized by Reddi[157] (1990).

Besides the choice of the model equations the choice of the numerical solution method is a major decision. Available methods comprise standard finite difference and finite element methods as well as method of characteristic (advection / dispersion; particles' tracking) and random walk techniques (particles' tracking). A *comparison by Kinzelbach*[158] *(1987)* of the several methods implemented into different models have shown that under certain conditions all models show flaws: standard finite difference and finite element models yield unreliable results if discretization conditions are not met. Particle tracking methods on the other hand show oscillations and demixing problems. Commercially used random walk models are usually based on simplified random walk equations which are efficient in calculation but yield unrealistic results in regions of strong velocity gradients. The random walk techniques should be used only for rough estimates as far as local concentrations are concerned. Another difficulty of random walks lies in the simulation of particle capture by wells. The random walk model can easily accommodate chemistry if only linear terms are involved. Calculation with random walk method is very time consuming.

> *Recent research has demonstrated the frequent failure of classic dispersion equation in describing the mass transport phenomena. The conclusion was that the velocity field should be described in greater detail by either a deterministic or a stochastic approach.*

Indeed, *three-dimensional spreading* can significantly affect the configuration of a contaminant plume.[159] The commonly-used types of two-dimensional simulation models cannot accurately depict the complete dispersion. In a 2-D vertical plane model, the transverse mass loss out of the section is neglected and the advance of the simulated plume will therefore be greater (i.e. to high a concentration) than that of the real plume. This effect already occurs at very low transverse dispersivities. *In a predictive mode, a 2-D vertical simulation can be regarded as conservative.* On the contrary, in a vertically-integrated horizontal plane simulation, the distribution of contaminants throughout the thickness of the aquifer is neglected, thus resulting in wrong information. A 3-D simulation eliminates these difficulties by representing the basic dispersion process correctly.

The reliability of groundwater model may be investigated by computing confidence intervals for model parameters and by computing confidence and prediction intervals for model output. Such reliability interval may be computed by using a general Lagrangian technique that determines extreme values of the variable in question subject to the constraint that the parameters lie in a joint confidence region with specified probability level. Three classes of reliability intervals must be considered, depending on: (a) observation, (b) manually calibrated models, and (c) nonlinear regression models. Cooley and Vecchia[160] (1987) have concluded that the intervals based on calibrated models generally should be more accurate than intervals based on uncalibrated models, and intervals based on nonlinear regression models will, in general, be more accurate than intervals based on manually calibrated models.

3.4. Water Quality Assessment

3.4.1. INTRODUCTION

Solving water pollution problems today involves a multidisciplinary approach in which the required water quality is related to agricultural, municipal, recreational and industrial requirements. In many cases a cost-benefit ratio must be established between the benefit derived from a specified water quality and the cost of achieving that quality.

Waste waters emanates from four primary sources:

(1) municipal sewage, (2) industrial waste waters, (3) agricultural runoff, and (4) stormwater and urban runoff.

As municipal and industrial wastewaters receive treatment, increasing emphasis is being placed on the pollutional effects of urban and agricultural runoff. In the future, water-quality management will have to consider storm-water as a major pollutant. Agricultural runoff is a major contributor to eutrophication in lakes and other natural bodies of water. Runoff of pesticides is also receiving increasing attention.

Guidelines for drinking-water quality assessment have been published by WHO[161] (1987), NAS[162] (1986), and NAS[163] (1987).

A procedure for the development of an effective water-quality management program is shown in *Figure 3.3*.

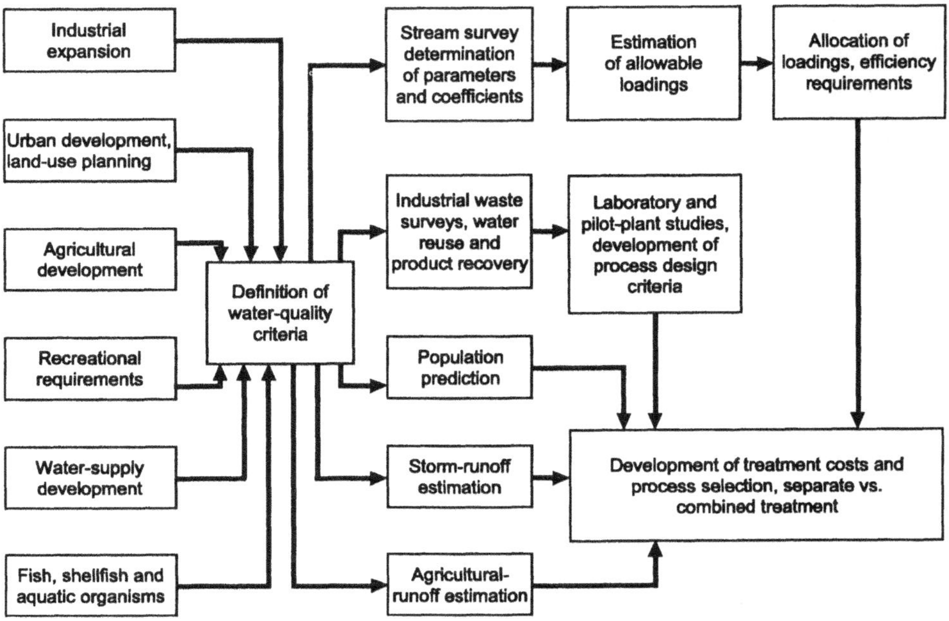

Figure 3.3. Development of a water quality management program (reproduced from Water Quality Engineering for Practicing Engineers[164])

3.4.2. WATER QUALITY STANDARDS AND INDEX

Water Quality Standards

Water Quality Standards are usually based on one of two primary criteria: *stream standards* or *effluent standards*.

▪ *Stream Standards.* These will usually be based on a system of classification of the water quality based on the intended use of the water. Although stream standards are the most realistic in light of use of the assimilation capacity of the receiving water, they are difficult to administer and control in an expanding industrial and urban area. On variation of stream standards is the specification of a maximum concentration of a pollutant (respective to BOD, for instance) in the stream after mixing at specified low flow condition.

▪ *Effluent Standards.* These may be based on the maximum concentration of a pollutant (mg/l) or the maximum load (kg/day) discharged to a receiving water.

Theses standards are usually related to a stream classification. The usual water quality parameters, usually, do not apply to shellfish area and aquatic reserves because they are non-specific as to detrimental effects on aquatic life (no differentiation between toxic and non-toxic organics/inorganic).

In this cases, a *species diversity index* has been employed as related to either free-floating or benthic organisms.

> ***Species Diversity Index***
> *The species diversity index, KD is an indication of the overall conditions of the aquatic environment.*
> $KD = (S - 1) / \log_{10} I$, *in which S is the number of species and I the total number of individual organisms counted.*
> *It is related to the number of species in the sample. The higher the species diversity index, the more productive is the aquatic system.*

The use of water quality index can bridge the gap between water quality monitoring and reporting methods. The use of a continuous scale provided by a water quality index allows changes in river water quality to be highlighted and the benefits resulting from investment in capital schemes to be evaluated. A *Water Quality Index (WQI)* can be defined as :

> ***Definition of a Water Quality Index***
> *A water quality index (WQI) can be defined as: "... a form of averaged derived by relating a group of variables to a common scale and combining them into a single number. The group should contain the most significant parameters of the data set, so that the index can describe the overall position and reflect change in a representative manner." (SDD[165], 1976; House and Newsome[166], 1988; Tyson and House[167], 1988).*

WQI scores are derived mathematically and lend themselves to computerization. The use of the WQI allows all of the data collected for an annual or longer time series to be used in the assessment of water quality at a particular site. A summary WQI score may be obtained for each site by calculating the 5 percentile WQI for the time series based on the individual sample WQI scores. The continuous scale provided by the use of the WQI facilitates the quantification of "good" and "bad" water quality. This is particular important where water quality approaches "class boundaries" or "water use" thresholds.

The WQI index can be sub-divided to provide four simple categories of water quality (I = water of high quality, II = water of reasonable quality, III = polluted water, and IV = badly polluted water) and possible water use. The WQI is based on nine physico-chemical or micro biological determinands:

- Biochemical Oxygen demand
- Ammoniacal Nitrogen
- Dissolved Oxygen
- Temperature
- pH
- Nitrates
- Chlorides
- Total Coliforms

Because it is not possible to develop a single index which would cover all possible water uses and contain information on toxic determinands directly within its structure, three other independent water quality indices were developed: the *Aquatic Toxicity Index (ATI)*, the *Potable Sapicity Index (PSI)*, and the biotic index. Information on legally adopted water quality standards and criteria was incorporated into the WQI via the development of determinand transformations. Rating curves were used to transform information on the individual determinand concentrations to the same scale. It is generally accepted that some determinands are considered more important indicators of water quality change than others. Consequently weighting coefficients are used to multiply the determinands. Finally the WQI index is calculated using the following formula:

$$WQI = \frac{1}{100}\left(\sum_{i=1}^{n} q_i w_i\right)^2 \tag{3.8}$$

where

WQI	=	water quality index
q_i	=	rating for the ith determinand
w_i	=	weighting factor for the ith determinand
n	=	number of determinands

A scale of 10 to 100 was selected (with a score of 100 for high quality water, and of 10 for highly polluted water). The WQI has now undergone extensive operational validation studies in the UK. Some other governmental classifications are heavily biased towards chemical determinands with the biotic indices being given credence where the chemical picture is confused.

Discovering Trends in River Water Quality
By determining the biological communities in a river, biologists can assess the impact of pollution on aquatic ecosystems, and thereby monitor the quality of the abiotic environment. Since changes in the quality of a river induces changes to biological community structure, observation of these can be used to indicate trends in river water quality.

Diagnosis of river quality may be reported in term of a proposed biological classification scheme. Ecological knowledge on a reference set of indicator organisms is then stored in the form of subjective probability distributions, representing the occurrence of the organism in each of four discrete states of abundance across the quality classes. These distributions provide a support leading to a classification of river water quality at a sampled site.

Expert Systems. A reasoning system based on artificial intelligence approach has been developed by Boyd[168] et al., (1993), who have used an application of Dempster-Shafer reasoning (D-S) for the biological surveillance of river water quality.[169]

Bayesian decision methods remain the primary numerical approach for representing and manipulating uncertainty, and have been employed in expert systems such as *PROSPECTOR*.

The Dempster-Shafer theory of evidence is another method for reasoning with uncertainty, due to its flexibility in the representation and aggregation of evidence. Its advantage include the ability to closely model diagnostic reasoning, and to distinguish between uncertainty, or lack of knowledge, and indifference.[170] Both the Bayesian and the Dempster-Shafer systems provide a ranked order of belief in each quality class for a site, but for the Bayesian model, a point value for each quality class is obtained, while for the D-S model, the belief is given as an interval.

Biota changes are obviously reflecting much longer trends in quality but can also pickup intermittent discharge problems that might otherwise be missed. Biological problems to be addressed would include toxicity data. The sub-lethal, synergistic effects of toxic chemicals must be considered alongside straight toxicity testing. Trophic effects such as bioaccumulation and biota community effects such as species diversity changes are required for a meaningful pollution profile. Macro-invertebrate community analyses with regard to seasonal change are required and more work on the integration of biotic indices with physico-chemical indices of water quality (Bascombe[171] et al., 1989).

International Water Quality Standards
- *Use-specific EC directives (EC[172],1975)*
- *Guidelines for Drinking Water Quality, (WHO[173], 1993)*
- *US Environmental Protection Agency (USEPA, 1976)*
- *European Inland Fisheries Advisory Commission (EIFAC, 1964 to 1983).*

3.4.3. WATER QUALITY ASSESSMENT AND WASTE LOAD ALLOCATION MODELS

Properly applied models are useful tools to study flow processes, assist in problem evaluation, design remedial strategy, recognize limitations in available data, and provide additional information to decisions' makers. *Generic models are useful tool for initial screening, but can never be used as a replacement for site-specific models.* Results of computations usually appear more certain than they really are. The use of contaminant transport models is sometimes necessary for long-term prediction of the migration potential of wastes. A model that is appropriate for one problem may not be applicable to another problem. As a result governmental Agencies cannot specify a list of government-approved models. One of the key requirement in successfully applying flow or contaminant transport models is good-quality, site-specific data. Good documentation of groundwater models is essential because of the complexities involved. The documentation should include a set of test problems, and the model should have been validated with field data.

Most groundwater models are structured to solve a partial differential equation, or a set of partial differential equations (basic equations of groundwater flow and transport). Among these models one distinguishes *four different application categories:*

- *Prediction models*, which simulate the behavior of the groundwater system and its response to stress
- *Resource management models,* which integrate hydrologic prediction with explicit management decision procedures
- *Identification models,* which determine input parameters for both of the above
- *Data manipulation and storage procedures,* which process and manage input data for all the above.

During the 1970's, a variety of mathematical codes were developed for general use in water quality assessment and waste load allocation. In several cases portions of the model codes were altered to reflect different assumptions of water quality interactions in rivers. In 1978-79, a survey of models was prepared on behalf of the *National Council of the Paper Industry for Air and Stream Improvement (NCASI)*[174] and the following models, listed below, were retained for practical use:

- NCASI, 1979: Use of Mathematical Modeling in the Development of Areawide 208 Plans, NCASI Technical Bulletin No. 326
- NCASI, 1979: A Review of the Mathematical Water Quality Model DOSAG, and Guidance for its Use, NCASI Technical Bulletin No. 327.
- NCASI, 1980: A Review of the Mathematical Model QUAL1E and Guidance for its Use, NCASI Technical Bulletin No. 331.
- NCASI, 1980: A Review of the Mathematical Model QUAL2 and Guidance for its Use, NCASI Technical Bulletin No. 338.
- NCASI, 1982: A Study of the Selection, Calibration, and Verification of the Mathematical Water Models, NCASI Technical Bulletin No. 367.
- NCASI, 1984: Modifications to the QUAL2 (SEMCOG) Water Quality Model and User Manual, NCASI Technical Bulletin No. 437.
- NCASI, 1985: A Review of the Mathematical Model WREGEV and Guidance for its Use, NCASI Technical Bulletin No. 452.
- NCASI, 1985: Supplemental User's Guide for Applying the WASP4 Model Program, NCASI Technical Bulletin No.(pending).

The first significant enhancement incorporated to water quality models in the 1970's (Zison[175], 1978; Zison[176] et al., 1978), concerned the modeling of *kinetic reactions in algal/nutrient cycle.* In the 1980's, the increased concern for water quality in impoundment and other deep river systems resulted in multi-dimensional models that provide for separate simulation of water quality in the water columns versus that in the sediment phase (Ambrose[177], 1988). More recently these models have been extended to include simulation of specific organic chemicals and possible food chain processes (Connolly[178], 1985).

The first generation of groundwater quality models was based on classical techniques, like finite difference or finite element methods. Such code produce poor results at high Peclet numbers. Nowadays three dimensional mass transport models based on the random-walk method are being developed, but when dispersion contrasts occur, random walk models suffer from particle accumulation in low dispersive regions. Several authors have already suggested algorithms to overcome this problem. An assessment of these new algorithms has shown that they are not efficient enough. A *new method has been presented by Semra*[179] *et al., (1993).* This method is based on a particle tracking technique: the random-walk method, which is based on the analogy between the advection-dispersion equation and the Fokker-Planck equation. The velocity field is calculated by a mixed hybrid finite element method. To avoid particle accumulation in regions with dispersion contrasts, a new method is proposed to allow conservation of the fluxes between high and low dispersive regions. This new code has been validated by on the basis of a 3D experiment made on a laboratory model.

Water quality models may also be classified in terms of the following characteristics:

a) research / management tool
b) deterministic /stochastic approach
c) deterministic approach (black-box) / stochastic approach (process-oriented)
d) simple / complex process
e) steady-state / dynamic simulation
f) lumped /spatially distributed parameter values

Water quality is deteriorating mainly due to the increase in anthropogenic loading placed on natural water bodies including rivers. Haith[180] (1981) has reported watershed loading functions for nonpoint sources.

A *probabilistic approach for water quality prediction and estimation* has been taken by Paal[181] et al., (1989). Most of the river or river water quality determinands, i.e. hydrodynamical and hydro-chemical regimes, the conditions for waste water discharge and its chemical composition, have time- and space variable nature. Thus the deterministic approach to river water quality prediction and estimation might appear too complicated. Therefore it may be reasonable to consider both the concentration of some constituents defining the water quality and some parameters of the river hydrological regime as random quantities, which are determined by their probability distribution or probability density function. The concentration probability distribution densities can be derived, based on balance relations for a particular contaminant. The river runoff and the pollutant background concentration are considered random quantities which can be described by their joint probability distribution. In that case the comparison of water of water quality characteristics to corresponding standards can be assigned a probabilistic content, which also means that the maximum pollutant loading released to a river can be estimated from the existing water quality standards and the risk of exceeding the legal limits.

Water Quality Models

Some water quality models are presented in this section and are briefly described below. Further literature reviews on such models have been compiled by Onishi[1] et al., (1981),

EPA[2] (1979). The idea is that the reader should gain an overview on the subject and get acquainted with different types of models:

EPA Code. The EPA code[182] (1988) *is a hydrodynamic and water quality model* developed at Environmental Research Lab. The mathematical model includes a hydrodynamic module, and a transport diffusion module.

QUAL2/SEMCOG. This public domain model is used for *water quality predictions.* It predicts both temporal and spatial quantities of the following water quality variables: temperature, conservative materials, carbonaceous BOD, algae as chlorophyll A, phosphorus, nitrogen as ammonia, nitrite, nitrate, dissolved oxygen, coliform, and an arbitrary non-conservative. *QUAL2*[183] is a steady state model with respect to head water flow and point source loads. The primary use of the dynamic load is to simulate the system response to solar radiation and temperature. Reactions' coefficients can be spatially variable, but cannot vary with time.

WASP4+ System. This system *is a proposed family of interrelated water quality modeling tools,* which was developed and is supported by the *US Federal Environmental Protection Agency (EPA CEAM).* It provides generic transport and transformation process options that users can configure to mathematically describe the water quality variables (advection, dispersion, liquid-gas exchange, point mass loading, diffusive mass loading, and transformation) of most water bodies. Transformation processes are incorporated into the model via a kinetic subroutine or user subroutines.

Seftran2 Code. The *SEFTRAN2* code[184] is based on numeric resolution of two-dimensional equations of *fluid motion in a porous media.* A finite element model on the basis of triangular elements and linear interpolation with forward differentiating in time (Huyakorn[185], 1984) has been adopted. The numeric resolution employs the Galerkin method. The spatial discretization grid uses both quadrilateral and triangular elements of varying dimensions. The SEFTRAN2 code can represent both flow and transport problems. The transport problems can take into account both conservative and first-order decay of pollutants. SEFTRAN is superior in terms of accuracy to the models *MOC* and *RANDOM WALK*, but the latter two models are preferable.

MT3D Model. It is a (commercial) modular *three-dimensional solute transport code* for simulation of advection, dispersion and chemical reactions of contaminants in groundwater systems (Papadopulos[186], 1990).

SWMM4.0 Model. It is one of the most sophisticated models in the water resource field. The *EPA Storm Water Management Model (SWMM),* of the University of Florida, is running on high speed computers (last release: 1988).

Vertical Horizontal Spread (VHS) Model. This model is used by the *US Environmental Protection Agency (EPA)* to determine when solid wastes need to be treated as hazardous wastes

MODFLOW Model. It is a groundwater flow model developed by McDonald and Harbaugh (1984) for *USGS (US Geological Survey).* The model is using three-dimensional finite-difference method.[187] The code is written in a modular format which allows the user to tailor the model to the specific area under consideration. The acceptance of this model is good.

STROP Model. It is a *horizontal two-dimensional groundwater flow model* that uses an approximate method for determining streamlines in a three-dimensional space as well as a method for modeling the actual transport of dissolved contaminants in the neighborhood of recoveries and injections (Burger[188], 1986).

Solute Transport Model of Konikow and Bredhoeft (1978): USGS (US Geological Survey) has developed a solute transport model (Konikow and Bredhoeft[189], *1978)* which can be used to calculate, for example the flow rate of mineralized water into a potable aquifer. The localized area of high chloride concentration can be represented by a source term.

GEOTHER/VT4 Code. This code has been developed by Bian et al.,[190] (1989) at Pacific Northwest Laboratory (PNL). It is an improved version of GEOTHER (Faust and Mercer[191], 1983). This code has been used for a realistic *simulation of the geothermal conditions in a high-level radioactive waste repository.* It is a three-dimensional, two-phase groundwater simulation code, which is based on a general mathematical model of a waste repository system that describes the three-dimensional flow of groundwater (vapor and liquid water) and heat transport from nuclear waste in a porous medium. The code solves the mass, momentum, and energy balance equation for porous/nonporous materials (e.g., host rock and waste nuclear container). Using Darcy's law in place of the momentum equations, the conservation equations are reduced to three nonlinear partial differential equations in terms of the dependent variables, pressure, enthalpy, and rock/solid temperature.

FEMWASTE Code. It is a *finite element model* of waste transport through saturated-unsaturated porous media (Yeh and Ward[192], 1981).

TRIPM Code. It is a *two-dimensional finite element model* for the simultaneous transport of water and reacting solutes through saturated and unsaturated porous media (Gureghian[193] ,1983).

3.4.4. MODELS FOR AGRICULTURAL LAND MANAGEMENT

A description of different environmental models, including water models, can be found in "Transferring Models to Users"[194], (1990).

CREAMS Model. It is an older scale model for chemicals, runoff, and erosion from agricultural management systems.[195]

GLEAMS Model. It is a more recent model, which was developed as an extension to CREAMS. *Both models simulate soil water balance and surface transport of sediments and chemicals from agricultural field management units.* GLEAMS in addition simulates chemical transport in and through the plant root zone. Several other features were added such as irrigation/chemigation options, pesticide metabolite tracking, and software to facilitate model implementation and output analysis. The GLEAMS model was designed for relative predictions and for comparative analysis.[196]

3.4.5. MODELS FOR ESTIMATING NITRATE AND PHOSPHORUS CONTAMINATION

Nitrogen Leaching Models

Numerous models describing nitrogen cycling in crop, pasture, and rotation systems of agriculture have been published. (Tanji[197], 1982). Most predicted plant growth but some focus on the relatively small nitrogen flux which is leached through the soil. The model of Burden (1982) is a simple model for grazed pastures. It gives a steady-state annual water and nitrogen balance in the root zone and uses only five unmeasured parameters.

BICHE Model. This models is of intermediate complexity (Thiery and Seguin[198], 1985). It has two unsaturated zone storages for water and nitrogen, takes account of mobile and immobile fractions of water, and introduces temporal variations. Monthly rates for nitrogen transformations and crop growth are superimposed on a five-day time step for rainfall. This model which uses ten mass transfer parameters and five hydrological parameters, requires considerable time series data for calibration.

LEACHM.N Model. One of the most rigorous nitrogen leaching models, LEACHM.N by Hutson and Wagenet[199] (1985), uses Richard's equation to describe unsaturated flow of water in a layered soil profile and separate storage of organic, ammoniacal, and nitrate nitrogen within each layer. This model accounts for plant and root growth in detail, and more than fifty parameters must be specified.. Finite difference in space and time is employed with variable time steps from a few minute to 12 hours, depending on the incidence of rain fall or irrigation.

NITWIT Model. A new simpler model, *NITWIT* (Nitrogen and water in transit) was developed by Dillon[200] (1989). This latter model follows the general form of the BICHE model and has monthly time steps. Nitrogen inputs to the pasture system are symbiotic fixation associated with clovers, nitrogen contained in rainfall, and nitrogen fertilizers. Outputs include animal products (meat, milk, and wool), volatilization of ammonia, and leaching. Losses through erosion, run-off, and denitrification are not considered in this model. Although the model assumes no long-term change in the inorganic content of soils, the organic nitrogen storage is allowed to increase with time by immobilization. In NITWIT, water is routed through two soil stores, neglecting lateral flow.

DIVAST Model. It is a diffuse source vertical analytical solute transport model (Dillon[201], 1986), and it is applied as *management model* for preliminary regional evaluation of potential groundwater contamination due to diffuse sources and for prediction of the future distribution of solutes in the aquifer under alternative land management strategies. It discriminates concentration changes through the aquifer thickness and models an unconfined aquifer with uniform thickness and recharge rate on a vertical slice aligned with the prevailing direction of regional groundwater flow. The use of flux-averaged concentrations allows account of flow in dual porosity media. DIVAST may be used together with the NITWIT model for groundwater quality management and for assessing the risk of nitrate pollution.

NITSOL Model. The NITSOL / PHOSOL *model* (see below) were develop for use on a regional scale. These models are coupled to a hydrologic model, *DEMGEN*. NITSOL is a numerical model (FORTRAN) that computes adsorbed and dissolved ammonium, dissolved nitrates and solid organic nitrogen concentrations in the soil.

RENLEM Model. It is a regional nitrate leaching model for estimating nitrate leaching from nitrogen input by atmospheric deposition, fertilizers, and manure (Kragt and de Vries[202], 1987).

Phosphorus Leaching Models

In the *planing of a future reservoir*, the question whether the water quality will satisfy the foreseen water use and how the conditions for good-quality water can be maintained is of great importance. *Usually the estimation of eutrophication related to water quality characteristics based on an OECD eutrophication modeling approach is used.* Phosphorus is the key element controlling reservoir eutrophication. The total annual phosphorus load can be estimated on the basis of instream phosphorus concentrations and flow measurements (C-Q relationship). The second methodology is based on the phosphorus export coefficients, land use patterns, population and livestock densities in the watershed. *Various sources of data on phosphorus watershed export coefficients are available* (Rast and Lee[203], 1983; Reckow et al.,[204] 1980). Some authors (Kauppi[205], 1984; Gillion[206], 1984) have developed equations for phosphorus emission calculations with land use in watershed as variable.

SPNM (Phosphorus and Nitrogen) Model. *SPNM* is a model for predicting sediment, phosphorus, and nitrogen yields from agricultural basins.[207]

PHOSOL Model. This model computes ortho-phosphate, mobile and solid organic phosphorus and solid calcium-magnesium phosphate *in the soil.* The adsorption of ortho-phosphate on Fe and Al components in the soil is, however, dominating over microbial processes (Hopstaken[208] et al., 1989).

REPTRAM Model. It is a *regional phosphate-transport model* predicting the impact of phosphate application through animal manure on phosphate transport in the unsaturated zone (Schoumans[209] et al., 1986).

RESAM Model. It is a *regional soil-acidification model* relating soil-water quality (e.g., aluminum concentrations) to atmospheric deposition of SO_2, NO_x, and NH_3 (de Vries[210], 1987).

3.4.6. MODELS DEALING WITH DEPOSITION RATE OF TRACE POLLUTANTS

Such models calculate the deposition rate of pollutants (heavy metals) transported in air and deposited into a water shell. Some of these models have been compared and are mentioned below:[211, 212]

EMEP Model. (European Monitoring and Evaluation Program) is two-dimensional receptor-oriented trajectory model, in which the transport layer corresponds to the planetary boundary layer (Peterson[213] et al., 1989). There exists also an *EMEP2* model.

HHLRT Model. The Hamburg Long-Range Transport Model is a source term oriented, three-dimensional, Monte-Carlo multi particles simulation model (Krell and Roeckner[214], 1988).

TREND Model. This model is a receptor-oriented, climatologic, statistical trajectory model with three transport layers (v. Jaarsveld[215] et al., 1986; v. Jaarsveld[216], 1990), which is combined with a Gauss model.

These models *(EMEP, HHLRT, and TREND)* have been used with success for the dispersion simulation of sulfur or nitrogen.(Eliassen and Saltbones[217], 1983; Eliassen[218] et al., 1985; Lehmhaus and Roeckner[219], 1986). The above mentioned models are of the Lagrangian type and make use of the K-theory for calculation of the turbulence. The advection is calculated separately.

How good such models are for assessing transport of heavy metals in air and deposition into water is not yet known with enough certainty. Comparisons of atmospheric transport models by Klug[220] et al., (1985) and Martens[221] et al., (1987) indicate that there is no known model which is generally better than the others.

3.5. Guidelines for Drinking Water Quality

3.5.1. RISK TO THE DRINKING-WATER IN THE CASE STUDY AREA

Concerning drinking-water quality at the regional/local level, the aim is the protection of public health. National as well as international standards are meant to ensure "the safety of drinking-water supplies through the elimination or reduction to a minimum concentration, of constituents of waters that are known to be hazardous to health".[222]
The judgment of an acceptable level of risk to public health due to the quality of the drinking-water is a matter in which society as a whole has a role to play. The final judgment is based on cost-benefit calculations for any particular circumstances. In general, guideline values for drinking-water quality have a certain degree of flexibility

and enable a judgment to be made regarding the provision of drinking-water of acceptable quality.

There are some "golden rules" regarding the quality of drinking-water at regional / local level:

a) Protection of water supplies from contamination is the first line of defense; source protection is the best method of ensuring safe drinking-water
b) Efforts should be made to achieve a drinking-water quality as high as practicable
c) Once the risk to health has been recognized, the availability to alternative supply sources must be considered
d) To the highest extent possible, water sources must be protected from contamination by human and animal waste
e) The category of the population at highest risk are infants and young children, people who are debilitated or living under unsanitary conditions, the sick, and the elderly
f) Regarding sources of drinking -water, the control of microbial contamination must always be of paramount importance; never make compromises.

Insufficient epidemiological evidence makes difficult the assessment of the risks associated with variations in microbial quality of the drinking-water. According to the international practice, "microbial risk can never be entirely eliminated".

Toxic chemicals in drinking-water induce health risk which differs from that caused by microbiological contaminants. Chemical standards for drinking-water are of secondary consideration in a supply subject to severe bacterial contamination. Contaminants that have cumulative toxic properties (e.g., heavy metals, carcinogenic substances, etc.) are of particular concern due to their ability to cause adverse health effects after prolonged periods of exposure.

Chemical by-products in drinking-water are potentially hazardous. The risk to health from these by-products are extreme small in comparison with the risks associated with inadequate disinfection.

The presence of naturally occurring radionuclides in drinking water could induce radiological health risk.

Parameters that must be considered when assessing the water quality are of various types: e.g., source protection, treatment efficiency and reliability, protection of the distribution network (corrosion control).

It is widely recognized that "infectious diseases caused by pathogenic bacteria, viruses and protozoa or by parasites are the most common and widespread health risk associated with drinking-water". Pathogenic agents have several properties that distinguish them from chemical pollutants:

i) Pathogens are discrete and not in solution
ii) Pathogens are often adherent to suspended solids in water; their average concentration in water is not a relevant risk indicator

iii) The invasiveness and virulence of the pathogen combined with the immunity of an individual give relevance to the likelihood of a successful challenge by a pathogen
iv) The dose response of pathogens is not cumulative
v) There is no tolerable lower limit for pathogens; water for various consumptions modes should contain no agent pathogenic for humans.

Next, a number of recommendations are considered for regional risk assessment and safety improvements:

- Water intended for drinking and household purposes must not contain waterborne pathogens
- Drinking-water must essentially be free of human enteroviruses; this would ensure negligible risk of transmitting viral infection (e.g., any drinking supply subject to fecal contamination presents a risk of viral disease to consumers)
- Drinking-water treatment can considerably reduce the levels of viruses but may not eliminate them completely from very large volumes of water
- The water must meet guideline criteria for turbidity and pH, bacteriological quality, and parasitological quality
- The monitoring function is an integral part of the responsibilities of the water supply agency; an independent body should verify that the waterworks correctly fulfills its duties
- Sampling frequencies for treated water leaving the waterworks depend on the quality of the water source and the type of treatment (e.g., One sample every two weeks for groundwater source, one sample every week for surface water source)
- A sound surveillance program is a continuous and vigilant public health assessment and overview of the safety and acceptability of drinking-water supplies.

The *TDI indicator* (Tolerable Daily Intake) is an estimate of the amount of a substance in food or drinking-water, expressed on a body weight basis (mg/kg, μg/kg of body weight), that can be ingested daily over a lifetime without appreciable health risk.

The TDI indicator is given by the following relation:

$$TDI = (A \text{ or } B) / UF \tag{3.9}$$

where

A - no observed adverse effect level
B - lowest observed adverse effect level
UF - uncertainty factor.

A guideline value (*GV*) derived from *TDI* has the following expression:

$$GV = (TDI * bw * P) / C \tag{3.10}$$

where

bw - body weight (60 kg for adults, 10 kg for children, 5 kg for infants)

P - fraction of the TDI allocated to drinking-water

C - daily drinking-water consumption (2 liters for adults, 1 liter for children, 0.75 liter for infants).

Chemical contaminants of drinking-water supplies are present together with other inorganic and organic constituents (mixtures). Guideline values must be calculated separately for individual substances, without taking into consideration their interaction. Large margin of safety are usually incorporated in the majority of the guideline values.

Concerning *radiological aspects*, a number of recommendations were recently outlined by WHO (World Health Organization), namely:

- The recommended reference level of committed effective dose is 0.1 mSv from 1 year's consumption of drinking-water; below this level the drinking-water is acceptable for human consumption and any action to reduce the radioactivity is not necessary;
- The recommended guideline activity concentrations are 0.1 Bq/liter for gross alpha and 1 Bq/liter for gross beta activity.

Regarding *acceptability aspects of drinking-water*, a number of different constituents should be considered:

- Physical parameters (color, taste and odor, temperature, turbidity)
- Inorganic constituents (aluminum, ammonia, chloride, copper, hardness, hydrogen sulfide, iron, manganese, dissolved oxygen, pH, sodium, sulfate, total dissolved solids, zinc)
- Organic constituents (toluene, xylenes, styrene, ethylbenzene, monochlorobenzene, dichlorobenzenes, trichlorobenzenes, synthetic detergents)
- Disinfectants and disinfectant by-products (chlorine, chlorophenols).

Safety Measurements

It is of relevance to remember that it is always better to protect water from contamination than to treat it after it has been contaminated. The watershed should be protected from human activities; dealing with risk assessment and safety management in the region, provisions must be taken in order to reach such target.

When water treatment is necessary, the intensity of the treatment must depend on the degree of contamination of the source water. A number of treatment processes could be considered, namely:

- pre-treatment
- coagulation, flocculation, and sedimentation
- rapid and slow sand filtration
- disinfection
- fluoride removal.

Regarding *corrosion control of pipelines for drinking water*, the main strategies include:

- the control of environmental parameters affecting corrosion rate
- electrochemical measures

– considerations of system design.

Contingency plans should be considered in case of emergency due to natural disasters (e.g., earthquakes, floods, damage to electrical equipment by lightning strikes), accidents (spills in the watershed), damage to treatment plant and distribution system, and human actions (e.g., sabotage).

In an emergency situation the regional/local public health authorities must be consulted about appropriate action.

3.6. Resource Prediction and Management Models

Groundwater management models have been developed to systematically relate the behavior of a flow system to the cost of utilizing scarce aquifer supplies. Such management models incorporate economic, technological, political, and institutional aspects. They can perform both management and optimization and indicate course of actions which will be consistent with stated management objectives and constraints.

This is accomplished by coupling physical principles of groundwater flow with optimization theory. Optimal flow and solute transport management strategies can be developed by including equations of solute transport and flow as linear or nonlinear constraints in an optimization model (review of the methods by Gorelick[223], 1983). A second category of models use gradient control and velocity influence coefficients (Colarullo[224] et al., 1984; Gorelick and Lefkoff[225], 1985). Such models may be too restrictive if some contaminant movement is acceptable, or impractical for regional use if the area of contamination is large. A third method makes use of influence coefficients describing the effect of a change in potentiometric head on steady state contamination (Datta and Peralta[226], 1986). This approach is also too restrictive if groundwater quality constraints must be considered for many locations. A groundwater quality model that can be embedded as linear constraints in steady state groundwater management models has been proposed by Prathapar and Peralta[227] (1989). This special equation accounts for initial level of contamination, change in contamination due to pumping, steady state flow and molecular diffusion. The model estimates contaminant density at the end of a specified time period in each cell (of uniform concentration). The applicability of this mathematical model as a water quality constraint has been demonstrated. The calibration took place by numerical simulation, using a finite element solute transport model.

A key part of the planning effort in water management is the capability to make accurate water demand projections and to match demands with existing and possible future supplies over time. Bishop[228] et al., (1988), have developed a color-graphics based water demand forecasting model based on a graphic interface (GIS) coupled with the model.

*The **International Groundwater Modeling Center (IGWMC)** was established in 1978 at the Holcomb Research Institute, Indianapolis, Indiana, USA, to advance the use of modeling methodologies by regulatory and management agencies in the development of effective groundwater management procedures. The center deals also with acquisition and distribution of models, supporting software, and documentation, etc. Another office is established in Delft (NL), and is supported by the Institute of Applied Geoscience of the Dutch research organization TNO.*

There exist special models (*BALANCE*[229], *CERES-Maize*[230]) *for assessing the water demand for irrigation purpose in a region.* CERES-Maize is an corn growth simulation model widely used in the US BALANCE. It is a water balance program to compute the maximum irrigation surface the stream flow can sustain. The daily minimum and maximum temperature and precipitations are generated for a full year using WGEN.

Software companies like *Advanced Computer Applications* (ACA)[231] are developing customized integrated programs and interfaces for specific applications and regions of interest. For instance a river basin simulation model, e.g., *MITSIM* (Strzepek and Lenton[232], 1978), *TOXSCREEN* (Hetricks and McDowell-Boyer[233], 1984), a site specific implementation for the river Rhine *(XSPILL)*, a prototype groundwater contamination model *FEFLOW/FEMCAD* (Fedra and Diersch[234], 1989).

■ *MITSIM.* It is a variable time step Monte Carlo, hydro-economic simulation model that operates on the basis of hydrological catchments or river systems, including the groundwater system. Its purpose is a stochastic analysis of the hydrologic and economic performance of water resource development projects in the context of the river basin system.

■ *TOXSCREEN.* An adaptation of the *river model* component of TOXSCREEN simulates pollutant dispersion in an arbitrary river segment (Weighkricht and Winkelbauer[235], (1987).

■ *XSPILL.* It is a *dynamic analytical model* that simulates the propagation of an accidental spill of a chemical, represented by its initial mass, time pattern, of the spill, and a first-order decay rate of the chemical substance.

Such models can be integrated with a GIS system and coupled to a rule-based expert system (e.g., *MEXSES*[236]) for performing environmental impact assessment of river development projects at a screening level.
Heatwole[237] (1986) has developed a *basin scale model* for the evaluation of the best management practices implementation programs. Another computer simulation model for *territorial systems of intensive agriculture* has been presented by Roosaare[238] (1984).

3.7. Waste Loads Management

3.7.1. RISK-BASED DECISION MAKING REGARDING MIXED-WASTE DISPOSAL SYSTEMS

An efficient approach for risk-based decision making regarding mixed-waste disposal systems has been developed by Roberds[239] *(1990)*. The approach followed by Roberds accommodates a system which consists of multiple interacting sites, each containing an inventory of mixed wastes, with multiple pathways to a variety of potential receptors, a variety of options (e.g., for remedial actions, barriers, operational controls), necessary tradeoffs among multiple and possibly competing objectives, and inherent uncertainties regarding the various parts of the system and how they behave.

▪ The *general approach* consists of defining and quantifying the set of objectives and criteria in terms of system attributes, through scoring/decision rules, and relating the system attributes to various system parameters through models. Specific system configurations are then evaluated and ranked. Finally the system configuration is optimized.

▪ The *utility of the system* can be defined for any configuration as the degree to which that configuration satisfies the set of objectives, considering the utility of achieving various levels of each of the objectives and the relative importance of the various objectives, as defined by scoring and decision rules (mini-max problem). In such a way the utility of a specific system configuration can be quantified.

▪ *Conceptual models* must be defined for the system, which qualitatively describe the major components and activities of the system, and the possible conditions and processes involved, as they affect the relevant system attributes (e.g., radioactive dose). Some examples of conceptual models are: conceptual dose model (based on the barrier approach), conceptual chemical concentration model (considering only the groundwater pathway), conceptual cost model. The conceptual model may be expressed either as a transport pathways scenario or in terms of hierarchy of functional relationships conducting to the final event (e.g., individual dose). Hence, the conceptual model can be described in terms of a system model, i.e., the top-level of the hierarchy, and detailed performance models, i.e., the lower levels of the hierarchy, where these performance models may also consists of several levels.

▪ *Possible and credible scenarios*, based on event and fault-tree analysis, must be identified and described. The parameters which quantify the description of the system must be defined and their values assessed for each significant scenario and configuration. For instance these parameters may include different types of wastes (hazardous or radioactive), waste matrix/packaging release factors, types of retaining barriers, natural pathway transport factors, biotic transport factors, (and in case of radioactive substances, the isotope-to-dose conversion factors), potential demographic receptors.

■ The *scenarios and parameters*, with their uncertainty, must be assessed and evaluated based on available information. The uncertainty of a scenario occurrence can be best expressed in terms of probabilities. For a comprehensive and mutual exclusive set of possible scenarios for a given configuration, the sum of their probabilities of occurrence must equal one. The uncertainty in parameter values for any scenario/configuration can best be expressed in terms of probability density functions (pdf) which express the relative likelihood of what the actual value will be, or cumulative distribution functions (cdf), which express the probability that the actual value will be less than some value. This is applicable to parameters with unique values, or the characteristics (e.g., mean, standard deviation) of a population of values. Parameters may be related (or correlated), either spatially, temporally, or with other parameters. If a correlation is observed, it can be described in terms of a covariance function or in terms of a conditional probability. If a large enough representative data set is available, then the assessment of a probability distribution for a parameter value can be developed directly from the statistics of the data set. Subjective probability assessment can be developed to represent individual opinions or to represent the consensus of a group (Roberds[240], 1988).

■ *Mathematical models*, which quantify the conceptual models, must be developed in order to relate system attributes to the system parameters. Because of the inherent simplifications and approximations in the mathematical model, the relationship between the system attribute and system parameters will not be exact. A model error term may be added, which expresses the relationship between the actual and predicted attribute values for a given set of system parameters' values. This model error term can be described probabilistically, with a mean value corresponding to the general degree of conservatism (or bias) and a standard deviation corresponding to the degree of approximation (or uncertainty). Uncertainties in the parameter values, as well as the scenarios and the models themselves, result in uncertainties in the attribute values. Using Monte Carlo simulation or second moment approximation technique allows to relate the uncertainties in the attribute values with the various components of uncertainty, in terms of *pdf's*. In order to focus on the important factors, the sensitivity of the attributes to those various factors should be determined. Sensitivity can best be discussed in terms of partial derivatives, so that the larger the partial derivative of y with respect to a given factor x_i, the more sensitive y is to x_i or the more effect that changes in x_i have on y. The sensitivity of each of the attributes to each of the parameters can be determined in terms of partial derivatives. Similarly, the sensitivity of each of these parameters to more detailed parameters can be determined. Thus, through the chain rule the sensitivity of the system utility to any of the most detailed parameters can be determined. More importantly, the sensitivity of the moments of the *pdf* for an attribute with respect to the moments of the *pdf's* for the various factors can also be determined.

Using the above-mentioned methodology, a specific system configuration can be evaluated and optimized with respect to the degree to which it satisfies the set of objectives/criteria (including costs).

3.7.2. RISK-BASED APPROACH TO THE EVALUATION OF GROUNDWATER CONTAMINATION FROM LAND-BASED WASTE DISPOSAL

Waste treatment prior to land disposal is an obvious technological approach to reduce groundwater contamination potential. Three problems must be considered in reducing groundwater risks from land disposal:

- the siting and location of land disposal systems
- the design, operation, and maintenance of each facility
- the characteristics of wastes to be disposed.

The health-based water quality limits will be determined from constituent-specific reference doses or from a specified risk level for carcinogens. The risk must be related to waste leachate concentrations as defined by a standard leach test. The release, pathways of exposure, and risk from toxic substances in land disposal systems are multimedia in scope. They involve air, soil, plants, surface water, and groundwater. *Comprehensive sets of scenarios can be found in EPA's MULTIMED modeling system (USEPA[241], 1988).*

The *HELP model* (Schroeder[242] et al., 1984) can be used to simulate leachate volumetric flux. The saturated zone solute transport model described by Huyakorn[243] et al., (1987) has been implemented in the multimedia exposure assessment model *MULTIMED*, which can be used for evaluating the land disposal by hazardous wastes.

3.7.3. MODELS FOR ESTIMATING GROUNDWATER CONTAMINATION FROM IMISCIBLE ORGANIC CHEMICALS

Analyses of multiphase flow pertinent to water resources problems have been discussed by a number of authors (Faust[244], 1985; Osborne and Sykes[245], 1986; Parker[246] et al., 1987), Kuppusamy et al.[247], 1987). Corapliogiu and Baehr[248] (1987), and Baehr and Corapliogiu[249] (1987) have presented a model for multiphase, multicomponent organic transport under conditions in which no convection occurs in NAPL and gas phases. *The analysis of coupled fluid flows and component transport in three fluid phase systems with mobile water and NAPL phases* has been discussed by Abriola and Pinder[250] (1986). A model following the approach of the latter authors has been presented by Parker[246] et al., (1989). This new model for multiphase flow and component transport is easier to use, due to some simplifications which has been introduced to it.

A model for *transient contaminant transport from single component non-aqueous phase liquid pool dissolution* has been presented by Chrysikopoulos[251] (1993), for two-dimensional, saturated, homogeneous porous media.

- *3DFEMWATER*. It is a three-dimensional finite element model of water through saturated-unsaturated media.[252]
- *Micro-Fem*. It is a Micro-computer multilayer steady state finite element program for groundwater modeling.[253].[254]

3.7.4. SITE CHARACTERIZATION PROCESS AT HAZARDOUS WASTE MANAGEMENT SITES

A central issue in the siting and design of hazardous waste disposal facilities is the *groundwater travel time beneath and adjacent to the facility.* The geologic environment is determining this factor. Differences in hydraulic conductivity of several orders of magnitude within a single geologic unit are not uncommon. With limited hydrogeological data on the repository site, there is no way to avoid the uncertainty that underlies any prediction of groundwater flow patters and contaminant transfer rates. To provide a framework for making engineering judgments in the face of such uncertainty, hydrogeologic modelers have adopted two methodologies:

(a) a geostatistical methodology, based on the theory of spatial stochastic processes, that can be used to quantify the uncertainty in medium properties, and
(b) methods of stochastic simulation that can be used to assess the impact of uncertainties in input parameters on the uncertainties associated with the predicted output variables.

One key objective of site characterization is to provide reliable indication of the pathways and rates of transfer of contaminants. This evaluation is carried out through a combined program of geologic mapping using borehole observations, and by taking measurements, both of hydraulic conductivity and hydraulic head Such evaluation allows to reduce the uncertainty but has its cost. A Bayesian approach may be adopted, whereby the subjective estimates of the medium properties are combined and updated with field measurements. A corresponding methodology has been proposed by Smith[255] et al., (1990) and is summarized below.

Prediction of the Fate of a Heavy Metal from a Municipal Sanitary Landfill
A feasibility study for a proposed sanitary landfill was prepared by Doran and Thresher[256] (1987). The study applied contaminant transport techniques to predict how one selected compound, chromium, contained in leachate generated from a proposed natural attenuation landfill would cause a given regulatory standard to be exceeded. Modeling was performed in three steps using a leachate generation model, an unsaturated zone transport model, and a groundwater transport model. Long term meteorological data were utilized to estimate a water balance for the landfill. Leaching and Landfill infiltration were modeled using a program supplied by *WDNR* (Scharch[257], 1985). Saturated zone contaminant transport was simulated by the *Random Walk Model* (Prickett[258] et al., 1981). This model requires groundwater flow velocity input, which was simulated by PLASM, the Prickett-Lonnquist Aquifer Simulation Model (Prickett and Lonnquist[259], 1971). *PLASM* is a quasi-3D finite difference groundwater flow code which has been widely tested and accepted in simulating groundwater flow systems.

Methodology for Landfill Site Characterization

1. *Selection of the form of the spatial stochastic process for hydraulic conductivity: the assumption of statistical homogeneity requires that within each terrain unit, the spatial variability can be characterized by a single probability distribution, with constant moments. Hydraulic conductivity has frequently been represented by a log normal probability distribution.*
2. *Determine a Bayesian before any field measurements are taken, and calculate a normal distribution with expected values and the corresponding uncertainty values.*
3. *Incorporate scale effects using variance reduction concepts: the statistics of the measured values are not immediately compatible with the scale of the finite element mesh that is used in solving the groundwater flow equation.*
4. *Perform prior unconditional uncertainty predictions: use a Monte Carlo technique on a subset of the prior values to make an unconditional uncertainty prediction of travel time.*
5. *Use field measurements to update estimates of uncertainty: the statistics describing the variability in hydraulic conductivity can be updated using both hydraulic conductivity measurements and hydraulic head measurements. The former approach is based on Bayesian updating, the latter on inverse simulation.*
6. *Perform posterior conditional uncertainty prediction on travel time*
7. *Use a decision-theory framework to assess alternative courses of action and data worth (Massman and Freeze*[260]*, 1987): for each of the alternatives, an objective function can be defined as discounted stream of benefits, costs, and risks. Here risks are defined as the expected costs associated with the probability of failure.*

Another paper by Massmann and Asce[261] (1990) summarizes recent developments that have been made for *assessing risks and costs of groundwater contamination.*

Bayesian Simulation and Inverse Simulation Bayesian Simulation

- The Bayesian updating equations use hydraulic conductivity measurements to reduce the uncertainties in describing the spatial pattern of hydraulic conductivity variations.

- Hydraulic head measurements also carry information on the spatial variability in hydraulic conductivity. A goal of inverse simulation is to use that information to construct a model of the hydraulic conductivity variations that has a smaller uncertainty than an approach based solely on hydraulic conductivity data.

3.7.5. NETWORK SIMULATION

The simulation of networks can help assessing municipal sewage loads. The Danish computer model *MOUSE*[262] (1990) may be used for *detailed dynamic simulation of networks.* The model contains three modules describing respectively the runoff processes (hydrologic model), the flow propagation through the system (by integrating the complete de Saint Venant equations, i.e., an hydraulic model), and the waste load discharges (base on the qualitative model *SAMBA*). MOUSE performs a detailed flow simulation based on design rainfall events providing time series of water level (piezometric heads), discharges and velocity in all manholes and pipes, as well as storage results through time.

The Belgian model *GLAS*[263] can be used for *quantitative and qualitative impact calculations*. The sewer network is modeled as a series of reservoirs (parallel or in series) linked by throttle pipes with a maximum throughput and delay times as only parameters. General data about the waste load concentration and sedimentation process in the network are also required for the different subsystems. Overflow results can also be treated statistically. Results from the program are overflow frequencies, volumes and mean discharges, and absolute and relative waste loads.

3.8. Commercial Software for Water Models

The *Hydrologic Engineering Center in California (HEC)*[264] provides technical services and is distributing ca. 100 computer programs (of which 43 are implementations for PC's), user's guides, and study method guides, for application in hydrologic engineering and planning analysis (Corps of Engineers[265], 1989).

Another source for technical software are the *US and foreign governmental Agencies*, and specialized institutions doing research and development work in the field of hydrology. Information on software is also available from *Electronic Bulletin Board Systems (BBS)*: for instance, from the *Center for Exposure Assessment Modeling (CEAM)*.[266]

Center for Exposure Assessment Modeling (CEAM)

The Center for Exposure Assessment Modeling (CEAM) was establish to meet the scientific and technical exposure assessment needs of the United States Environmental Protection Agency's (EPA) Programs and Regional Offices and the various state environmental agencies.

To support environmental risk-based decisions concerning protection of air, water, and soil, CEAM provides predictive exposure assessment techniques for aquatic, atmospheric, terrestrial, and multimedia pathways for organic chemicals and metals.

A wide range of analyses' techniques is provided by CEAM, ranging from simple desk-top techniques suitable for screening analysis through computerized steady-state models to sophisticated, state-of-the-art continuous simulation models. CEAM has developed different environmental simulation models (12): for urban and rural nonpoint sources, conventional and toxic pollution of streams, lakes and estuaries, tidal hydrodynamics, geochemical equilibrium, and aquatic food chain bioaccumulation.

CEAM is also supporting a large user software-base of CEAM models.[9]

The following remarks concern the availability of computer software/codes for water modeling:

■ *Models Developed at Universities.* The models are often still under development, and usually older versions only are made available to the user; the documentation may be scarce or missing; the models may not have been tested or follow a theoretical approach along the research line pursued at the university in question.

■ *Models from Governmental Agencies.* Not all of them are made available to the public. Usually the models may be tide up to a database containing regional hydrogeologic data.

■ *Commercial Software.* This is usually available through specialized software companies, like *Scientific Software*, or from hydrologic engineering centers, like HEC in California.

3.8.1. DESCRIPTION OF SELECTED CODES FROM "SCIENTIFIC SOFTWARE"

■ *MODFLOW^{PC}*. It is a *modular three-dimensional finite-difference groundwater flow model,* by McDonald and Harbaugh. The code consists of an input processor (MODINP), an output processor MODOUT, and a special utility program MODREC. A routine package (STR1), preconditioned conjugate gradient solver package (PCG2), and the BCF2 package, which permits the wetting of dry cells, have been included. MODFLOW^{PC} represents the most up-to-date IBM/AT PC-version code available. The code includes a velocity vector package which provides for the computation of Darcy velocity at block boundaries and subsequent superposition on head contours for use with MODGRAF^{PC}. Complementary programs to MODFLOW are available:

■ *MODVEL.* It is a new program which uses unformated files of two-dimensional groundwater velocity vectors for each model layer. MODGRAF, MODOUT, MODCELL, TECVIEW are utility routines.

■ *MODLOCAL.* It consists of two programs, MODGROW and MODSTRES. MODLOCAL can be used to construct local MODFLOW models from an existing regional model. MODGROW expands all 2-D arrays including IBOUND arrays. MODSTRES modifies all stress package files. The maximum resultant size of the local grid is 200 x 200 respectively (400 x 400). It is used to modify the regional MODFLOW well, drain, river, general head boundary and stream flow routine.

■ *MODGRAF.* It was designed to provide graphics capability to MODFLOW, and includes groundwater velocity vector capability. It can create contour plots.

■ *MODINVEM.* It is a 386/486 extended memory version of MODINV. It contains MODFLOW and is a complete groundwater modeling tool kit.

■ *MODPATH.* It is a particle tracking program to MODFLOW (Pollock[267], 1990).
TABLE 3.4 gives an overview of the codes available through the company *Scientific Software*[268]:

TABLE 3.4. Software for water modeling (available from "Scientific Software")

Model Name	Description / Author	Price ($)
AIRFLOW	*2-D radial symmetric model; simulation of vapor flow in the unsaturated zone;* It computes vapor pressure, pathlines, vapor velocities under steady-state conditions for an ideal gas in a vertical section; single phase vapor flow; applicable to different gases	650.-
AQUA^{EM}	*2-D groundwater flow and Transport equations,* with inhomogeneous and anisotropic flow conditions; steady-state and transient transport of contaminants and heat with velocity-dependent dispersion, conversion decay and adsorption; Galerkin finite element method	2,300.- to 2,500.-
AQUASEA^{EM}	*Tidal flow problems* in estuaries, coastal areas, lake circulation; mass, heat and suspended solids' transport.	6,000.-
AQUIFEM	*For analyzes of 2-D horizontal groundwater flow;* anisotropic, heterogeneous, phreatic or confined, leaky or non leaky aquifers under transient or steady-state conditions; time varying boundary conditions. *Application:* pumping/recharge wells, springs, drains, excavation dewatering, groundwater discharge, evapo-transpiration effects of geologic faults, and exchange of water between the aquifer and surface-water bodies.	1,100.-
ARMOS	*Numerical model to simulate areal flow of water and separate phase light hydrocarbon* in unconfined aquifers under natural gradients during or after spills or leak from pipelines, or underground tanks, and or induced gradients associated with well or trench free product recovery systems. Assumes vertical equilibrium pressure distributions of water / hydrocarbon in saturated and unsaturated zones associated with changing water and oil tables. Needs SOILPROP	990.-
BALANCE	Calculates *mass transfer* for geochemical reactions in groundwater; redox reactions; isotopic compositions	250.-
BIO1D	*Biodegradation and sorption in contaminant transport;* 1-D code, uniform flow field and material properties for both substances; constant velocity; advective and dispersive transport of a hydrocarbon and an electron acceptor (e.g., oxygen); aerobic biodegradation using modified Monod function; anaerobic biodegradation for both substances; Linear, Freundlich, Langmuir adsorption isotherms for both substances; Dirichlet, Neumann, and Cauchy boundary conditions modified to include first-order degradation; cumulative mass balance report; microbial density assumed constant.	250.-
BIOPLUME-II	*US/EPA code; 2-D transport of dissolved hydrocarbons under the influence of oxygen-limited biodegradation.* It computes two plumes which are then combined using superposition; three sources of oxygen are provided: initial dissolved oxygen in the unconfined aquifer, natural recharge of oxygen across the boundaries, and vertical exchange of oxygen from the unsaturated zone.	300.-

TABLE 3.4 Software for water modeling (continued)

Model Name	**Description / Author**	**Price ($)**
CTRAN/W	*Finite-element contaminant transport analysis through soil and rock;* uses SEEP/W; finite-element solve function; steady-state and transient water flow conditions; transient hydraulic boundary conditions; anisotropic and heterogeneous ground conditions	2,050.-
FEMSEEP	*2-D steady / non-steady groundwater flow & Contaminant Transport;* finite element with particle tracking. Advective / dispersive transport in a 2-D horizontal plane, a vertical cross-section, 3-D axi-symetric system	
FLOWTHRU	*2-D anisotropic vertical section through water bodies.* Needs AQUIFEM-N	300.-
FLONET	*2-D cross-sectional steady-state groundwater flow;* it computes potential streamlines, groundwater velocities in a vertical section through a confined / unconfined aquifer	650.-
FLOWCAD	*2-D transient-state groundwater flow model for the calculation of time-variant hydraulic heads and drawdowns;* water balance computation	650.-
FLOWPATH	*Complete 2-D groundwater steady-state flow and wellhead protection analysis model;* used for hydraulic heads, groundwater-velocities, time-related pathlines, capture zones, water balances, and steady-state draw-down distributions; uses finite-difference method	650.-
GEOVIEW	*Desktop contouring and analysis for the Macintosh;* Creates contour maps from random X, Y, Z data; performs analysis on those maps	600.-
Groundwater/ DMS	*Groundwater data management system:* monitoring well data; lab and field results; historical data summaries.	10,000.-
HSPF	*US/EPA Hydrologic simulation program for watershed hydrology and water quality for both conventional and toxic organic pollutants.* Uses time-history of rainfall in urban / or agricultural watershed, temperature and solar radiation; land use characteristics / patterns. Flow rate, sediment load, and nutrient and pesticide concentrations are predicted (FORTRAN).	350.-
HST3D	*3-D Heat and Solute Transport;* runs on AT 386/486 (expanded memory version) AU: K. L. Kipp	450.-
HYDROLOGY System (EDSC)	*Watershed modeling;* water surface profiles; storm sewers; sanitary sewers	895.-
INTERSAT	*3-D Groundwater flow;* runs on AT 386/486 (expanded memory version)	
INTERTRANS	*3-D Solute transport;* particle tracking; contaminant advection / convection	500.-

TABLE 3.4 Software for water modeling (continued)

Model Name	Description / Author	Price ($)
MINTEQA2	*US / EPA geochemical equilibrium speciation model* capable of computing equilibria among the dissolved, adsorbed, solid, and gas phases in an environmental setting.	350.-
MOC	*2-D solute transport and dispersion in groundwater* AU: Konikow & Bredehoeft	
MODFLOW	*Groundwater flow model;* 3-D Finite difference AU: McDonald & Harbaugh	200.-
MODFLOWP	*USGS computer program for estimating parameters of a transient or steady-state, 3-D groundwater flow model using non-linear regression.* Parameters are estimated by minimizing a weighted-least squares objective function by the modified Gauss-Newton method or by a conjugate-direction method.	400.-
MODINV	*Groundwater flow model.* Based on MODFLOW; 386/486 expanded memory version	800.-
MOCDENSE	*2-D solute transport and dispersion in groundwater;* (386/486 expanded memory version) AU: Sanford & Konikow	300.-
MODRET	*Infiltration of stormwater retention ponds;* uses MODFLOW; modified Green & Ampt equation	325.-
MULTIMED	*Multi media exposure assessment model (US / EPA program);* simulates transport and transformation of contaminants released from a waste disposal facility into the multimedia environment. It includes releases to air or soil, incl. the saturated or unsaturated zone. For risk assessment of land disposal of chemicals in a multimedia environment.	350.-
PESTRAN	*Pesticide transport (US / EPA) program used for evaluating the transport of organic pollutants through soil to groundwater.* It calculates organic movement based on a linear isotherm, first-order degradation and hydrodynamic dispersion.	200.-
PHREEQE	*Model for geochemical reactions;* based on ion pairing aqueous model; Calculates pH, redox potential, mass transfer as a function of reaction process; composition of solutions in equilibrium with multiple phases	350.-
PROCESSING MODFLOW	AU: W.H. Chiang, W. Kinzelbach	
QUAL2EU	*Enhanced Stream water quality model with uncertainty analysis;* (US / EPA steady-state model for conventional pollutants in branching streams and well mixed lakes); as steady-state or dynamics model; for use as water quality planning tool; for studying impact of waste loads on instream water quality; quality characteristics and magnitude of non-point waste loads on instream water quality	350.-

TABLE 3.4 Software for water modeling (continued)

Model Name	Description / Author	Price ($)
SEEP/W	*Finite element seepage analysis;* Analyzes; saturated steady-state / unsaturated time-dependent flow problems; migration of a wetting front; dissipation of excess pore water	3,550.-
SOILPROP	*Estimation of unsaturated soil hydraulic properties,* and their uncertainty from particle size distribution data (saturated hydraulic conductivity; parameter in the van Genuchten and Brooks-Corey models)	200.-
SPACE	*Data-base management system (for spatial data),* which uses the relational data model and the script / SQL language; map display capability using MS-Windows	2,000.-
SPILLVOL	*Estimation of free, recoverable and residual hydrocarbon spill volumes from monitoring well fluid level data;* Calculations are based on a physically-based model for vertical equilibrium three-phase fluid distributions (oil entrapment in saturated zone and non-entrainable residual oil in unsaturated zone)	300.-
SPILLCAD	*Hydrocarbon spill site assessment and remedial design;* Needs PROPS, SPILLVOL, ARMOS, and SPILLTRANS	230.-
SWIM	*Soil-water infiltration and movement;* soil-water flow equation (Richards' equation). Simulation of infiltration, redistribution, deep drainage, simultaneous evapo-transpiration; four types of vegetation, transient surface-water storage and runoff.	550.-
SWMM	*Storm Water Management Model (US / EPA computer model);* analysis of quantity and quality problems associated with urban runoff; both single-event and continuous simulation; all aspects of urban hydrologic and quality cycles, incl. rainfall and snowmelt, flow routine through the drainage network, storage and treatment.	350.-
WASP4	*Water quality analysis simulation program;* US / EPA generalized modeling framework that simulates contaminant fate in surface water. Flexible compartment modeling approach in 1-D, 2-D, or 3-D.	350.-
SWIFT/486	*3-D model to simulate groundwater flow, heat (energy), brine, and radionuclide transport in porous and fractured geologic media.* The primary equations for fluid flow, heat and brine are coupled by fluid density, viscosity or porosity; transient analysis or steady-state (for coupled flow and brine). The equations are solved using central or backward spatial and time weighting approximations by the finite difference method. In addition to Cartesian, cylindrical grids may be used. Contaminant transport includes advection, dispersion, sorption and decay, including chains of constituents. Both dual, and discrete fracture representation along with rock matrix interactions may be simulated.	800.- to 1,800.-
SLOPE/W	Uses the limit equilibrium theory for computing the *factor of safety of earth and rock slopes* for a variety of methods;	3,550.-
SUTRA^{EM}	*2-D Saturated-Unsaturated Transport;* (386/486 expanded memory version, 1991)	

■ *MODFLOW^{PM}*. It is an *easy to use program* developed by Chiang and Kinzelach. PM is a combination of preprocessing and post-processing modules for the popular model MODFLOW. It can be used for simulation and analysis of flows.

■ *MOCEM.* MOC is a *computer model of two-dimensional solute transport and dispersion in flowing groundwater* (El Kadi[269], 1987). The model is both general and flexible, and it can be applied to a wide range of problem types. It is applicable to one or two dimensional problems involving steady state or transient flow. The model computes changes in concentration over time caused by the processes of convective transport, hydrodynamic dispersion, and mixing (or dilution) from fluid sources. The model assumes that gradients of fluid density, viscosity, and temperature do not affect the velocity distribution. However, the aquifer may be heterogeneous and/or anisotropic. The model is based on a rectangular, block centered, finite-difference grid. It allows the specification of injection of withdrawal wells and of spatially varying diffuse recharge or discharge, saturated thickness, transmissivity, boundary conditions and initial heads and concentrations. The model incorporates:

- First-order irreversible rate-reaction
- Reversible equilibrium-controlled sorption with linear, Freundlich, or Langmuir isotherms
- Reversible equilibrium-controlled ion exchange for monovalent or divalent ions
- Decay.

MOC executes in a batch processing shell similar to MODFLOW. The utilities comprise MOCOUT, MOCGRAF, MOCTIME and MOCREC. MOC and RANDOM WALK both use a particle tracking technique.

■ *MOCDENSE^{EM}*. It is the 386/486 PC version of MOC with extended memory. This model simulates solute transport and dispersion of either one or two constituents in groundwater where there is two-dimensional, density-dependent flow. It can simulate flow in a cross-sectional plane rather than in an areal plane. Since the problems of interest involve variable density, this model solves for fluid pressure rather than hydraulic head in the flow equation. Density is considered to be a function of the concentration of one of the constituents. Use of the model depends on the following assumptions:

- Flow is two-dimensional, with one principal axis being parallel to gravity
- constituents are non-reactive (conservative)
- Density and viscosity are a function of concentration and not of other factors such as pressure and temperature
- Grids of 72 x 60 are permitted.

Submodules packages comprise MOCDINP, MOCREC, TECVIEW, MOCDSIP, MOCDR4. Utilities include MOCDGRAF^{EM} ($ 250.-).

3.8.2. MODELING AND SIMULATION TOOLKITS

Other recently developed *modeling and simulation tools* are:

■ *SAM-Set*. This model was developed by Vancso-Polacsek[270] (1990), and *ModelWorks* by Ulrich[271] (1987), and Fischlin et al.,[272] (1990). With both tools, models can easily be formulated by short definition programs written in a common programming language (modula-2). SAM-Set further allows interactive modification of model structures. The latter program which is based on concepts of modeling and simulation theory, is very versatile and allows hierarchical system definition with different model types (discrete time, continuous time, etc.). Both tools provide user-friendly interfaces including pull-down menus, and graphical display of calculation results.

■ Another software called *AIDE Interaction Development Environment*[273] provides a methodology for implementing new interactive interface systems for environmental software. AIDE combines a toolkit of utility subroutines for building individual interactive screens with instructions for developing two parallel products: a file containing all text, questions, and messages used in interactive communications, and a FORTRAN program containing the control strategy and sequencing instructions for interactions.

■ The *US Environmental Protection Agency (USEPA)* and the *US Geological Survey (USGS)* have been developing an integrated framework for environmental models called the Watershed Data Management (WDM) System (Lumb and Kittle[274], 1986). This framework has been developed to facilitate the integrated use of environmental models and relevant databases required for model application.

■ The user interface, called *ANNIE* (Lumb[275] et al., 1990) is organized as a toolkit of FORTRAN subroutines that allow manipulation of large data bases (storage, retrieval, listing, checking, updating) and screen information and perform many of the functions required to support model operation (analyze spatial, parametric and time-series data. for hydrological models, graphics input/output). Many hydrologic and water-quality models and analysis tools developed by the *USGS* and the *USEPA* currently use the WDM file, ANNIE, or both.

■ A second user interface toolkit, *EMIFE*, the Environmental Model Input File Editor (General Sciences Corporation[276], 1988), was developed in a joint project between *EPA's Office of Research and Development,* and *Office of Toxic Substances.* EMIFE is based on user interaction capabilities developed for the Graphical Exposure Modeling System (GMS) and the microcomputer version PCGMS (General Services Corporation[277], 1988), which were designed primarily to deal with models and data needed to evaluate fate and exposure of chemicals covered by the *Toxic Substances Control Act.*

EMIFE is used to develop interfaces to the environmental databases and models in GEMS and PCGMS. It is complementing ANNIE. The ANNIE and EMIFE capabilities are now integrated into the consistent user interaction framework AIDE.

AIDE has been applied to the existing environmental model QUAL2E-UNCAS (Brown and Barnwell[278], 1987), to a soil properties analysis program, DBAPE (Imhoff[279] et al., 1989), and to other EPA and USGS models, including RRM, J407, MULTIMED, EXAMS-II, FGETS, and DYNHYD5.

Summary (Chapter 3)

This chapter deals with models and risk assessment methods for the hydrological cycle. The aquatic environment is divided into a number of media submitted to different sub-regimes, each requiring a different kind of model e.g., surface waters, subsurface waters), either simple aquatic models (box models, two-dimensional models), either simple aquatic models or complicated flow models (1-D, 2-D, 3-D models). River, lake, and subsurface models are also summarized and presented in this chapter. Risk assessment methodology for groundwater pollution control is included in the economy of the chapter. Special attention is paid to the risk assessment method, potential exposure pathways, quantifying exposure and risk from groundwater, aquifer restoration techniques, the development of expert systems for groundwater contamination. Groundwater models are reviewed and classified for the case of flow transport and numerical geochemical models. Water quality assessment plays an important role in the risk assessment process. The present chapter includes water quality standards and index models for agricultural land management, for estimating nitrate and phosphorus contamination. The chapter includes also guidelines for risk to the drinking water and quality, resource prediction and management models, waste loads management (risk based models regarding mixed-wastes disposal systems, contamination from land based waste disposal, contamination from imiscible organic chemicals, etc.), and from adequate commercial software for water models.

References (Chapter 3)

[1] Onishi, Y.; Serne, R.J.; Arnold, E.M., et al., Critical Review: Radionuclide Transport, Sediment Transport, and Water Quality Mathematical Modeling; and Radionuclide Adsorption/Desorption Mechanisms, Battelle Pacific Northwest Laboratories, January 1981.

[2] Environmental Protection Agency (EPA), Environmental Modeling Catalogue, Abstracts of Environmental Models, May 1979.

[3] Grismud, G.P.; Finnemore, E.J.; Owens, H.J., Evaluation of Water Quality Models, A Management Guide for Planners, U.S. Environmental Protection Agency, Report EPA-600/5-76-004, July 1976.

[4] Orlob, G.T., Mathematical Modeling of Surface Water Impoundments, Vol. I and II, Resource Management Associates, Lafayette, California, (1977).

[5] Bolten, J.G.; Morrison, P.F.; Solomon, K., Risk-Cost Assessment Methodology for Toxic Pollutants from Fossil Fuel Power Plants (R-2993-EPRI, June 1983).

[6] Methodology for Evaluating the Radiological Consequences of Radioactive Effluents Released in Normal Operations, Joint Report by the United Kingdom National Radiological Protection Board and the Commissariat à l'Energie Atomique, Doc. No. V/3865/79-EN, Commission of the European Communities, Luxembourg (July 1979).

[7] General Principle of Calculation for the Radiation Exposure Resulting from Radioactive Effluents in Exhaust Air and in Surface Water (Guidelines Under Sect. 45 of the Radiological Protection Ordinance), Translations - Safety Codes and Guides, Edited by: Gesellschaft für Reaktorsicherheit, Cologne, FRG (Nov. 1980).

[8] Rossi, G., Modelling Lake Pollution, Environment Institute, Joint Research Center, Commission of the European Communities, Luxemburg, EUR 13998 EN (1991).

[9] Ambrose, R.B.; Barnwell, T.O., Environmental Software at the U.S. Environmental Protection Agency's Center for Exposure Assessment Modeling, Environmental Software, 4(2) (1989).

[10] Cohen, Y.; Tsai, W.; Chetty, L.S.; Mayer, G.C., Dynamic partitioning of organic chemicals in local environments: a multi-media screening level modeling approach, Environ. Sci. Technol. 24(10) (1989) 1549-1558.

[11] Jury, W.A.; Focht, D.D.; Farmer, W.J., Evaluation of pesticide groundwater pollution potential from standard indices of soil chemical adsoprtion and biodegradation,. J. Environmental Quality 16(4) (1987) 422-428.

[12] Reichert, P.; Wanner, O., RIVERSIM - Modeling of Hydraulics, transport and transformation processes in rivers, User Manual, EAWAG, Dübendorf, Switzerland, (1990).

[13] Jorgensen, S.E., (ed.) Modelling in Ecotoxicology, Developments in Environmental Modelling, Vol. 16 (1990), Elsevier, Amsterdam.

[14] Burns, L.A., Models for Predicting the Fate of Synthetic Chemicals in Aquatic Ecosystems, Special Technical Testing Publication 865, (1985). American Society for Testing and Materials, 1916 Race Street, Philadelphia, PA 19103.

[15] Exposure Analysis Modelling System, EXAMS (1990). User Manual and System Documentation. National Technical Information Service (NTIS), 5285 Port Royal Road, Springfields, VA 22161, USA.

[16] Matthies, M.; Trapp, S., Environmental Models for Exposure and Hazard Assessment of Chemicals, 1st Eoropean Conference on Ecotoxicology, Copenhagen, Denmrk, Oct. 1988, pp. 420-438.

[17] Ulrich, M., Modelling of Chemicals in Lakes - Development and Application of User-Friendly Simulation Software (MASA & CHEMSEE) on Personal Computers. Thesis: Diss. ETH No. 9632 (1991). Swiss Federal Institute of Technology Zurich (ETHZ), Switzerland.

[18] Ulrich, M., CHEMSEE II, Programm-Anleitung. User's Manual (1989). Swiss Federal Institute of Technology Zurich (ETHZ), Switzerland.

[19] MacMETH. A Fast Modula-2 Language System for the Apple Macintosh. User Manual. Institut für Informatik (1988), Eidgenössische Technische Hochschule (ETH), CH-8096 Zürich, Switzerland.

[20] Fischlin, A., The "DialogMachine" for the Macintosh. Internal Report, Project Center IDA, Swiss Federal Institute of Technology Zurich (ETHZ), Switzerland.

[21] Mackay, D.; Joy, M.; Paterson, S., A quantitative water, air, sediment interaction (QWASI) fugacity model for describing the fate of chemicals in lakes, Chemosphere 12 (1983) 981-997.

[22] Mackay, D.; Paterson, S.; Joy, M., A quantitative water, air, sediment interaction (QWASI) fugacity model for describing the fate of chemicals in rivers, Chemosphere 12 (1983) 1193-1208.

[23] Mackay, D.; Paterson, S.; Schroeder, W.H., Model describing the rates of transfer processes of organic chemicals between atmosphere and water, Environ. Sci. Technol. 20 (1986) 810-816.
[24] Mackay, D.; Paterson, S., Evaluating the multimedia fate of organic chemicals: A level III fugacity model, Environ. Sci. Technol. 25 (1991) 427-436.
[25] Mackay, D., Multimedia Environmental Models, The Fugacity Approach, Lewis Publishers, (1991).
[26] Neely, W.B., Chemical in the Environment, M. Decker, New York, N.Y., (1980).
[27] Neely, W.B.; Blau, G.E., Environmental Exposure from Chemicals, Vols I and II. CRC Press, Boca Raton, FL, (1985).
[28] Conway, R.A., Environmental Risk Analysis for Chemicals. Van Nostrand-Reinhold, New York, 1982.
[29] Hutzinger, O., (ed.), : The Handbooks of Environmental Chemistry (various editions and years). Springer-Verlag, Berlin.
[30] Handbook of Environmental Data on Organic Chemicals, Van Nostrand Reinhold, New York, NY. (2nd edition 1983).
[31] Southwood, J.M.; Harris, R.C.; Mackay, D., Modeling the fate of chemicals in an aquatic environment: the use of computer spreadsheet and graphics software, Environ. Toxicol. Chem. 8 (1989) 987-996.
[32] Thomann, R.V., Bioaccumulation model of organic chemical distribution in aquatic food chains, Environ. Sci. Technol. 23 (1989) 699-707.
[33] Schramm, K.W.; Reischl, A.; Hutzinger, O.; et al., A multimedia compartment model to estimate the fate of lipophilic compounds in plants, Chemosphere 16 (1987) 2653-2663.
[34] Paterson, S.; Mackay, D., A steady-state fugacity-based pharmakokinetic model with simultaneous multiple exposure routes, Environ. Toxicol. Chem. 6 (1987) 395-408.
[35] Paterson, S.; Mackay, D., A model illustrating the environmental fate, exposure and human uptake of persistent organic chemicals, Environ. Toxicol. Chem. 47 (1989) 85-114.
[36] Boethling, R.S.; Sabijic, A., Screening-level model for aerobic biodegradability based on a survey of expert knowledge, Environ. Sci. Technol. 23 (1989) 672-679.
[37] Lange, W.J., A method to assess the risk and admissibility of one or more activities in the inflow area of a receptor, in G. Jousma, J. Bear, Y.Y. Haimes and F. Walter (eds.), Groundwater Contamination: Use of Models in Decision-Making, International Groundwater Modeling Center (IGWC), Kluwer Academic Publishers, Dordrecht (1989), pp. 47-57.
[38] Bolten, J.G.; Morrison, P.F.; Solomon, K; Wolf, K., Alternative models for risk assessment of toxic emissions, RAND / N-2261-EPRI, April 1985.
[39] Foster, S.A.; Chrostowski, P.C., Integrated household exposure model for use of tap water contaminated with volatile organic chemicals. Paper 86-12.3 presented at the 79th Annual APCA Meeting, Minneapolis, Minnesota (1986).
[40] Guidelines for Groundwater Classification under the EPA Groundwater protection Strategy, U.S. Environmental Protection Agency (USEPA), Office of Groundwater Protection, Office of Water. Washington, DC, (1986).
[41] Anderson, E.L.; Chrotoswski, P.C.; Vreeland, J.L., Risk assessment for use in groundwater management, in W.F. McTernan and E. Kaplan (eds.), Risk Assessment for Groundwater Pollution Control, Published by the American Society of Civil Engineers, New-York. (1990), pp. 111-125.
[42] Walton, W.C., Practical Aspects of Groundwater Modeling. Second Edition. National Water Well Association, Worthington, Ohio (1985).
[43] Users Manual for the Pesticide Root Zone Model (PRZM), Release 1 Prepared by Carsel, R. F., et al., for Development and Application Branch, Environmental Research Laboratory, Athens, Georgia, Dec. 1984, Environmental Protection Agency (USEPA), Report EPA-600/3-84-109, (1984).
[44] U.S. Environmental Protection Agency (USEPA). Superfund Public Health Evaluation Manual (EPA), Prepared by ICF., for Office of Emergency and Remedial Response, Washington, DC, October 1986. Report EPA 540/186/060.
[45] U.S. Environmental Protection Agency (USEPA), Remedial Action at Waste Disposal Sites (Revised) , USEPA Report EPA/625/6-85/006 (1985).
[46] American Petroleum Institute: Cost Model for Selected Technologies for Removal of Gasoline Components from Groundwater, API Publication No. 4389 (1985).
[47] van Ree, C.C.; Kabos, R., et al., Environmental risk assessment using stochastic modelling, in G. Jousma, J. Bear, Y.Y. Haimes and F. Walter (eds.), Groundwater Contamination: Use of Models in Decision-

Making, Kluwer Academic Publishers, Dordrecht. International Groundwater Modeling Center (IGWC), (1989), pp. 63-72.
[48] Loxham, M.; Van Ree, D., Risk analysis as a rational basis for soil protection and contaminated land clean-up, Environmental Technology Conference, June 22-26, 1987.
[49] REFCON: A ground- and surface water pollution response model for the design and optimalization of clean-up and soil protection strategies for petro-chemical complexes, in G. Jousma, J. Bear, Y.Y. Haimes, and F. Walter (eds.), Groundwater Contamination: Use of Models in Decision-Making, Kluwer Academic Publishers, Dordrecht. International Groundwater Modeling Center (IGWC), (1989), pp. 599-606.
[50] Shih, Chia Shun; Riojas, A.H., Risk and its acceptability for groundwater contamination by hazardous wastes, in National Research Council (ed.), Groundwater Models, Scientific and Regulatory Applications, National Academic Press, Washington, DC, (1989).
[51] Vesely, W., et al., Fault Tree Handbook, U.S. Nuclear Regulatory Commission, NUREG-0492 (1980).
[52] Hart, P.; Duda, R.; Enaudi, M., PROSPECTOR: a computer based consultation system for mineral exploration. Mathematical Geology 10(5) (1978).
[53] Osterkamp, G.; Richter, B.; Skala, W., An expert system for groundwaterrisk assessment, in G. Jousma, J. Bear, Y.Y. Haimes and F. Walter (eds.), Groundwater Contamination: Use of Models in Decision-Making, Kluwer Academic Publishers, Dordrecht. International Groundwater Modeling Center (IGWC), (1989), pp. 59-62.
[54] Ludwigsen, P.J.; Sim, R.C.; Grenneg, W.J., A demonstration expert system to aid in assessing groundwater containation potential by organic chemicals, in W.T. Lenocker (ed.), Computers in Civil Engineering, Proceedings of the Fourth Conference, American Society of Civil Engineers, Boston, Mass. (1986), pp. 687-698.
[55] Law, K.H.; Zimmie, T.F.; Chapman, D.R., An expert system for inactive hazardous waste site characterization, in C.N. Kostem and M.L. Maher (eds.), Expert Systems in Civil Engineering, American Society of Civil Engineers, Seatle, Wash. (1986), pp. 159-168.
[56] Mc Clymont, G.L.; Schwartz, F.W., Development and application of an expert system in contaminant hydrogeology, National Hydrology Research Institute, Environment Canada, (1987).
[57] Hipel, K.W., Fuzzy sets techniques in decision making, in IFAC Theory and Aplication of Digital Control. Pergamon Press, London, (1982).
[58] Kindler, J.K., et al., Imprecision in groundwater modelling and management, in G. Jousma, J. Bear, Y.Y. Haimes and F. Walter (eds.), Groundwater Contamination: Use of Models in Decision-Making, Kluwer Academic Publishers, Dordrecht. International Groundwater Modeling Center (IGWC), (1989) pp. 139-148.
[59] Groundwater Models, Scientific and Regulatory Applications. Water Science and Technology Board; Committee on Groundwater Modeling Assessment; Commission on Physical Sciences, Mathematics, and Resources; National Research Council. National Academy Press, Washington, DC, (1990).
[60] Sloan, F.G.; Moore, I.D., Modelling Subsurface Storm-flow on Steeply Forested Watersheds. Water Resources Research, 20 (1984) 1815-1822.
[61] Whelan; Brown (1984).
[62] Bachmat, Y.; Andrews, B.; Holtz, D.; Sebastian, S., Utilization of Numerical Groundwater Models for Water Resource Management, U.S. Environmental Protection Agency, Report EPA 600/8-78-012, Environmental Research Laboratory, Ada, Oklahoma, (1978).
[63] Anderson, M.P., Using models to simulate the movement of contaminants through groundwater flow systems, CRC Critical reviews in environment control, 9(2), (1979) 97-156.
[64] Knox, R.C.; Canter, L.W., Summary of groundwater modeling in the United States, National Center for Groundwater Research, NCGWR 80-20, Norman, Oklahoma, (1980).
[65] Lappala, E.G., Modeling of water and solute transport under variably saturated conditions: State-of-the-art, in Modeling and Low-Level Waste Management: An Interagency Workshop, Oak Ridge National Laboratory, ORO-821, Oak Ridge, Tennessee, (1980).
[66] Science Applications, Inc. (SAI), Tabulation of Waste Isolation Computer Models, Office of Nuclear Waste Isolation, Battelle Memorial Institute, ONWI-78, Columbus, Ohio, (1981).
[67] Koines, A., Technical review of Groundwater Models, Office of Solid waste, U.S. Environmental Protection Agency, (1982).

[68] Oster, C.A., Review of Groundwater Flow and Transport Models in the Unsaturated Zone, Battelle Pacific Northwest Laboratories, PNL-4427, NUREG/CR-2917, Richland, Washington, (1982).
[69] Nelson, R.W., A Summary of Subsurface Fluid Flow and Contaminant Transport Models Useful in Waste Isolation Assessment, Prepared for the Waste Isolation Safety Assessment Program, Batelle Pacific Northwest Laboratories, Richland, Washington, (1982).
[70] Kincaid, C.T.; Morrey, J.R.; Rogers, J.E., Geochemical Models for Solute Migration: The Selection of Computer Codes and Description of Solute Migration Processes, Vol. I, Electric Power Research Institute, Palo Alto, California, (1983).
[71] van der Heijde, P.K.M.; Bachmat, Y., et al., Groundwater management: The use of numerical models. Water Resources Monograph 5, American Geophysical Union, Washington, DC, 2nd ed. (1985).
[72] Kinzelbach, W., Groundwater modelling - An introduction with sample programs in Basic. Elsevier Science Publishers, Amsterdam, (1986).
Kinzelbach, W.; Frind, E.O., Accuracy criteria for advection-dispersion models, in Finite Elements in Water Resources, Proc. of the 6th International Conference, Lisboa, (1986), pp. 489-504.
[73] van der Heijde, P.K.M.; El-Kadi, A.I.; Williams, S.A., Groundwater Modeling: An Overview and Status Report. International Groundwater Modelling Center, Holcomb Research Institute, Butler University, Indianapolis, Report No. GWMI 88-10, (1988).
[74] Cederberg, G.A.; Street, R.L.; Leckie, J.O., A groundwater mass transport and equilibrium chemistry model for multicomponent systems, Water Resour. Research, 21 (1985) 1095-1104.
[75] Miller, C.W.; Benson, L.V., Simulation of solute transport in a chemically reactive heterogeneous system: model development and application, Water Resour. Research, 19 (1983) 381-391.
[76] van Rijn, L.C., Mathematical modelling of suspended sediment in non-unoform flow, J. Hydr. Engineering, ASCE, 12(6) (1986) 443-455.
[77] Clik, I.; Rodi, W., Modeling of suspended sediment transport in non-equilibrium situations. J. Hydr. Engrg, ASCE, 114(10) (1988) 1157-1191.
[78] Christodoulou, G.C., A model for the dispersion of suspended matter in surface water bodies, in L.C. Wrobel, C.A. Brebbia (eds.), Water Pollution II, Modelling, Measuring and Prediction, Computational Mechanics Publications, Southhampton Boston, (1993), pp. 209-216.
[79] Ouillon, S.; Le Guennec, B., Development of a 3D non cohesive suspended sediment transport model for free surface flows, in L.C. Wrobel, C.A. Brebbia (eds.), Water Pollution II, Modelling, Measuring and Prediction, Computational Mechanics Publications, Southhampton Boston, (1993), pp. 201-208.
[80] Harleman, D.R.F., Hydrothermal analysis of lakes and reservoirs, J. Hydr. Div. ASCE 108 (HY3), (1982), pp. 302-325.
[81] Rodi, W., Turbulence models and their application in hydraulics - A state of the art review, Int. Assoc. Hydr. Res. (IAHR), Delft, The Netherlands, (1980).
[82] Owens, E.W.; Harleman, D.R.F.; Parsons, R.M., A one-dimensional turbulence model for vertical transport in stratified lakes, in L.C. Wrobel, C.A. Brebbia (eds.), Water Pollution II, Modelling, Measuring and Prediction, Computational Mechanics Publications, Southhampton Boston, (1993), pp. 125-139.
[83] Swensson, U., PROBE - An instruction manual Rep. Oceanogr. No. 10 (1986), Swedish Meteorological and Hydrological Institute (SMHI), S-60176 Sweden.
[84] Huyakorn, P.S.; Pinder, G.F., Computational Methods in Subsurface Flow, Academic Press, New York, (1983).
[85] Remson, I.; Hornberger, G.M., Molz, F.J., Numerical Methods in Subsurface Hydrology, Wiley-Interscience, New York, (1971).
[86] Wang, H.F.; Anderson, M.P., Introduction to Groundwater Modeling, Freeman, San Francisco, (1982).
[87] Haverkamp, R.; Vauclin, M., et al., A comparison of numerical simulation models for one-dimensional infiltration, Soil Science Society of America Journal, 41 (1977) 285-294.
[88] Pruess, K.; Wang, J.S.Y., The numerical modeling of isothermal and nonisothermal flow in unsaturated fractured rock - A review in Flow and Transport Through Unsaturated Fractured Rock, in D.D. Evans and T.J. Nicholson (eds.), Geophysical Monograph 42, American Geophysical Union, Washington, DC, (1987).
[89] Bachmat, Y.; Bredehoeft, J.; et al., Groundwater Management: The Use of Numerical Models, Water Resources Monograph 5, American Geophysical Union, Washington, DC, (1980).

[90] Bachmat, Y.; Andrews, B.; Holtz, D.; Sebastian, S., Utilization of Numerical Groundwater Models for Water Resource Management, U.S. Environmental Protection Agency, Report EPA 600/8-78-012, Environmental Research Laboratory, Ada, Oklahoma, (1978).
[91] International Groundwater Modeling Center (IGWMC), Indianapolis, Indiana, USA.
[92] Soil Conservation Service: A methods for estimating volume and rate of runoff in small watersheds, SCS-TTP-149, U.S. Department of Agriculture, Soil Conservation Service, (1973).
[93] Holtan, H.N., Concept for infiltration estimates in watershed engineering. USDA-ARS Paper No. 41-51, (1961).
[94] Green, W.H.; Ampt, G.A., Studies of Soil Physics - The Flow of Air and Water through Soils, J. Agri. Sci., 4(1) (1911) 1-24.
[95] Moore, I.D., Infiltration equations mofified for surface effects, J. of the Irrigation and Drainage Division, ASCE, Vol. 107, No. IRI:71-86 (1981).
[96] Illangasekare; T.H.; Reible, D.D., A simple model for transport of free-phase organics in unsaturated zones of aquifers, in Solving Groundwater Problems with Models, Vol.1, Conf. Febr. 10-12, (1987), Denver, Colorado, (1987), pp. 404-423.
[97] Pinder, G.F.; Abriola, L.M., On the simulation of nonaqueous phase organic compounds in the subsurface. Water Resources Research, 22(9) (1986) 109-119, (Supplement).
[98] Faust, C.R., Transport of immiscible fluids within and below the unsaturade zone: A numerical model. Water Resources Research, 21(4) (1985) 587-596.
[99] Hochmuth, D.P.; Sunada, D.K., Groundwater model of two-phase immiscible flow in coarse material. Groundwater, 23(5) (1985) 616-626.
[100] Osborne, M.; Sykes, J., Numerical modeling of immiscible organic transport at the Hyde Park landfill, Water Resources Research, 22(1) (1986) 25-33.
[101] Baehr, A.; Corapcioglu, M.Y., A predictive model for pollution from gasoline in soils and groundwater, in Petrolueum Hydrocarbons and Organic Chemicals in soils and groundwater, National Water Well Association, Washington, OH 43085. No. 225 (1984), American Chemical Society.
[102] Bredhoeft, J.D.; Pinder, G.F., Digital analysis of areal flow in multi-aquifer groundwater systems: a quasi-three-dimensional model. Water Resource Res, 6(3) (1970) 883-888.
[103] Javandel, I.; Witherspoon, P.A., A method of analyzing transient fluid flow in multilayered aquifers, Water Resource Res., 5(4) (1969) 856-869.
[104] Pinder, G.F.; Bredehoeft, J.D., Application of the digital computer for aquifer evaluation, Water Resource Res., 17(1) (1968) 1069-1093.
[105] Chorley, D.W.; Frind, E.O., An iterative quasi-three dimensional finite element model for heterogeneous multi-aquifer system, Water Resource Res., 14(5) (1978) 943-952.
[106] Premchitt, J., A technique in using integrodifferential equations for model simulation of multi-aquifer systems, Water Resource Res., 17(1) (1981), 673-685.
[107] Freeze, R.A., Three-dimensional, transient, saturated-unsaturated flow in a groundwater basin, Water Resource Res., 7, (2) (1971) 347-366.
[108] Gupta, S.K.; Tanji, K.K., A three-dimensional Galerkin finite element soultion of flow through multi-aquifers in Sutter Basin, California, Water Resource Res., 12(2) (1976) 155-162.
[109] Find, E.O; Verge, M.J., Three-dimensional modeling of groundwater flow systems, Water Resource Res., 14(5) (1978) 844-856.
[110] Yon; Little; Hahn, Applications of finite element groundwater model to Cherry Creek Basin, Colorado, Proc. of the National Water Well Association Conference on Practical Applications of Groundwater Models, August 19-20, (1985).
[111] Trescott, P.C., Iterative digital model for Aquifer Evaluations. U.S. Geological Survey Open-File Report 75-438, (1973); see also: Trescott, P.C.; Larson, S.P., Supplement to Open-File Report 75-438, Documentation of Finite-Difference Model for Simulation of Three-Dimensional Groundwater Flow, U.S. Geological Survey Open-File Report 76-591 (1976).
[112] Toth, J., A theoretical analysis of groundwater flow in small drainage basins, Journal of Geophysics Research 68 (16): 4795-4812.
[113] Engelen, G.B., Groundwater systems, methodology, and definitions, in G.B. Engelen and G.P. Jones (eds.), IAHS publication no. 163, Wallingford, (1986).
[114] Zeijl, W.; Brouwer, G.K., FLOSA: A tool for regional three-dimensional flow systems analysis, in G. Jousma, J. Bear, Y.Y. Haimes and F. Walter (eds.), Groundwater Contamination: Use of Models in

Decision-Making, Kluwer Academic Publishers, Dordrecht. International Groundwater Modeling Center (IGWC), (1989), pp. 541-552.

[115] Nawalany, M., Are reconstructors useful in groundwater protection decision-making, in G. Jousma, J. Bear, Y.Y. Haimes and F. Walter (eds.), Groundwater Contamination: Use of Models in Decision-Making, Kluwer Academic Publishers, Dordrecht. International Groundwater Modeling Center (IGWC), (1989), pp. 201-208.

[116] Dettinger, M.D.; Wilson, J.L., First order analysis of uncertainty in numerical models of groundwater flow. Part I. Mathematical development, WRR., 17(1) (1981) 149-161.

[117] Dillon, R.T.; Lantz, R.B.; Pahwa, S.B., Risk methodology for geologic disposal of radioactive wastes: The Sandia Waste Isolation Flow and Transport (SWIFT) model, (1978). Report SAND 78-1267.

[118] Huyakorn, P.S.; Broome, R.W., et al., SEFTRAN, Geotrans Inc., Herndon VA, USA, (1984).

[119] Konikow, L.F.; Bredehoeft, T.D., Computer model of two-dimensional solute transport and dispersion in groundwater, Techniques of Water Resources Investigation of the United States Geological Survey, Book 7, Ch. C2, (1978). United States Government Printing Office, Washington.

[120] Javandel, I.; Daughty, D.; Tsang, C.F., Groundwater transport: handbook of mathematical models. Water Resources Monograph 10, American Geophysical Union, (1984).

[121] Naymik, T.G., Mathematical modeling of solute transport in the subsurface, Critical reviews in Environmental Control, 17(3) (1987) 229-251.

[122] Anderson, M.O., Movements of Contaminants in Groundwater: Groundwater transport-advection and dispersion, in Groundwater Contamination, National Academy Press, Washington, DC, (1984), pp. 37-45.

[123] Cole, C.R; Foote, H.P.; Zimmermann, D.A.; Simons, C.S., Understanding, Testing, and Development of Stochastic Approaches to Hydrologic Flow and Transport through the Use of the Multi-Grid Method and Synthertic Data Sets. Preprints of the Symposium on the Stochastic Approach to Subsurface Flow, (1985) 126-140, Greco 35 Hydrogeologie Ecole des Mines de Paris, Fontainebleau, France.

[124] Sykes, J.F., Sleep, B.E., Thomson, N.R., A moment method for calculating groundwater flow uncertainty during the analysis of a landfill, in International Groundwater Modeling Center (IGWC), G. Jousma, J. Bear, Y.Y. Haimes and F. Walter (eds.), Groundwater Contamination: Use of Models in Decision-Making, Kluwer Academic Publishers, Dordrecht, (1989), pp. 219-227.

[125] Sykes, J.F., GWPGM3 Documentation and user guide, University of Waterloo, Waterloo, Ontario, (1987).

[126] Pepper, D.W.; Stephenson, D.E., Calculation of three-dimensiaonal groundwater transport using second-order moments, in G. Jousma, J. Bear, Y.Y. Haimes and F. Walter (eds.), Groundwater Contamination: Use of Models in Decision-Making, Kluwer Academic Publishers, Dordrecht. International Groundwater Modeling Center (IGWC), (1989) 313-328.

[127] Egan, B.A.; Mahoney, J.R., Numerical modeling of advection and diffusion of urban area source pollutants, Journal of Applied Meteorology 11 (1972) 312.

[128] Yanenko, N.N., The method of fractional steps: solutions of problems of mathematical physics in several variables, Springer-Verlag, Heidelberg, (1971).

[129] Sykes, J.F., GWPGM3 Documentation and user guide, University of Waterloo, Waterloo, Ontario, (1987).

[130] Lauren, J.M.; Teasdale, I., WOLFNET: A standalone calculation program for groundwater flow networks, in L.C. Wrobel, C.A. Brebbia (eds.), Water Pollution II, Modelling, Measuring and Prediction, Computational Mechanics Publications, Southhampton Boston, (1993), pp. 91-97.

[131] Selim, H.M.; Iskandar, I.K., Simplified model for the prediction of nitrogen behaviour in land treatment of waste water, Hanover, New Hampshire, CRREL Report 80-12, (1980).

[132] Benes, V.; Vrba, J., Solute transport in the unsaturated zone: Model verification and input data, in G. Jousma, J. Bear, Y.Y. Haimes and F. Walter (eds.), Groundwater Contamination: Use of Models in Decision-Making, Kluwer Academic Publishers, Dordrecht. International Groundwater Modeling Center (IGWC), (1989).

[133] Endangerment Assessment Handbook. U.S. Environmental Protection Agency (USEPA). Prepared by PRC, Environmental Management, Inc., for Office of Waste Programs Enforcement, Washington, DC, August, 1985; and Superfund Public Health Evaluation Manual U.S. Environmental Protection Agency (USEPA). Prepared by ICF, Inc., for Office Emergency and Remedial Response, Washington, DC, October 1986. Report EPA 540/186/060.

[134] Grove, D.B.; Wood, W.W., Predictions and field verification of subsurface-water quality changes during artificial recharge, Lubbock, Texas. Groundwater, 17 (1979) 250-257.

[135] Jennings, A.A.; Kirkner, D.J.; Theis, T.L., Multicomponent equilibrium chemistry in groundwater quality models, Water Resour. Res., 18 (1982) 1089-1096.
[136] Cederberg, G.A., The effects of geochemical processes on the transport of contaminants in multicomponent systems: A modelling perspective, in C.F. Tsang (ed.), Proc. Intern. Symposium on Coupled Processes Affecting the Performance of a Nuclear Waste Repository, Berkeley, USA, (1985) pp. 71-76.
[137] Read, D.; Williams, D.R.; Liew, S.K., A coupled approach to radioelement migration modelling in the United Kingdom, in C.F. Tsang (ed.), Proc. Internat. Symposium on Coupled Processes Affecting the Performance of a Nuclear Waste Repository, Berkeley, USA, (1985) pp. 109-114.
[138] Miller, C.W.; Benson, L.V., Simulationof solute transport in a chemically reactive heterogeneous sytem: model development and application. Water Resour. Res., 19 (1983) 381-391.
[139] Carnahan, C.L., Simulation of Uranium transoport with variable temperature and oxidation potential: the computer program THCC, Scientific Basis for Nuclear Waste Management X, J.K. Bate, and W.B. Seefeldt (eds.), Materials research Society, Pittsburgh, USA, (1987) pp. 713-721.
[140] Noy, D.J., Coupling of transport and geochemical models, CEC Contact Report No. 413-83-7 WAS.UK, (1983).
[141] Read, D.; Liew, S.K., Recent developments in coupling geochemical and hydrologic processes, in G. Jousma, J. Bear, Y.Y. Haimes & F. Walter (eds.), Groundwater Contamination: Use of Models in Decision-Making, Kluwer Academic Publishers, Dordrecht, International Groundwater Modeling Center (IGWC), (1989) pp. 383-393.
[142] Gupta, S.K.; Cole, C.R.; Monti, A. M., Finite-Element Three-Dimensional Ground-Water (FE3DGW) Flow Model: Formulation, Computer Source Listings, and User's Manual. Technical Report Oct. 1984, BMI/ONWI-548. Office of Nuclear Waste Isolation, Battelle Memorial Institute (505 King Avenue, Columbus, OH 43201-2693).
[143] Gupta, S.K.; Kincaid, C.T.; Meyer, P.; Cole, C.R., CFEST - Multi-Dimensional Finite Element Code for the Analysis of Coupled Fluid, Energy and Solute Transport, Pacific Nortwest Laboratory, (1982). Report PNL-4260.
[144] Kharaka, Y.K.; Barnes, I., SOLMNEQ: Solution-mineral equilibrium computations, (1973). NTIS Report PB214-899.
[145] Plummer, L.N.; Jones, B.F.; Truesdell, A.H., WATEQF- A Fortran IV version of WATEQ, a computer program for calculating chemical equilibrium in natural waters, USG.S. Water Resources Investigations, (1976), pp. 76-13.
[146] Westall, J.C.; Zachary, J.L.; Morel, F.M.M., MINEQL, a computer program for the calculation of chemical equilibrium composition of aqueous systems, M.L. Parsons Laboratory for Water Research and Hydrodynamics, Technical Note 18, Massachusetts Institute of Technology, Cambridge, Massachusetts, (1976).
[147] Wolery, T.J., Calculation of chemical equilibrium between aqueous solutions and minerals: the EQ3/6 software package, Lawrance Livermore National Laboratoy, Livermore, California, (1979). NTIS Report UCRL-52658.
[148] Parkhurst, D.L.; Thorstenson, DC; Plummer, L. N., PHREEQE - a computer program for geochemical calculations, USG.S. Water Resources Investigations, (1980) pp. 80-96.
[149] Reeves, M.; Johns, N.D., Cranwell, R.M., Theory and Implementation for SWIFT II, Sandia Waste-Isolation Flow and Transport Model for Fractured Media, Sandia National Laboratories, Release 4.84 (1986). NUREG/CR-3328 and SAND83-1159.
[150] Geotrans Inc., Data Input Guide for Swift III, Herndon, Virginia (1986).
[151] van der Heijde, P.K.M.; Huyakorn, P.S.; Mercer, J.W., Testing and validation of groundwater models. Proceedings, Practical Applications of Ground-Water Modeling, NWWA / IGWMC Conference, Columbus, Ohio, August 19-20, (1984).
[152] Stanford Geothermal Program. Proceedings, 6th Annual Workshop on Geothermal Reservoir Engineering., Stanford University, Stanford, California, (1980). SGP-TR-42.
[153] Hamilton, D.A., Groundwater modeling: Selection, Testing, and Use, Michigan Department of Natural Resources, Water Management Division, (1982).
[154] Kinzelbach, W., Methods for the simulation of pollutant transport in groundwater. NWWA/IGWMC Conference Proceedings, Solving Groundwater Problems with Models, Denver, Colorado, February 10-12, (1987).

155 INTRACOIN (1986) International nuclide transport code intercomparison study; Final report levels 2 and 3: model validation and uncertainty analysis. Swedish Nuclear Power Inspectorate, Stockholm, Sweden.

156 HYDRACOIN (1984) Progress Report No. 1, Swedish Nuclear Power Inspectorate, Stockholm, Sweden.

157 Reddi, L.N., Potential Pittfalls in Using Groundwater Models. Transferring Models to Users, Symposium Proceedings, AWRA, Denver, (1990).

158 Kinzelbach, W., Methods for the Simulation of Pollutant Transport in Groundwater. A Model Comparison, in Solving ground Water Problems with Models, Vol.1, Conf. Febr. 10-12, (1987), Denver, Colorado, pp. 656-674.

159 Burnett, R.D.; Frind, E.O., Simulation of contaminant transport in three dimensions: 1. The Alternating Direction Galerkin Technique; 2. Dimensionality effects, Water Resources Research, 23 (1987).

160 Cooley, R.L.; Vecchia, A.V., Reliability Analysis for Groundwater Models: An Approach Through the Use of Confidence and Prediction Intervals, in Solving ground Water Problems with Models, Vol.1, Conf. Febr. 10-12, (1987), Denver, Colorado, pp. 613-625.

161 WHO (1987) Guidelines for Drinking-Water Quality, WHO Geneva, (1987).

162 NAS (1986) Drinking Water and Health, Vol. 6, Safe Drinking Water Committee, National Research Council, National Academy Press, Washington, DC, (1986).

163 NAS (1987) Pharmacokinetics in Risk Assessment, Drinking Water and Health, Vol. 8, Subcommittee on Pharmacokinetics in Risk Assessment, National Research Council, National Academy Press, Washington, DC, (1987).

164 Eckenfelder, W.S.; Water Quality Engineering, Barnes and Noble, Inc. (eds.), New York (1970), pp. 4.

165 Development of a water quality index, Scottish Development Department. Applied Research and Development Engineering Division, HMSO Edinburgh, Report ARD 3, (1976).

166 House, M.A.; Newsome, D.H., Water quality indices for the management of surface water quality, Water Science and Technology, 21 (10/11) (1988) 1137.

167 Tyson, J.M.; House, M.A., The application of a water quality index to river management, Water Science and Technology. 21(10/11) (1988) 1149.

168 Boyd, M.; Walley, W.J.; Hawkes, H.A., Dempster-Shafer reasoning for the biological surveillance of river water quality, in L.C. Wrobel, C.A. Brebbia (eds.), Water Pollution II, Modelling, Measuring and Prediction, Computational Mechanics Publications, Southhampton Boston, (1993), pp. 349-356.

169 Shafer, G., A Mathematical Theory of Evidence, Princetown University Press, Princeton, N. J., USA, (1976).

170 Gordon, J.; Shortliffe, E.H., A method of managing evidential reasoning in a hierarchical hypothesis space, Artificial Intelligence, 26 (1985) 323-357.

171 Bascombe, A.D.; House, M.A., Ellis, J.B., The Utility of Chemical and Biological Monitotoring techniques for the Assesment of Urban Pollution. IAWPRC 5th River Basin Management Conference, Rovaniemi, Finland. Adv. Water Poll. Contr. 9 (1989) 59-69.

172 Commission For European Communities, Council directive of 16 June (1975) concerning the quality required of surface water intended for the abstraction of drinking water in member states, Official Journal No. L194/26 (1975).

173 Guidelines for Drinking-water Quality, Second edition, Vol. 1: Recommendations, World Health Organization, Geneva, WHO (1993).

174 National Council of the Paper Industry for Air and Stream Improvement (NCASI), USA.

175 Zison et al., Comparison of First Order Error Analysis and Monte Carlo Simulation in Time Dependent Lake Eutrophication Models, Water Resources research, 17(4) (1981) 1051-1059.

176 Zison, S.W.; Mill, W.B.; Deimer, D.; Chen, C.W., Rates, constants, and kinetics formulations in surface water quality modeling, USEPA Environmental Research Laboratory, Athens, (1978). GA. Report EPA-600/3-78-105

177 Ambrose, R.B., et al., WASP4, A Hydrodynamic and Water Quality Model. Model Theory, User's Manual and Programmer's Guide, Report EPA/600/3-87/039, (1988).

178 Connolly, J.P.; Thomman, R.V., WASTOX, A Framework for Modelling the Fate of Toxic Chemicals in Aquatic Environments. Part 2. Food Chain, PB85-214435.

179 Semra, K.; Ackerer, Ph.; Mosé, R., Three dimensional modelling in heterogeneous media, in L.C. Wrobel, C.A. Brebbia (eds.),Water Pollution II, Modelling, Measuring and Prediction, Computational Mechanics Publications, Southhampton Boston, (1993), pp. 3-11.

[180] Haith, D.A.; Tubbs, L.J., Watershed loading functions for nonpoint sources, J. Env. Eng. Div., Proc. Am. Soc. Civ. Eng. 107 (1981) 121-137.

[181] Paal, L.; Toompuu, A.; Lode, E., Probabilistic mode for water quality prediction and estimation, IAWPRC 5th River Basin Management Conference, Rovaniemi, Finland, Adv. Water Poll. Contr. 9 (1989) 125-132.

[182] EPA, A Hydrodynamic and Water Quality Model - Model, Theory, Users Manual and Programmers Guide, Environmental Research Lab., Athens, GA, (1988).

[183] Computer program documentation for the enhanced stream water model QUAL2E, U.S. EPA Environmental Research Laboratory, Athens, GA. Report EPA-600/3-85-065.

[184] GEOTRANS Inc. (ed.), SEFTRAN: a simple and efficient two-dimensioanl groundwater flow and transport model, Herdon, Virginia (USA), (1988).

[185] Huyakorn, P.S.; Broome, R.W., et al., SEFTRAN, Geotrans Inc., Herndon VA, (1984).

[186] Zengh, C.A., A modular three-dimensional solute transport model for simulation of advection, dispersion and chemical reactions of contaminants in groundwater systems, S.S. Papadopulos & Associate, Rockville, (1990).

[187] McDonald, M.G.; Harbaugh, A.W., A modular three-dimensional finite-difference groundwater flow model: U.S. Geological Survey Open-File Report 87-690 (1984); Note: MODFLOW Model.

[188] Burger, J., The calculation of pseudo-three-dimensional stream lines and contaminant transport in the saturated groundwater with the computer program STROP, Dienst Waterbeheer, Provincie Gerlderland, The Netherlands, (1986).

[189] Konikow, L.W.; Bredhoeft, J.D., Computer Model of Two-Dimensional Solute Transport and Dispersion in Groundwater. Techniques of Water Resources Investigations, Chap. C2, U.S. Geological Survey, Washington, DC, (1978).

[190] Bian, S.H.; Budden, M.J.; Yung, S.C., Application of GEOTHER/VT4 to assess groundwater transients in a high-level radioactive waste repository, in G. Jousma, J. Bear, Y.Y. Haimes and F. Walter, (eds.), Groundwater Contamination: Use of Models in Decision-Making, International Groundwater Modeling Center (IGWC), Kluwer Academic Publishers, Dordrecht , (1989), pp. 243-251.

[191] Faust, C.R.; Mercer, J.W., GEOTHER: a two-phase fluid flow and heat transfer code, Report ONWI-434 (1983), Battelle Memorial Institute, Columbus, Ohio.

[192] Yeh, G.T.; Ward, D.S., FEMWASTE: a finite element model of waste transport through saturated-unsaturated porous media. Report ORNL-5601 (1981), 137, Oak Ridge National Laboratory, Oak Ridge, Tennessee.

[193] Gureghian, A.B., TRIPM, a two-dimensional finite element model for the simulataneous transport of water and reacting solutes through saturated and unsaturated porous media. Technical Report ONWI-465 (1981), Office of Nuclear Waste Isolation, Columbus Ohio.

[194] Transferring Models to Users, in E.B. Janes; W.R. Hotchkiss (eds.), Symposium Proceedings, American Water Resources Association, Denver, (1990).

[195] Knissel, W.G., (ed.), CREAMS - A field scale model for chemicals, runoff, and erosion from agricultural management systems, Conservation Research Report No. 26 (1980), U.S. Department of Agriculture, Science and Education Administration.
User's Guide for CREAMS Computer Model: Washington Computer Center, Version U.S.DA-SCS Engineering Division Technical Release 72 (1984), U.S. Department of Agriculture, Soil Conservation Service, Washington, DC.

[196] Knisel, E.G. (ed.), CREAMS/GLEAMS: A Development Overview, Proceedings of CREAMS/GLEAMS Symposium, Sept. 1989, University of Georgia, Athens, GA., (1990).

[197] Tanji, K.W.; Modeling of the soil nitrogen cycle, in F.J. Stevenson (ed.), Nitrogen in Agricultural Soils. No. 22 (1982) in Agronomy Series, American Society of Agronomy, Madison, Wisconsin.

[198] Thiery, D.; Seguin, J.J., First model of nitrate transfer over a river basin with a lumped model, in Hydrogeology in the Service of Man, IAHS publication No. 154, (1985).

[199] Hutson, J.L.; Wagenet, R.J., LEACHM, leaching estimation and chemistry model, Agronomy mimeo 85-9, Department of Agronomy, Cornell University, Ithaca, New York, (1985) .

[200] Dillon, P.J., Models of nitrate transport at different space and time scales for groundwater quality management, in G. Jousma, J. Bear, Y.Y. Haimes and F. Walter (eds.), Groundwater Contamination: Use of Models in Decision-Making, Kluwer Academic Publishers, Dordrecht. International Groundwater Modeling Center (IGWC), (1989), pp. 273-284.

[201] Dillon, P.J., An analytical model for transport of contaminants from diffuse sources in saturated porous media, and DIVAST Users Manual, Internal Report, CSIRO Groundwater Reasearch Division, Perth, Australia, (1986).

[202] Kragt, J.F.; de Vries, W., Onderzook naar de effecten van mestbeperking op de nitratuitspoeling in waterwingebieden in Overijssel; 1 beschrijving van RENLEM: een nitratuitspoeling model op regionale schaal, STIBOKA, Wageningen, the Netherlands. Report No. 1935, (1987).

[203] Rast, W.; Lee, G.F., Nutrient Loading Estimates for Lakes, J. Env. Eng. ASCE, 109(2) (1983) 502-517.

[204] Reckow, H.M., et al., Modelling Phosphorus Loading in Lake Response under Uncertainty, A Manual and Compilation of Export Coefficients. Report USEPA-440/80-011, (1980).

[205] Kauppi, K., The Contribution of Agriculture Loading to Eutrophication in Finnish Lakes, Wat. Sci. Tech. 17(6/7) (1984) 1133-1140.

[206] Gillion, J.R., Estimation of Background Loading and Concentration of Phosphorus for Lakes in the Puget Sound Region. Wat. Resource Res., 17(2) (1984) 410-421.

[207] William, J.R., SPNM, A model for predicting sediment, phosphorus, and nitrogen yields from agricultural basins, Water Res. Bull. 16 (1980) 843-848.

[208] Hopstaken, C.F.; Ruygh, E.; Grashoff, P.; Menke, M., The environmental impact of agriculture on surface and groundwater quality and the use of NITSOL / PHOSOL in decision making, in G. Jousma, J. Bear, Y.Y. Haimes and F. Walter (eds.), Groundwater Contamination: Use of Models in Decision-Making, Kluwer Academic Publishers, Dordrecht. International Groundwater Modeling Center (IGWC), (1989).

[209] Schoumans, O.F.; Breuwsma, de Fries, W., Een fosfaattransport model voor toepassing op regionale schaal, STIBOKA, Wageningen, The Netherlands, Report No. 1951, (1986).

[210] de Vries, W., A conceptual model for analysing soil and groundwater acidification on a regional scale, International Symposium on Acidification and Water Pathways, Norwegian National Committee for Hydrology, Oslo, Norway, (1987).

[211] Krell, U., Vergleich von Modellen für den Eintrag von Spurenstoffen aus der Atmosphäre in den Nordsee - Beispiel Blei., Diss. vom Fachbereich Geowissenschaften der Universität Hamburg, (1991), GKSS-Forschungszentrum Geesthacht GmbH, Geesthacht.

[212] Derwent, R.G.; Hov, O.; et al., An intercomparison of long-term atmospheric transport models; The budgets of acidifying species for the Netherlands, Atm. Env., 23 (1989) 1893-1909.

[213] Petersen, G.; Weber, H.; Grassl, H., Modelling the atmospheric transport of trace metals from Europe to the North Sea and the Baltic Sea, in J. M. Pacyna and B. Ottar, (eds.), Control and fate of atmospheric trace metals, NATO ASI Series C, 268 (1989) 57-84.

[214] Krell, U.; Roeckner, E., Model simulation of the atmospheric input of lead and cadmium in the North Sea, Atm. Env. 22 (1989) 375-381.

[215] van Jaarsveld, J.A.; van Aaalst, R.M., Onderlinden, D., Deposition of metals from the atmosphere into the North Sea: model calculations, RIVM, Bilthoven, Report No. 842015002, (1986).

[216] van Jaarsveld, J.A.; Onderlinden, D., TREND: An analytical long term deposition model for multi-scale applications, RIVM, Bilthoven, Rep. No. 228603009, (1990).

[217] Eliassen, A.; Saltnones, J., Modeling of long-range transport of sulphur over Europe: A two year model run and some experiments. Atm. Env., 17 (1983) 1457-1473.

[218] Eliassen, A.; How, Q., et al., Estimates of airborne transboundary transport of sulphur and nitrogen over Europe, (1988). EMP / MSC-W Rep. 1/88, Norwegisches Meteorologisches Institut, Oslo.

[219] Lehmhaus, J.; Roeckner, E., Entwicklung von Ausbreitungsmodellen zur Analyse und Prognose von Emissionsepisoden beim mittel- und grossräumigen Transport von Luftverunreinigungen, F+E Vorhaben Nr. 104 02 618, Umweltbundesamt, Berlin, (1986).

[220] Klug, W.; Builtjis, P.J.H., et al., A comparison between four different inter-regional air pollution model, in S. Zwerfer and J. Ham (eds.), Interegional air-pollution modelling, NATO Challenges of Modern Society, 9 (1985), Plenum Press, New York.

[221] Martens, R; Massmeyer, W.; et al., Bestands-Aufnahme und Bewertung der derzeit genutzten atmospherischen Ausbreitungsmodelle, Gesellschaft für Reaktorsicherheit mbH, Köln (BRD), (1987).

[222] Guidelines for drinking-water quality, Vol 1, Recommendations, World Health Organization, Geneva, (1993).

[223] Gorelick, S.M., A review of distributed parameter groundwater management modeling methods, Water Resources Research, 19(2), (1983) 305-319.

[224] Colarullo, S.J.; Heidari, M.; Maddock, T., Identification of an optimal groundwater management strategy in a contaminated aquifer, Water Resources Bulletin, 20(5), (1984) 747-760.
[225] Gorelick, S.M.; Lefkoff, L.J., Design and cost analysis of rapid aquifer restoration systems using flow simulation and quadratic programming, Groundwater, 24(6), (1985) 777-790.
[226] Datta, B.; Peralta, R.C., Optimal modification of regional potentiometric surface design for groundwater contaminant containment, Transactions of ASAE, 29 (6), (1986).
[227] Prathapar, S.A.; Peralta; Peralta, Peralta, R.C., Development of a quality model for optimal groundwater management, in G. Jousma, J. Bear, Y.Y. Haimes and F. Walter (eds.), Groundwater Contamination: Use of Models in Decision-Making, Kluwer Academic Publishers, Dordrecht. International Groundwater Modeling Center (IGWC), (1989), pp. 170-187.
[228] Bishop, A.B.; Hughes, T.C.; et al., A Color-Graphics Based Water Demand Forecasting Model, in Water-Use Data for Water Resources Management, M. Waterstone and R. J. Burt, eds., American Water Association, Bethesda, MD, August 1988, pp. 223-238.
[229] Chansheng, He; Edens, Th.C.; Bartholic, J.F., A Water Resource Management Model for Irrigation in the Sginaw Valley, Michigan, Transferring Models to Users, Symposium Proceedings, AWRA, Denver (1990).
[230] Jone, C.A.; Kiniry, J.R., CERES-Maize: A Simulation Model of Maize Growth and Development, Texas A&M University Press, College Station, Texas, (1986).
[231] Advanced Computer Applications (ACA), International Institute for Applied Systems Analysis (IIASA), A-2361 Laxenburg, Austria.
[232] Strzepek, K.M.; Lenton, R.L., Analysis of Multipurpose River Basin Systems: Guidelines for Simulation Modelling, Ralph M. Parsons Laboratory. MIT, Cambrige, USA, Technical Report No. 236, (1978).
[233] Hetrick, D.M.; Mc Dowell-Boyer, L.M., User's Manual for TOX-SCREEN: A Multi-media Screening-level Program for Assessing the Potential Fate of Chemicals Released to the Environment, Oak Ridge National Laboratory, Report ORNL-6041, (1984); and USEPA, Washington, DC, Report EPA-560/5-83-024.
[234] Fedra, K; Diersch, H.J., Interactive Groundwater Modeling: Color Graphics, ICAD and AI, in A. Andreu et al., (eds.), Proceedings of the International Symposium on Groundwater Management: Quantity and Quality, October 2-5, (1989), Benidorm, Spain, IAHS Publication No. 188 (1989), pp. 305-320.
[235] Fedra, K.; Weighkricht, E.; Winkelbauer, L., A Hybrid Approach to Information and Decision Support Systems: Hazardous Substances and Industrial Risk Management, International Institute for Applied Systems, A-2361 Laxenburg, Austria, Report RR-87-12, (1987).
[236] MEXSES: Environmental Guidelines for Selected Industrial and Power Development Projects, Environment Unit, Asian Development Bank, (1988).
[237] Heatwole, S.D., Basin scale model for evaluation of best management practices implementation programs, Trans. Am. Soc. Agric. Eng., 29 (1986) 436-447.
[238] Roosaare, J., A computer simulation model for territorial systems of intensive agriculture, Acta Univ Tartu, USSR., 676 (1984) 87-103.
[239] Roberds, W.J., Risk-Based Decision Making Regarding Mixed-Waste Disposal Systems, in W.F. McTernan and E. kaplan (eds.), Risk Assessment for Groundwater Pollution Control, Published by the American Society of Civil Engineers, New-York, (1990), pp. 187-265.
[240] Roberds, W.J., Manual for Conducting Subjective Probability Assessment, draft internal report by Golder Associates Inc., Redmond, Washington, April 1988.
[241] MULTIMED, The Multimedia Exposure Assessment Model for Evaluating the Land Disposal by Hazardous Wastes, Environmental Protection Agency, USEPA, Athens, GA, (1988).
[242] Schroeder, P.R.; Morgan, Wolski, J.M.; Gibson, A.C., The Hydrologic Evaluation of Landfill Performance (HELP) Model, Vol. I, Users Guide for Version 1, Report USEPA/530-SW-84-009, (1984).
[243] Huyakorn, P.S.; Ungs, M.J.; et al., A three-dimensional analytical method for predicting leachate migration, Groundwater, 25(5) (1987).
[244] Faust, C.R.; Transport of immiscible fluids within and below the unsaturated zone: a numerical model, Water Resources Research, 21 (1985) 587-596.
[245] Osborne, M.; Sykes, J., Numerical modeling of immiscible organic transport at the Hyde Park landfill, Water Resources Research, 22(1) (1986) 25-33.
[246] Parker, J.C.; Lenhard, R.J.; Kuppusamy, T., A parametric model for constitutive properties governing multiphase fluid flow in porous media, Water Resources Research, 23 (1987) 618-624.

[247] Kuppusamy, T.; Sheng, J. ; Parker, J.C.; Lenhard, R.J., Finite element analysis of multiphase immiscible flow through solids, Water Resources Research, 23 (1987) 625-632.
[248] Corapliogiu, M.Y.; Baehr, A.L., Groundwater contamination by petroleum products, 2. Numerical solution, Water Resources Research, 23 (1987) 191-200.
[249] Baehr, A.L.; Corapliogiu, M.Y., Groundwater contamination by petroleum products, 2. Numerical solution, Water Resources Research 23 (1987) 201-213.
[250] Abriola, L.M.; Pinder, G.F., A multiphase approach to the modeling of porous media contamination by organic compounds, 1. Equation development, Water Resources Research, 21 (1985) 11-18.
[251] Chrysikopoulos, C.V., Transport contaminants from non-aqueous phase liquid pool dissolution in subsurface formations, in L.C. Wrobel, C.A. Brebbia (eds.), Water Pollution II, Modelling, Measuring and Prediction, Computational Mechanics Publications, Southhampton Boston, (1993), pp. 29-36.
[252] Yeh, G.T., 3DFEMWATER: A three-dimensional finite element model of water through saturated-unsaturated media, Environmental Science Division, Oak Ridge National Laboratory, Publication No. 2904, (1987).
[253] Hemker; C.J; van Elburg, H., Micro-Fem: MicroComputer Multilayer Steady State Finite Element Groundwater Modeling, Elandsgracht 83-1016-TR, Amsterdam, The Netherlands, User's Manual, Version 2.0, (1987).
[254] Yeh, G.T., 3DFEMWATER: A three-dimensional finite element model of water through saturated-unsaturated media, Environmental Science Division, Oak Ridge National Laboratory, Publication No. 2904, (1987).
[255] Smith, L.; Freeze, R.A.; Massmann, J., Geostatistical approach to site characterization and risk assessment related to groundwater contamination at hazardous waste management sites, in W.F. McTernan and E. kaplan (eds.), Risk Assessment for Groundwater Pollution Control, Published by the American Society of Civil Engineers, New-York, (1990), pp. 281-311.
[256] Doran, F.J.; Thresher, J.E., Prediction of the fate of a chromium fom a municial sanitary landfill, in Solving ground Water Problems with Models, Vol. 1, Conf. Febr. 10-12, (1987), Denver, Colorado.
[257] Scharch, J.F., Water Balance Analysis Program for the IBM-PC Micro-computer, Wiss. Det. of Nat'l Res., Bur. of Solid waste Management, (1985).
[258] Prickett, T.A.; Naymik, T.G., Lonnquist, C.G., A Random-Walk Solute Transport Model for Selected Groundwater Quality Evaluations, Illinois State Water Survey, Bulletin 65, (1981)
[259] Prickett, T.A.; Lonnquist, C.G., Selected digital computer techniques for groundwater resources evaluation, Illinois State Water Survey Bulletin, 55 (1971).
[260] Massmann, J.; Freeze, R.A., Groundwater contamination from waste management sites: The interaction between risk-based engineering design and regulatory policy, 1, Methodology, Water Resour. Res., 23(2), (1987) 351-367.
ibid: 2, Results. Water Resour. Res., 23(2), (1987) 368-380.
[261] Massmann, J.; Asce, A.M., Risk Assessment and Groundwater Contamination: Methods and Relationships, in W.F. McTernan and E. kaplan, (eds.), Risk Assessment for Groundwater Pollution Control, Published by the American Society of Civil Engineers, New-York, (1990), pp. 331-368.
[262] OUSE User's Guide and Technical Reference, Danish Hydraulic Institute, Edition 3 0, (1990).
[263] Troch, P.A.; Van de Velde, J.; De Troch, F.P., A stochastic point rainfall generator based on the modified Neyman-Scott rectangular pulses model, Application to two Italian raingauge stations at the Lago Maggiore: Pallanza and Ispra, Laboratory of Hydrology and Water Management, State University of Gent, (1991). See also: GLAS User's Manual, Labo Hydrologie, Gent, Belgium, (1989).
[264] Hydrologic Engineering Center in California, 609 Second Street, Davis, California 95616, USA.
[265] Corps of Engineers, U.S. Army Hydrologic Engineering Center, Annual Report, (1989).
[266] Center for Exposure Assessment Modeling (CEAM), ERL, U.S. Environmental Protection Agency, Athens, GA 30613, U S A.
[267] Pollock, W.D., Documentation of computer programs to compute and display pathlines using results from the USGS modular three-dimensional finite-difference groundwater flow model, U.S. Geological Survey Open-File Report, (1990), pp. 89-381.
[268] Scientific Software, P.O.Box 23041, Washington, DC 20026-3041.
[269] El Kadi, A.I., Application of the USGS two-dimmensional mass-transport model MOC to remedial actions by recovery wells, Groundwater, 25 (1987).

[270] Vancso-POLASCEK, K., Theory and Practice of Computer Assisted Simulation and Modelling on Professional Workstations, Swiss Federal Institute of Technology Zurich (ETHZ), Switzerland, ETH Dissertation No. 9104, (1990).

[271] Ulrich, M., ModelWorks - An Interactive Modula-2 Simulation Environment, Pilot Project CELTIA, Institute of Automatic Control and Industrial Electronics, Swiss Federal Institute of Technology Zurich (ETHZ), Switzerland, Postgraduate Thesis, (1987).

[272] Fischlin, A.; Roth, O; Gyalistra, D; Ulrich, M.; Nemecek, Th., ModelWorks - An interactive simulation environment for personal computers and work stations, Systems Ecology Group, Swiss Federal Institute of Technology Zurich (ETHZ), Switzerland, Report No. 8, (1990).

[273] Kittle, J.L.; Barnwell, T.O.; Hummel, P.R.; Imhoff, J.C., AIDE - A system for developing interactive user interfaces for environmental models, in Transferring Models to Users, American Water Resources Association, Bethesda, MD., Symposium Proc., Denver, Nov. 4-8, (1990), pp. 253-262.

[274] Lumb, A.M.; Kittle, J.L., Preliminary Design of a Data Storage and Retrieval System for Hydraulic Models, Watershed Data Management System, (WDMS), U.S. Geological Survey, Reston, VA, (1986).

[275] Lumb, A.M.; Kittle, J.L.; Flynn, K.M., User's Manual for ANNIE, A Computer Program for Interactive Hydrologic Analyses and Data Management, U.S. Geological Survey Water-Resources Investigations Report No. 89-4080, Reston, VA, (1989).

[276] General Sciences Corporation. EMIFE-Environmental Model Input File Editor FORTRAN Utilities - Preliminary Documentation, Office of Pesticides and Toxic Substances, USEPA, Washington, DC, (1987).

[277] General Sciences Corporation, PCGMS User's Guide, Draft Report prepared for Office of Pesticides and Toxic Substances, USEPA, Washington, DC, (1987).

[278] Brown, L.C.; Barnwell, T.O., The Enhanced Stream Water Quality Models QUAL2E and QUAL2E-UNCAS: Documentation and User's Manual Environmental Research Laboratory, USEPA, Athens, Ga., 30613, Report EPA/600/3-87-007, (1987).

[279] Imhoff , J.C.; Carsel, R.F.; Kittle, J.L.; Hummel, P.R., Data Base Analyzer and Parameter Estimator (DBAPE) Interactive Computer Program User' Manual., Environmental Research Laboratory, USEPA, Athens, Ga., 30613, Report EPA/600/3-89-083, (1989).

CHAPTER 4

MODELS AND ASSESSMENT METHODS FOR ENVIRONMENTAL EXPOSURE AND HEALTH IMPACTS

4.1. Assessment of Continuous Emissions Impacts to the Environment

Assessment of environmental impacts is more complex than that of human health impacts, because of the large variety of species and physical entities involved, availability of toxicological data on only a few, and the need to consider competition, predation, and other ecological interactions. *To make the task manageable, effects are usually addressed in terms of aggregate indices (total biomass or species diversity) or assessment is limited to key species.*

Four key issues in such ecological risk assessment program are relevant:

■ *Which ecological resources are at risk?* What are the characteristics of these ecosystems and how do they respond to pollution? What are the best indicators and endpoints to determine the condition of these ecosystems? What are the best methods for screening and characterizing pollutants in these ecosystems?

■ *What is the condition of the environment and how is it changing?* What are the baseline characteristics that define a healthy ecosystem against which to measure change? How are the affected ecosystems changing? Which pollutants are contributing to ecosystem deterioration? How accurately can ecosystem exposure and effects models predict reality?

■ *To what levels of pollutants are the ecosystems exposed?* What pollution levels exist in the environment? What biological, chemical or physical processes form and transform complex pollutants and how are they taken up in the environment? What are the most accurate and sensitive biomarkers of pollution exposure?

■ *How do pollutant exposures affect ecosystems?* What structural properties of chemicals predispose them to be biologically active and what are the best methods for predicting their effects? How can we predict effects of long term, indirect, or cumulative exposures of ecosystems to pollutants? How can laboratory data be extrapolated to ecosystem effects? How can effects recorded for one species, population, or community be extrapolated to others?

4.1.1. ASSESSMENT METHODS

The following assessment methods are widely used and can be recommended:

Matrix Approach. The matrix approach is essentially a *checklist approach,* widely used in environmental impact assessment. It helps assure completeness, but, in itself, is not sufficient for environmental risk assessment.

Thresholds. There are often specific levels of environmental quality parameters (e.g., pH in lakes) that represent the *threshold of ecological change.* A series of such thresholds can represent a progression of stages in ecological decline. *Thresholds may also be toxicological benchmarks such as* LC_{50}.

Functions. Where decline is continuous, without clear thresholds, *continuous damage functions* may be available. Such functions have been developed for the impact of air pollution on agricultural crops, for example. Ideally, functions include extrapolation error, and the appropriate uncertainty factors associated with extrapolation from laboratory test organisms to field observations.

Simulation Models. Thresholds or formulae, while often useful for environmental assessment, are generally limited in the kinds of impact they can represent and are always a simplification. Analysis of complex ecosystems usually requires more *detailed models* that can integrate the combined effect of multiple relationships.

Assessing the Impacts of Different Policy Options
Models have often proven to be powerful guides for studying and understanding ecological relationships, but less useful as predictors of future effects. Nonetheless, they can serve as a useful comparative measure of the possible impacts of different policy options.

Added detail in highly complex models may improve predictability, although too much detail can introduce crippling problems of parameter estimation and error propagation. In some cases, *simplified models may be preferred* even when detailed data are available.

Status of Ecological Models
The status of ecological models has been summarized, according to (DOE, 1987), as follows:

- Long-term predictability remains an elusive target.

- There are severe limits to how long into the future the behavior of complex systems with many feedbacks can be projected.

- What must replace these are short-term predictive schemes coupled with monitoring and adaptive management approaches that incorporate both modeling and monitoring.

Probabilistic Models. They explicitly take into account the variability and uncertainty in natural systems. These include analysis of extrapolation error (mentioned above), fault-tree analysis applied to elucidate causal linkages between pollutants and endpoints, and ecosystem uncertainty analysis using Monte Carlo simulation and ecosystem models to extrapolate from laboratory toxicological data to estimate risks to populations and ecosystems (Barnthouse et al.[1], 1982).

Expert Judgment. In the future, it will not be possible to cover satisfactorily all the needs for environmental risk assessment with ecological models. *While these models may provide valuable guidance, they must be used in an integrated way with expert judgment.* Methods of eliciting expert judgment are described by Morgan and Morris[2] (1981). *Judgments can be applied to policy making using techniques such as decision theory or analytical hierarchy method.* Environmental applications of the latter are described by Barnthouse et al.[1] (1982).

Remarks Regarding Endpoints
The diversity of possible endpoints requires that they be divided into classes.

■ **Biome:** *Biomes represent large types of environment like tundra, deciduous forest, grassland, and desert.* Many specific processes of environmental impact and ecological models through which those processes are quantified to explain impacts or estimate risk are biome specific.

■ **Structure and Function:** *The health of ecological communities can be evaluated by parameters measuring their structure and function.* Species diversity is a frequently used measure. The greater the diversity of species in a community, the stronger that community is ecologically.

■ **Physical support entities:** Air, water, and soil are the basis of the environment. While *environmental impact assessment* often focuses on the biosphere, the biological inhabitants of the physical world, the term *environmental quality* itself usually refers to the physical state of air, water, and soil. While in health effects assessment, *degrading environmental quality* is addressed in terms of potential risks to human health, in environmental assessment it is addressed both in terms of its own merit and for its implications for the biosphere. In some cases the latter relationships are sufficiently well established *that physical environmental parameters can be used as indices of ecological damage.* For example, the relationship between acidification of freshwater lakes and the resulting impacts on aquatic life is sufficient that the latter is often indexed solely on the basis of predicted changes in pH of the lake water.

4.2. Environmental Exposure Assessment Methods for Chemicals

An OECD workshop on "Practical Approaches to the Assessment of Environmental Exposure" took place in Vienna in April 1986. As a result a *Compendium of Environmental Exposure Assessment Methods for Chemicals*[3] was prepared. It is a compilation of some of the methods used by government, industry and research institution in OECD Member countries to assess:

a) the sources of release of chemicals to the environment

b) the fate and pathways of chemicals in the environment.

> ***Integrated Multi-media Transport and Contaminant Fate***
> *In the past, contaminant assessment generally focused on only one environmental pathway, such as groundwater, surface water, or atmosphere. However, in the real world, contaminants released into one medium eventually migrate through and interact with several media. Thus to more accurately portray the complexities of the interactions of contaminants and the environment, scientifically based assessment methodologies must consider integrated multimedia transport and contaminant fate.*

A summary of these methods and models has been prepared and is presented below. The references for the different methods, as well as the contact addresses of the "models owners/developers", can be found in the original compendium and have not been reported here.

The models have been classified into :

a) Source assessment methods
b) Fate and Pathways assessment methods
c) Surface water methods
d) Soil (ground water methods)
e) Air methods
f) Multi-media methods
g) Miscellaneous methods

Some of these methods are used to assess the sources of release, or fate and pathways, of a large number or a broad class of chemicals. Other are designed to be more chemical-specific. Most of the methods described in this compendium have been developed and/or used in the areas of chemicals control, air and water pollution control, or waste management.

TABLE 4.1. Classification of simplified models

Category	Level I	Level II
Models	Simple	Complex
Data	Limited /aggregated	Extensive/detailed
Required knowledge	Familiarity	Expert
Sensitivity analysis	Extensive	Limited
Use of results	Screening/bounds	Final answers

This section discusses the use of simplified models for calculating long-term average pollutant concentrations and intermedia transfer rates for the air, surface soil, surface water, and groundwater. Depending on the analysis level I or II we can have the following classification (TABLE 4.1).

Basic Information and Data

Before proceeding to some calculation, employing regional maps and preliminary information, the user should define the *initial boundaries of the analysis*. This definition should include, among other things, the (1) study area, (2) overall time period and intervals, (3) environmental media, (4) pollutant species, (5) surface water network, and (6) potential exposure routes. The definition should be used to structure the analysis and initial data collection activities, but can later be changed if necessary.

Data requirement depend on the specific type of analysis. In general data will be needed to describe regional population distributions (urban and rural repartition based on census information), water sources and consumption, atmospheric conditions, land uses, hydrologic characteristics, concentrations of background pollutants, and other variables. As an example, data needed for a typical hydrologic analysis might include:

- Runoff quantity and quality.
- Stream flows (low, mean, high, critical) and water quality parameters (e.g., temperature, pH, dissolved solids, pollutant concentrations).
- Impoundment/lake residence times, inflows/outflows, stagnant regions, thermal stratification, internal flow patterns, rates of sediment accumulation (for impoundment), and water quality.
- Esturial tidal prism, flushing times, and water quality.

Most of the data should be available from governmental organizations such as environmental agencies. If runoff data are not available, this information can be estimated from the simplified procedures or more complex models. Basic climatological data will then be needed, including precipitation rates and patterns. Statistical analysis may be necessary to convert raw material data to the form required by the analysis and models. It may be also necessary to determine flows, water quality, and other characteristics during critical periods. Waste loads can be separated into *point sources*, according to different types of industries: chemical processing plants, nuclear and power plants thermal, and their associated facilities, and discharges from cities and municipal waste treatment; and *nonpoint sources* represented by runoff, agricultural drainage, and urban meteor water. All of the above discharges are contributing to the existing water quality, and can be incorporated into the analysis by using current pollutant background concentrations in water bodies. Surface water consumption's data should include a description of the type of users that withdraw surface waters from potentially affected sources, and the withdrawal rates.

If atmospheric transport and inhalation exposure will be included in the study, the area should be partitioned into subregions. These should be based on eight equal angular sectors and up to five radial divisions of such sector. If inhalation exposure is not considered, the sub-region definition can be more flexible and the subregions will be probably based on regional population, water consumption pattern, and the surface water network.

Atmospheric Pollutants Transport. The analyst should use an atmospheric transport model to calculate the regional distribution of average pollutant concentrations and deposition rates for different emission rates and environmental conditions. These results should be used to estimate inhalation health risks, if appropriate, and to provide necessary deposition rates for the overland runoff analysis.

In general, plume dispersion will be affected by:

- Wind speed, direction, and velocity gradients
- Atmospheric stability (responsible for vertical motion)
- Source parameters (velocity, temperature, and stack height)
- Regional topography

To successfully calculate plume dispersion, one must normally simplify the process by separating the transport calculations into initial plume rise and subsequent dispersion and deposition. Plume rise is caused primarily by the condition of release. Although atmospheric turbulence and stability affect the plume immediately after discharge, their influence is generally secondary. After the initial phase, atmospheric conditions govern the dispersion and deposition process.

Overland Pollutants Transport. A level I or II overland transport model should be used to calculate pollutant concentrations in surface soil layers and the distribution of runoff into surface water streams and lakes. Soil concentrations can be used with irrigation data and other information to estimate exposure rates associated with consumption of animal and vegetable products.

Groundwater Transport. Simplified or detailed groundwater flows and contaminant transport models should be used with well location and geologic information to determine the transport of pollutant through the soil and regional aquifers. Unsaturated soil models will probably be required for transport down through the soil from pollutant sources into local aquifers. From there saturated soil models should be used to estimate groundwater flow patterns and rates, the transport of pollutants to regional wells, and the effects of interactions with surface water bodies.

Surface Water Transport. For such analysis, simplified or complex models can be used to calculate pollutant concentrations in stream segments and lakes of the surface water network. Thermal stratification, sediment accumulation, and pollutant decay should ideally be considered when determining water quality in lakes and impoundments. Pollutant concentrations can be calculated for both rivers and lakes using background water quality data, mass balance equations, and estimates of decay constants, given pollutant loading from point and nonpoint sources. The primary use of the results of such analysis is in calculating exposure rates associated with consumption of surface water and aquatic organisms (fishes, etc.). Pollutant concentrations can also be compared with the appropriate water quality standards or criteria, to determine potential problems and areas for more detailed analysis.

Lakes and Impoundments. Pollutants may enter an impoundment with inflowing water, through deposition from air, or in runoff from the surrounding area. These pollutants will be mixed through the current unstratified, non-stagnant volume and may decay and settle out to the bottom. The resulting concentration within the active volume of the impoundment will reflect the dynamic balance between these processes. By making certain simplifying assumptions, one can develop usable water quality models for lakes and impoundments. The models can easily be used for sensitivity analysis to estimate the uncertainty associated with the results of an analysis to place upper and lower bounds on these results.

The water quality in lakes and impoundments depends strongly on the occurrence and strength of thermal stratification. When stratification does not occur, lakes mix well and pollution concentrations are more uniform throughout the water body. Stratification, on the other hand, may trap pollutants in the various water layers, causing large concentration gradients and temporarily high concentrations during some periods of the year.

Sedimentation in an impoundment may also affect pollutant concentrations. Reservoirs and lakes frequently collect suspended particulate material carried in by streams and rivers. If toxic pollutants are associated with suspended particulates, total concentrations can be significantly reduced through settling processes. Polluted sediments settled on the bottom of impoundments may also act as a source of pollution in the future if conditions in the water change.

To calculate concentrations in a lake or impoundment, one must know its hydraulic residence time, or the mean time that a given volume of water resides within the impoundment. The residence time normally depends on the basin inflow, discharge, depth, shape, and other hydraulic characteristics.

4.2.1. SOURCE ASSESSMENT METHODS

Source Assessment Methods (Methods 101 to 110). They start from an understanding of the manufacturing, processing, use and disposal patterns of chemicals. Based on information relevant to their release to the environment and using expert judgment, an estimation is made of the amount released to the environment. This estimation can serve as input to fate and pathways assessment methods, which are used to estimate the environmental concentration of a chemical. The source assessment methods can be grouped according to scenarios for the release of chemicals to surface water, soil, (groundwater, and air). The methods are described in TABLE 4.2.

The *facility-specific approach* generally require the assessor to collect input data on a site-specific basis in order to complete the source assessment. On the other hand, in most of the *industry/use category* approaches reported here, data or assumptions concerning the average and range of important facility characteristics for a particular process and/or use category are used along with information on the chemical.

Input data for source assessment methods are derived from literature, existing data bases, information supplied by chemical manufacturers or users, or reports from the facility (chemical or industrial plant, nuclear plant, regional incinerators, sewage plants, etc.).

TABLE 4.2. Source assessment methods

Source Assessment Method										
Method No.:	**101**	**102**	**103**	**104**	**105**	**106**	**107**	**108**	**109**	**110**
A. Release to surface water										
- Direct releases from industrial facilities										
Industry / Use category Approach	X	X	X	X	X					
Facility-specific approach						X	X	X		X
- Indirect releases from sewage treatment plants										
Industry / Use category Approach	X	X	X	X	X					
Facility-specific approach						X			X	
B. Release to soil (ground water)										
- Direct deliberate application										
Industry / Use category Approach				X						
Facility-specific approach						X				
- Landfill disposal of wastes										
Industry / Use category Approach	X	X	X		X					
Facility-specific approach						X		X		X
C. Release to air										
- Point source discharge										
Industry / Use category Approach	X	X		X	X					
Facility-specific approach						X				X
- Diffuse release										
Industry / Use category Approach										
Facility-specific approach				X	X					

TABLE 4.2. Source assessment methods (continued)

Source Assessment Method										
Method No.:	**101**	**102**	**103**	**104**	**105**	**106**	**107**	**108**	**109**	**110**
- Hardware / Software requirements						X				X
Manual operation	X		X							
Not yet computerized							X			
Desk calculator	X		X							
Spreadsheet / Lotus123									X	
Micro-computer						X			X	
IBM-PC / compatible (MS-DOS)				X		X		X	X	
Workstation		X							X	
Mainframe computer (V = Vax-VMS; U = Unix)					V			U		
Mainframe computer (S= Siemens)				S						
Cray computer				X						X
Fortran77 / Basic / Pascal				F				P		
DBMS / Informix / C-ISQL					X			X		
- Assessment Effort										
Training (h / d / w / month)	h	d	h	d/w	d	w	h/d	d	h	
Data collection and preparation (h / d / w / month/Year)	w/m	m	w/m	w	h/d	m/y	h/d	h/d	d	m/y
Assessment (h / d / w / month)	h/d	h	h/d	h	d	M	h	h	h	m
- Validation / Review										
not applicable			X		X					
limited experience	X	X		X						
validated with model experiments								X	X	
validated with field experiments / data	X	X				X			X	
validated through reviews/experts							X	X	X	X
used for regulatory purpose						X			X	
- Availability C= Code/Model; P= Procedure	P	P	P	C	C	No	P		C/P	

Method 101: Product Category Approach
The "product category approach" is used to estimate point source and/or multiple point source emissions during processing, use and/or disposal/recycling into the environmental compartments of air, water and soil, and is applicable for the initial exposure assessment of new chemicals.

The product category approach (M-101) is based on the assumption that a chemical under consideration will behave in a similar way, under similar conditions, as other chemicals in the same product category (considering similar types of applications and use patterns). Based on product category-derived data, the emission of the chemical under consideration can be quantified.

From the implementation of the *German Chemical Act* until now, about 100 notified chemicals have been assessed using this approach.

Method 102: Assessment of the Environmental Release of New Chemicals
The method is used to estimate emissions and discharges of new chemicals in certain applications from all processes which they go through. Based on general data, an estimation is made of emissions and discharge of new chemicals.

A calculation model has been developed for the behavior of an organic substance in an activated sludge water treatment system (with powdered activated carbon , PAC). The method (M-102) assumes that data are available on only a limited number of characteristics of the process and of the chemicals. As output the amount of chemical which is released with waste water or waste or into the air, or which is removed by degradation is calculated.

Method 103: Source Assessment Using Simple Dilution Models
The method is based on appropriate scenarios developed for each chemical to estimate source discharge to the aquatic and terrestrial environment at the point where the chemical (in solution, or suspension, or as solid waste) enters the aquatic or terrestrial environment.

The method (M-103) may be applied at the point of manufacture, use, or disposal. The initial data are usually sufficient to allow a screening estimation to be made of environmental concentration. The method is based on the development of relevant scenarios and is performed by making a series of simple calculations. The data available and any assumption made are used to create "worst-case" scenarios. The output of the assessment will comprise an estimation of the number of sites where the discharge or disposal will occur, frequency of discharge, and quantity of the chemical released. The quantity and quality of the data will affect the estimation. Worst case situations may or may not be realistic. The model has been applied to the estimation of surfactant concentrations in UK sewage and rivers.

Method 104: RLTEC Method (Release from the Technosphere)
This method is used to estimate area-wide releases to air, water and soil from production, manufacturing and use. The purpose is to screen existing chemicals. The results are used as input for models which estimate the fate of chemicals in the corresponding media.

The method (M-104) is also part of a system of models and methods (Exposure and Ecotoxicity Estimation for Environmental Chemicals, E4CHEM) for assisting priority-setting exercises regarding potential environmental hazard of chemicals. As a result the relative flows into the media air, water and soil relative to the marketing volume in the geographic area of interest is obtained. The method should only be applied where some information on the use pattern of the chemical, but where the relative flows into the media are unknown.

> ***Method 105: Environmental Information and Planing Model RIM***
> *RIM is an interactive tool which decision-makers can use to evaluate environmental and financial implications of abatement measures, primarily at a national level. It is intended to be an integrated system giving a simultaneous overview of different polluting sectors.*

In principle the model RIM (M-105) can be applied to all chemicals, but the data base contains actually data on acidifying substances. It functions as an information system and as a tool for generating emission scenarios. The complex system is user friendly. Emissions and costs are calculated by multiplying levels of activities by emission and cost factors. Emissions of pollutants in all compartments can be evaluated in order to forestall the transfer of emissions from one compartment into another when abatement measures are taken. Presently all relations are linear. Sub-models are being developed for a more detailed analysis of important sectors (traffic, agriculture, solid wastes etc.). Data on economic development are derived from publications of the Dutch Central Planing Office, etc.

> ***Method 106: Assessment of Industrial Reports of Estimated Releases for the Toxic Chemical Release Inventory***
> *The method evaluates site-specific release data for all media as reported by plants and facilities under various US Environmental Protection Agency or US State reporting requirements. The purpose of this screening method is to assess releases of specific industrial chemicals from all operations at a facility. Exposure assessment based on reported data will generally be site-specific. The method uses data reported as the total annual releases from each reporting facility. In some instances peak or short-term releases are also reported.*

The assessment (M-106) will report daily releases to each media of concern, based on annual releases and days of operation; the treatment and disposal practice used by the facility; the reported basis for estimate of each release. At present substantial errors may be expected in fugitive air emission estimates and some waste water estimates.

> ***Method 107: Industrial Sources of Water Releases***
> *The method is a case-by-case estimation of point-source discharges to water from industrial facilities. The method is used when data on discharges are not available for the chemical in question.*

It should be noted that end-of-pipe reporting of aggregate waste water discharges (after any on-site treatment) makes it difficult to apply monitoring data for a single chemical to an analog. Since the treatment step vary from site to site, end-of-pipe data for one facility may not be applicable to another. Process flow diagrams or descriptions, batch volumes and amounts of water used in the process at various step, as well as the estimated water solubility or partition coefficient should be known (M-107). The water release points must be identified. The prevailing conditions should be estimated (temperature, chemical concentration, etc.). As output one obtains the amount of chemical per day released to the water, etc.

> ***Method 108: Waste Water and Waste from Metal Plating***
> *A model for calculating amounts of heavy metals and other compounds in waste water after a treatment step is available. The model shows how discharges and costs of treatment are related.*

This method (M-108) offers a basis for effluent guidelines an for a policy for prevention of hazardous waste. The model deals only with metal finishing processes in aqueous solutions.

> ***Method 109: SIMPLETREAT Box Model***
> *SIMPLETREAT is a steady-state box model of the removal of micro-pollutants in a communal sewage treatment plant (primary clarify / aeration tank / settling tank). It is based on a general purpose box modeling system SIMPLEBOX. It is meant to be a diagnostic method of analyzing the fate of micro-pollutants after they enter a sewage treatment plant.*

A computation is done of the distribution of a chemical which is subject to reaction, advection, and interphase transport under steady-state, non-equilibrium conditions in the separate systems of which a communal sewage plant is composed. A multiple mass balance computation at steady state is performed with the built-in box modeling system SIMPLEBOX (M-109). From the produced steady-state concentrations, the emission rates are calculated and represented in tables and diagrams. In screening mode only the steady loading of the system and the partition coefficient are needed. In the diagnostic mode, operation parameters of the sewage plant may be redefined. Steady-state concentrations in the effluent and fate of the chemical in terms of the percentage removed by degradation, volatilization, exported sludge and exported water.

> ***Method 110: Environmental Risk Analysis***
> *This method can be used to analyze point-source discharge to water, air and soil from single plants and from incidental sources related to the transport of chemicals by railways or trucks. The method is intended to yield information on chemical releases during routine operation, and information on potential accidental releases.*

The method (M-110) includes a review of relevant site-specific information (using check-lists) on processing, storage, and transport of chemicals. Probable amount of release is calculated by multiplying the probability of failure by the amount of the chemical in containers. Occasionally mass balance or multiplication of the actual concentration by the amount of effluent are also used to calculate long-term environmental risks.

4.2.2. FATE AND PATHWAYS ASSESSMENT METHODS

The fate and pathway assessment methods are categorized according to the environmental media for which the prediction is made, and are designed to make assessment for single chemicals (with the exception of Methods 209 and 210, which deal with several chemicals). Four methods deal with products degraded from parent chemicals (Methods 209, 210, 306, and 511). They start by defining a scenario for the release, fate and pathway of a chemical. The scenario facilitates the task of the assessor since it limits the collection of input data. The fate and pathway assessment methods require as input the release rate of chemicals from sources, data on chemical properties, and data on the environment to which chemicals are released. Many of the above mentioned methods have been programmed for personal computers. A few refined methods require a mainframe computer. Since adequate experimental data on chemicals are not always available, a number of methods therefore employ data estimation techniques for physical-chemical properties.

4.2.3. ASSESSMENT METHODS FOR SURFACE WATER

The methods in this group range from simple dilution model (Methods 201 to 203), which yield rough estimates, to highly refined models (Methods 209 to 213) in which hydrological and meteorological influences on the fate and pathway of chemicals are taken into account. The methods are described in TABLE 4.3.

Chemicals' Concentrations in Rivers

■ **Methods 201 to 208** are used for predicting concentrations of chemicals in rivers. Methods 204 to 208 are analytical fate models or compartment models, which require not only chemical release data, but also data regarding physical-chemical properties and river characteristics' data.

■ **Methods 209 to 213** are hydrodynamic simulation models. They all require the input of data on chemical release, physical-chemical properties and hydrology, meteorology, and other environmental characteristics. They predict concentrations in more than one compartment of the system simultaneously.

Fresh Water Systems

■ **Methods 209 and 210** are for fresh water systems (river, lake), whereas

Sea Water Systems

■ **Methods 211 to 213** are used to predict concentrations in sea waters (estuaries, coastal waters, and seas).

TABLE 4.3. Fate and pathway assessment methods for surface water

Source Assessment Method													
Method No.:	**201**	**202**	**203**	**204**	**205**	**206**	**207**	**208**	**209**	**210**	**211**	**212**	**213**
Models / Methods for:													
- Fresh water													
Simple dilution model	X	X	X										
Simple analytical fate model				X	X			X					
Simple compartment model						X	X						
Refined hydrodynamic simulation model									X	X	X	X	X
- Sea water													
Refined hydrodynamic simulation model											X	X	X
Hardware / Software requirements													
Manual operation													
Not yet computerized													
Desk calculator	X												
IBM-PC / compatible (MS-DOS)		X	X	X	X		X	X	X	X			
Workstation									X	X	X	X	X
Mainframe computer (V = VAX-VMS; U = Unix)		V			V			V	V, U	V	V		
Mainframe computer (S = Siemens; C = Cyber)							S		S			C	C
Fortran /Basic / Pascal		F	P		F	B	F	F	F	F	F	F	F
Assessment Effort													
- Training (h / d / w / month)	d	d	h	h		d	h	d/w	d/w	w	w	m	m
- Data collection and preparation (h / d / w / month)	w	d	m		d		d/w	d/w	d	d/w	w/m	w/m	d/m
- Assessment (h / d / w / month)	w	d	d		d	d	d	h/d	d	w/m	w/m	m	d/m
Validation / Review													
not necessary	X												
limited experience							X					X	
validated with model experiments		X	X	X	X	X			X	X			
validated with field experiments / data					X	X	x	X	X	X			X
validated through reviews/experts		X	X	X					X	X			X
used for regulatory purposes		X							X	X	X		
Availability C= Code/Model; P= Procedure	P	C	C	C	C	C	C	P (C)	C	C	C	C	C

Method 201: River dilution model
This simple stream dilution calculation method is used to calculate the surface water concentration of a chemical in a river.

The method (M-201) can be applied for single and multiple-point releases. Calculations represent rough estimations. The method is applicable to all chemicals which are emitted into the water compartment during their life cycle. Dilution factors are selected under reasonable worst case assumptions. Transfer and degradation processes are not taken into account. The most important data are source related and must be worked out for product categories. Based on these specific data the concentration of the chemical in waste water is estimated.

Method 202: Dynamic Toxicity Model (DYNTOX)
DYNTOX is a waste load allocation computer program that uses probabilistic dilution techniques to estimate concentrations of toxic substances or fractions of whole effluent toxicity.

DYNTOX (M-202) performs three types of simulation: continuous, Monte-Carlo, and log-normal. Based on probabilities, they can aid in analyzing the frequency and duration of toxic concentrations from a waste discharge. DYNTOX considers dilution and net first-order loss, but not sorption or benthic exchange. The net loss rate must be determined empirically on a case-by-case basis and should not be extrapolated to different conditions off-flow, temperature, solids, pH or light. DYNTOX requires input of flow and concentration information for a stream and a waste source. Mean, variance and lag-one correlation coefficients are required for Markow process (four inputs). The log-normal methodology requires mean and variance for the log-normal distributions of the four inputs. A toxicity measure (e.g. LC_{50}) for the contaminant is required for evaluating the output. The output of the assessment is the frequency and duration of daily average contaminant in the receiving water body.

Method 203: Probabilistic Dilution Model PDM3 (Version 3)
It is a river dilution model which takes into account the variability in the stream flow. The primary use of this methodology is to calculate the surface water concentration in rivers/streams receiving point discharge and to calculate how frequently a specified concentration of concern (COC) would be exceeded. The model can deal with intermittent and continuous releases and can be applied either to site-specific situations or release scenarios that occur in an industrial sector.

The model PDM3 is a mass balance calculation system which assumes the chemical is non-reactive and remains in the dissolved state (M-203). A feature of this system is the treatment of surface water concentration as a probability distribution rather than as a single steady-state concentration corresponding to a mean or low-flow condition. The model assumes that discharges take place randomly over the year. Stream flow may also violate the assumption of log-normality (for small flows or where flows are controlled by dams). An actual draw-back of the implemented correlations (for computing the coefficient of variation of stream flow) is that they only apply to hydrological sub-basin located in the United States. A similar equation would have to be implemented for a location outside the United States.

Method 204: ReachScan
This a river dilution model which uses stream flow and river connection data. The program combines data base search functions for understanding the reach system of rivers designations with a limited surface water modeling approach to predict surface water concentrations of an introduced pollutant. The model is applied to continuous point sources.

The program (M-204) make use of *US data bases* and was developed for the site specific conditions prevailing in this country.

Method 205: SARAH, a Surface Water Assessment Model for Back Calculating Reductions in Abiotic Hazards
SARAH is steady-state mixing zone model for the back-calculation of acceptable concentrations of hazardous wastes discharged to land disposal or waste water treatment facilities. It is designed for rapid screening analysis of the potential hazard due to disposal of waste in a land disposal facility.

Beside different loading scenarios, the following exposure scenarios can be evaluated (M-205):

- Human exposure through drinking water
- Human exposure through fish consumption
- Toxicity to aquatic organisms.

Method 206: Modified QWASI Model
This model is used to predict the distribution of chemicals in aquatic environments. It has four compartments: water, bottom sediment, suspended sediment and biota.

The model QWASI (M-206) assumes that mass transfer coefficients and the diffusion coefficients have the same values for all chemicals, and that concentrations are homogeneous. in each layer of water and bottom sediments. The output includes: concentrations, time for steady-state, transfer between compartments, fugacities of compartments, advection, and reaction in compartments, and evaporation rate.

Method 207: EXWAT
This steady-state model which takes into account the major transport mechanisms, degradation and removal processes was developed for the characterization of chemical's fate in surface water bodies with a special view to comparative aspects. EXWAT is also part of system of models and methods called E4CHEM.

EXWAT (M-207) is a box model with two compartments per box, for fluid and sediment. The approach is similar to the QWASI model. The output of the model is a list of steady-state concentrations and mass flows. Transient behavior cannot be studied. Stratified lakes and ponds are not considered. The model was designed primarily for the comparative assessment of environmental hazards of existing chemicals.

Method 208: MEXWA, Model of Exposure Assessment in Water
The method is used to estimate the concentration of chemicals in the river downstream of a source.

MEXWA (M-208) includes the following features:

- Mixing process in the close vicinity of a source of chemicals (transverse and vertical mixing)
- Mixing process considering the transverse dispersion and volatility of chemicals
- Long-distance mixing processes considering longitudinal dispersion.

The output produces a plot of contours of plume, concentrations at points downstream, and calculates the evaporated fraction.

Method 209: EXAMS, Exposure Analysis Modeling System (Version 2.92)
It is a refined approach to water surface modeling. EXAMS is an interactive modeling system that allows a user to specify and store the properties of chemicals and aquatic ecosystems, to modify either these properties, and to conduct rapid evaluation and error analyses of the probable aquatic fate of synthetic organic chemicals.

EXAMS (M-209) includes process models of the physical, chemical and biological phenomena governing the transport and fate of compounds. EXAMS allows the simultaneous treatment of up to 28 molecular species of a chemical, including the parent uncharged molecule, and singly or other charged cations and anions, each of which can occur in a dissolved, DOC-(Dissolved Organic Carbon) complexed, sediment-sorbed or bio-sorbed form. EXAMS computes the kinetics of transformation attributable to photolysis, hydrolysis, biolysis, and oxidation reactions. As output tables and graphs describing exposure, fate and persistence are produced. EXAMS has been designed to evaluate the consequences of longer-term time-averaged chemical loading that ultimately result in trace-level contamination of aquatic systems.

Method 210: WASP4, Water Analysis Simulation Programs
WASP4 is a dynamic compartment model of the transport and fate of organic chemicals and metals in all types of aquatic systems.

This simulation program (M-210) combines hydrodynamic mass transport, sediment balance, and chemical transformation reactions adapted from EXAMS.

Method 211: WAQUA, Water Quality Model
It is a fully operational model for the calculation of horizontal, two-dimensional water elevation, velocities, and concentrations of water-quality parameters in rivers, estuaries and coastal seas.

The method (M-211) consists of a water-movement part and a water-quality part. Their computation can be carried out together or separately.

Method 212: XTEM, A Xenobiotics and Effect Transport Model
XTEM is a time-dependent, two-dimensional, layer-averaged surface water circulation and transport model coupled with an inhibition model. The code is based on the detailed simulation of hydrodynamics.

The model XTEM (M-212) simulates in situ concentrations of xenobiotics (e.g. chlorinated phenolic and resin acids) and estimates the inhibition of related heterotrophic bacterial activity. Different types of release can be handled: single-point, multiple point, or diffuse. Both intermittent and continuous releases can be modeled. The present prototype model is suitable for detailed analyses, but with the restriction that a first-order degradation / elimination rate is defensible.

Method 213: Three Dimensional Transport And Influence Model
It is a time-dependent, multi-layer horizontal distribution model for transport, dilution, interaction and environmental influence of chemicals in surface waters.

This model (M-213) is classified as a most refined level approach for lakes, rivers, estuaries, coastal waters, reservoir and seas. It is use to predict, estimate or explain the effects of different releases and of different management decisions (location of discharge, dams, embankments, etc.). Continuous and intermittent, single, multiple-point and diffuse releases can be taken into account. The method can be applied to chemicals whose interaction rates with and influence properties in the environment are well-known or can be estimated. The output of the model includes time series, field comparison, distribution and animation outputs. The method as such is generally applicable.

SSM (Simplified Stream Model). The SSM model is a one-dimensional, steady-state water quality model developed by Hydroscience[4] (1971) for use with a hand calculator. The SSM simulates dilution, advection, reaeration, and temperature in streams, rivers and shallow nonstratified lakes. BOD (Biological Oxygen Demand), DO (Dissolved Oxygen), conservatives, and non-conservatives with first order decay can be represented. Both coupled BOD/DO and uncoupled chemical reactions can be handled.

4.2.4. ASSESSMENT METHODS FOR SOIL (GROUNDWATER)

These methods are classified according to whether they deal with chemicals in soil zones saturated or unsaturated by water. The methods range from simple analytical techniques to refined numerical methods including transport equations for chemicals in soil. The input data are largely data on loading of chemicals to the soil, physical-chemical properties, soil-specific properties, terrestrial plant-specific properties, and meteorological data. The methods are described in TABLE 4.4.

Unsaturated Zone

- **Methods 301 to 305** deal with the *vertical movement* of chemicals in the unsaturated zone.
- **Methods 301, 302, and 304** assume vertically steady-state water flow.
- **Methods 303 and 305** simulate transient vertical water flow.

Saturated Zone

- **Methods 306 to 308** deal with the *transport of chemicals in the saturated zone.*
- **Methods 306 and 307** are used for *transport of chemicals through uniform aquifers.*
- **Method 308** simulates the *transport of chemicals in heterogeneous subsoils.*

Method 301: PSM, Pesticide Estimation Model
This vertical movement model calculates the amount of pesticide upon a certain area reaching the saturation zone of ground water, following several kinds of primary and secondary losses. It is characterized as a screening model and can generate, in a relative short time, a first estimation of the leaching characteristics of a pesticide.

The model PSM (M-301) can handle intermittent release of pesticide and can be applied to general or site-specific situations to characterize a specific deposition scenario. The model is capable of describing the decay of an organic pesticide in any of three ways:

- First-order kinetics
- Percentage decrease per unit of time
- A mass transfer equation for evaporation / transpiration and plant uptake.

The model acts as a series of well-mixed soil layers, for each of which a mass balance is calculated and is particularly suited to initial screening. PSM is based on the analytical solution of a first-order differential equation. Because several first-order differential processes can be added, the final first-order equation is a sum of individual processes. The soil structure can be defined as having up to ten horizontal layers of variable thickness. The output of the program is expressed as a diagram indicating the actual and cumulated flow of pesticide to the saturated zone of ground water and a tabulation of the concentrations remaining in the several soil layers. The model is one-dimensional and only vertical movement of water and of a substance can be taken into account.

TABLE 4.4. Fate and pathways assessment methods for soil and groundwater

Methods for Soil Ground Water									
Method No.:	**301**	**302**	**303**	**304**	**305**	**306**	**307**	**308**	
Models / Methods for:									
- Unsaturated zone									
Simple analytical fate model	X								
Refined numerical fate model		X	X	X	X				
- Saturated zone									
Simple analytical fate model						X	X		
Refined numerical fate model								X	
Hardware / Software requirements									
Manual operation									
Not yet computerized									
Desk calculator									
Apple II	X								
IBM-PC / compatible (MS-DOS)		X	X		X	X		X	
Workstation									
Mainframe computer (V = VAX-VMS; U = Unix)			V	V			V	V	
Mainframe computer (S= Siemens)		X							
Fortran		X	X	X	X		X	X	
Basic / Pascal	P					B			
Assessment Effort									
- Training (h / d / w / month)	h	h	d	d	w	d	d		
- Data collection and preparation (h / d / w / month)	w	w	d	h/d	w	w	h		
- Assessment (h / d / w / month)	h		d/w		d	d			
- Validation / Review									
not necessary									
limited experience	X								
validated with model experiments			X	X	X	X	X		
validated with field experiments / data		X	X	X	X	X	X	X	
validated through reviews/experts			X	X		X			
used for regulatory purposes			X	X		X			
Code / Model availability	Yes		Yes	Yes	Yes	Yes	Yes		

Method 302: EXSOL (Exposure of soil)
The model was developed for behavior assessment in the upper soil zone for a wide variety of organics after single, intermittent or continuous deposition onto soil surface. It is a refined multi-layer soil model (for unsaturated zone) which takes into account the major transport mechanisms, degradation and removal processes.

The dynamics of mobility, accumulation and degradation can be studied under various soil and climatic conditions, and the amount of volatilized or leached substances can be estimated. EXSOL (M-302) is also part of a system of models and methods under the E4CHEM code. The model considers the following processes:

- Advection with interstitial water flow
- Hydrodynamic dispersion
- Partitioning between soil water and soil air
- Sorption on organic matter
- Ionization equilibrium
- Gaseous diffusion in the soil air
- Volatilization
- Biotic and abiotic degradation
- Root uptake

Up to 77 horizontal sub-layers of variable thickness can be defined. The output of the assessment is a table giving mass balance (or concentration profiles) after various time intervals.

Method 303: PRZM Model (Pesticide Root Model)
The PRZM model simulates the vertical movement of pesticides in unsaturated soil, both within and below the plant root zone and extending to the water table.

The model PRZM (M-303) consist of hydrology and chemical transport modules that simulate the runoff, erosion, plant uptake, leaching, decay, foliar washout and volatilization of a pesticide. Predictions can be made daily, monthly, or annually. The output indicates the amount of organic pesticide leaving the root zone and entering the water table. The model is limited to estimation of the vertical movement of contaminants.

Method 304: SESOIL-4, Seasonal Soil Compartment Model (4-Layer version)
It is an unsaturated zone model for contaminant migration through soils, incorporating hydrologic and sediment budget estimates. The model is used to predict chemical migration of organics and metals through unsaturated soil zones, and releases to adjacent ground water and surface water and to the atmosphere.

The estimates of chemical concentrations in soils incorporate partitioning among dissolved, gaseous, adsorbed and pure states. The model SESOIL-4 (M-304) accepts time-varying chemical loading and generates time profile results. The model has an unusually large variety of built-in process sub-models. It is part of *GMS*, a larger computerized system *(OTS, Office of Toxic Substances, US Environmental Protection Agency)*. A three-layer version of SESOIL forms the terrestrial module of a multi-media fate and transport model *TOX-SCREEN*. The model provides mass balance and concentrations of chemicals in each layer as a function of time, and the mass of chemicals entering ground water as a function of time.

Method 305: ONZAT Model
ONZAT is a one-dimensional mathematical model for (un)saturated flow and transport of a solute, including equilibrium, sorption and first order decay. The code calculates evaporation and ground water recharge as well.

The ONZAT model (M-305) describes averaged horizontal flow and transport in a layered soil profile. Only non point-source of chemicals (e.g. herbicides) are considered. The calculation for the flow equations is based on finite difference method, half-implicit in time (Cranck-Nicholson scheme), using an iterative scheme. For the transport equations, finite difference method, implicit in time, backward in space (mixed cells), are used without iterative scheme. The program, at present, still has some numerical problems, and the interaction of two more chemicals cannot be simulated. The volatilization of the chemical is not modeled.

Method 306: SOLUTE Model
SOLUTE is a program package which contains the analytical solution of one-, two-, and three-dimensional advection dispersion equations. Retardation and decay options are included. It is a screening model, used to predict concentrations of conventional and organic contaminants in ground water and requires input of flow and concentration information about an aquifer and the contamination source.

Depending on the model selected, concentrations are calculated in one, two, or three dimensions for specified time periods and distance increments. The SOLUTE model (M-306) assumes a uniformly porous, confined aquifer, which is homogeneous, isotropic and infinite in area extent. The density and viscosity of the solute are assumed to be the same as those of the native water in the aquifer.

Method 307: AT123D (Analytical, Transient, 1, 2, and 3-D- Model)
AT123D is a ground water pollution model for plume migration through relatively uniform aquifer, and is used to predict chemical concentrations in ground water at specified times. It estimates time-varying concentrations within the saturated zone on a specified three-dimensional grid.

The model (M-307) accepts either constant or time-varying chemical releases to ground water. Hydraulic conductivity and hydraulic gradients are constant numbers. The flow is assumed to be nearly horizontal. The adsorption coefficients and degradation rates have constant values. AT123D accepts time-varying releases from an area or volume source.

Method 308: FLOSA, Flow Systems Analysis
It is a ground model with refined approach. The model is used to describe several independent subsurface flow systems and the flow paths and residence times of a chemical dissolved in the ground water. It can be used for both local and diffuse release and for either one-, two-, or three-dimensional calculations in a heterogeneous and anisotropic subsoil.

The FLOSA model (M-308) predicts ground water flow in the saturated zone. A finite difference method is used and mechanisms such as advection and adsorption (e.g. linear, Langmuir, or Freundlich type) are taken into account. This model differs from others in that the velocity in one-, two-, or three-dimensional flow field is calculated directly. A promising development is the combining of conventional modeling techniques, to predict hydraulic heads, with Kalman filtering technique in order to obtain reliable spatial interpolation of the water table. As output, maps of velocity vectors and iso-concentration maps are produced. The residence time and the front of contamination of a chemical are indicated. As a limitation, only the ground water flow is modeled, and no interactions excepts adsorption of a chemical are considered.

4.2.5. ASSESSMENT METHODS FOR AIR

Methods 401 to 405 predict *ground-level concentrations of a chemical at various distances from the source*. The methods are described in TABLE 4.5.

- **Methods 401 to 405** treat *aerial spraying of pesticides.*
- **Methods 402 to 404** are concerned with the *dispersion of chemicals* up to 50 - 100 km from the source.
- **Method 406** deals with the *global dispersion of airborne substances.*

Method 401: Pesticides Emission and Dispersion Model
This is an emission model coupled with a Gaussian plume model. The model is used to calculate pesticide concentrations in the air and depositions on the soil and foliage in the vicinity of an area where pesticides are applied by spraying.

The method (M-401) makes use of two emission models: one for droplet emission and the other for vapor emission.

TABLE 4.5. Fate and pathways assessment methods for air

Fate and Pathways Assessment Methods for Air						
Method No.:	**401**	**402**	**403**	**404**	**405**	**406**
Models / Methods for:						
- Local dispersion from point sources						
Simple Gaussian plume model		X	X	X		
Refined Gaussian plume model			X			
- Local dispersion from disperse sources						
Simple Gaussian plume, compartment model	X			X		
Refined numerical fate model					X	
- Global dispersion						
Refined compartment model						X
Hardware/Software requirements						
Manual operation						
Not yet computerized						
Desk calculator						
Apple II						
IBM-PC / compatible (MS-DOS)	X			X	X	X
Workstation			X			
Mainframe computer (V = VAX-VMS; U = Unix)			X		V	
Mainframe computer (S = Siemens; C = Cyber)		S		S	C	S
Fortran	X		X	X	X	X
Basic / Pascal						
Assessment Effort						
- Training (h / d / w / month)	w			d	w	d
- Data collection and prepn. (h / d / w / month)	d		d	d	w	d/w
- Assessment (h / d / w / month)	d		d	d	d	d/w
Validation / Review						
not necessary						
limited experience	X					
validated with model experiments			X			X
validated with field experiments / data	X	X	X	(X)		X
validated through reviews/experts			X			X
used for regulatory purposes		X	X		X	
Code / Model availability	Yes		Yes		Yes	Yes

The droplet emission occurring during spraying is determined by the spraying conditions, while vapor emission (after spraying) is determined by diffusion in a thin air layer near the foliage. The dispersion model is of the Gaussian plume type; but for droplets it accounts for the change in the deposition rate, which decreases during their transport owing to the evaporation of the volatile component (the dispersant). The shape of the sprayed area has to be represented as an approximate rectangle. The output gives the pesticide concentrations and deposition fluxes as a function of the distance from the area boundary during and after spraying.

> ***Method 402: TA Air Dispersion Model***
> *TA Air Dispersion Model, is a pollutant concentration distribution model to be used according to the technical guidance of the German Air Pollution Control Act.*

The *TA model* (M-402) is a climatological Gaussian plume model which estimates average concentrations of a pollutant downwind from a point source. It is a mesoscale model (250 m to 50 km) capable of treating single point sources as well as several emission sources by summing up their contributions.

The method is to be applied to continuous emissions, but through the use of default values it can be sued for screening purpose. Removal processes such as dry-fall deposition are included. The coordinates of the system consists of radial distances and the wind rose is divided into 26 sectors. Meteorological and site-specific data must be supplied to the model. The output consists of a listing of annual concentrations at defined circular distances as a function of wind directions. This dispersion model is an empirical model. The calculations are based on meteorological data monitored during a ten-year's period. Specific and incidental emissions cannot be considered.

> ***Method 403: UNAMAP, User's network for Applied Modeling of Air Pollution***
> *The system UNAMAP contains over 30 models, generally Gaussian dispersion models for continuous emissions sources. As a class, they are single and multi-source and include plume rise equations.*

The *(ISC) Industrial Source Complex Model* (M-403) is included in the package. It is a steady-state source model which can be used to assess pollutant concentrations from a wide variety of sources. ISC is appropriate for industrial source complexes, rural or urban areas, flat or rolling terrain, transport distances of less than 50 km, and one hour to annual averaging times.

Most of the models available on this network have been evaluated by the *US Environmental Protection Agency* and deemed appropriate for simulations in regulatory applications. Other models are still under development. Some of the models include provisions for down-wash, terrain considerations, deposition through settling, and trapping by inversion layers. For *air toxic assessments*, the UNAMAP models are appropriate for simulation of continuous or quasi-continuous releases of materials that behave in the atmosphere as neutrally buoyant tracers.

The source data required by the ISC model are: location, emission rate, stack height, stack gas exit velocity, stack inside diameter, and stack gas temperature. Optional inputs include source elevation, building dimensions, particle size distribution with corresponding settling velocities, and surface reflection coefficients. The meteorological data required for short term assessment are: hourly surface weather data; including hourly stability class, wind direction, wind speed, temperature, and mixing height. For long-term assessment, the corresponding averaged data are required. The output indicates the average concentration or total deposition calculated at each receptor for any desired combination of sources.

> ***Method 404: EXAIR, Exposure of Air in the Lower Troposphere***
> *EXAIR is an analytical model of the tropospheric mixing layer under different weather conditions, taking into account the major transport mechanisms, deposition, and degradation. The primary aim of this model is to assess the fate of airborne organic substances at distances of up to 100 km downwind of the source (single point and area source). EXAIR is also part of the code E4CHEM.*

Concentration profiles as well as mass flows are estimated. Using meteorological statistics, the mean and variance of the concentration and mass flow can be calculated. The release is assumed to be independent of the meteorological conditions. EXAIR (M-404) is also part of the code E4CHEM. The model assumes lateral dispersion and uniform vertical distribution within the atmospheric mixing layer. The movement downwind is governed by the vertically averaged wind speed. The state of airborne substances may be gaseous or adsorbed to aerosols. Both the physical-chemical properties of the pollutant and the properties of the aerosol are taken into consideration for calculating the partition between the two phases. The removal processes are wet and dry deposition of both aerosol and gas particles. The attack by OH radicals is treated as degradation. The advection-dispersion equation is analytically solved, taking into consideration the mass balance of the substance. EXAIR uses frequency distribution for wind speed, dispersion classes, and precipitation intensities. The output of the model is a table giving mobility, degradation and deposition for individual weather conditions as a percentage of the total mass of the air pollutant as a function of the time period.

> ***Method 405: Airborne Pesticide Particles Dispersion Model***
> *This air model calculates the concentration and dose of airborne non-volatile pesticide particles in the down wind area during agricultural spraying (using an airplane).*

Concentrations in this model (M-405) are calculated as a functions of the coordinates of points in the downwind area and of time. A number of instantaneous line sources adjacent in time and position can be considered. The influence of vortices induced by the aircraft may be considered. The model is of type mono-swath and consists of two sub-models (the cloud development model and the atmospheric dispersion model) which are successive in time. Evaporation of the droplets is taken into consideration. The quasi-stationary one-droplet model, including the ventilation and the temperature drop effect is used for this purpose. The dispersion of the initial cloud is based on an extension of the source depletion model.

Method 406: EXATM (Exposure of the Global Atmosphere)
EXATM is a global eight-compartment model which takes into account degradation and mass transfer processes between the troposphere and stratosphere and between four zone of latitude. It is used to assess the long-term, world-wide dispersion of air-borne substances. It is applicable to compounds with weak degradability and low water solubility. EXATM is part of the system E4CHEM.

The EXATM model (M-406) consists of four tropospheric and four stratospheric compartments. The compartment interfaces are positioned in the north-south direction at the latitude of 30° N, 0°, and 30° S, and in altitude at the earth's surface and in the tropopause. The transfer rates between the compartments are based on literature values, which were calculated from measured concentration profiles of radio nuclides. The values are assumed to be constant in time, although global atmospheric circulation changes with the seasons. The values of these transfer rates reflect the kinetics of tropospheric and stratospheric transport by the mean wind and by turbulent processes. The transfer rates are coefficients in a set of eight linear first-order differential equations solved numerically, which describe the time-dependent dispersion of the substance mass in the global atmospheric compartment system. The equations also contain source and sink terms. the output of the model is a list of the concentrations in the atmospheric compartments at consecutive time points.

4.2.6. ASSESSMENT METHODS FOR MULTI-MEDIA

The input data for the models are generally volume and density of the media in question, amount of emission (Level I complexity) or emission rate (Levels II to IV complexity).

■ **Methods 501 to 507** *use Mackay's fugacity modeling approach*, which describes the partitioning and transfer of chemicals between the media air, water, and soil, sediments and biota. They vary from level I to IV according to Mackay's classification of the complexity of the fugacity modeling.

■ **Methods 508-511** are *multi-media methods dealing with the chemicals in two or more environmental media*, but are not based on the fugacity modeling approach. Of these methods, method 510 simulates pesticide runoff and adsorption into the soil in small agricultural basins, another (Method 511) simulates watershed hydrology and water quality.

The methods are described in TABLE 4.6.

TABLE 4.6. Fate and pathways assessment methods for multi-media

Fate and Pathways Assessment Methods for Multi-Media											
Method No. :	**501**	**502**	**503**	**504**	**505**	**506**	**507**	**508**	**509**	**510**	**511**
Models / Methods for:											
- Air, Water, Soil, Biota											
Simple, Mackay's fugacity model	X	X	X	X	X	X	X				
Simple combination of media specific models								X			
Refined combination of media specific models									X		
- Surface water, Groundwater											
Refined hydrodynamic simulation model										X	X
Hardware / Software requirements											
Manual operation											
Spreadsheet						X					
Not yet computerized											
Desk calculator				X							
Apple II / Hewlett-Packard		HP									
IBM-PC / compatible (MS-DOS)	X		X	X	X	X	X	X		X	X
Workstation											
Mainframe computer (V = VAX- VMS; U = Unix)					V				V		V
Mainframe computer (S = Siemens; C = Cyber)								S			
Fortran / Basic / Pascal	B	B	P	B	F	P	B	F	F	F	F
Assessment Effort											
- Training (h / d / w / month)	d	h/d	h	h	h/d	h	h	h	d	d/w	w/m
- Data collection and preparation (h / d / w / month)	w	d/w		h	h	d	d/w	h	d/w	d	w/m
- Assessment (h / d / w / month)	d/w	d		h	h	h	d/w	h	h/d	d	w/m
Validation / Review											
not necessary											
limited experience											
validated with model experiments				X						X	X
validated with field experiments / data	X	X		X		X	X				X
validated through reviews/experts		X	X		X				X	X	X
used for regulatory purposes										X	
Code / Model availability		Yes		Yes	Yes	Yes	Yes		Yes	Yes	Yes

> ***Method 501: Multi-media Fugacity Models***
> *The purpose of the models is to predict the relative or absolute amounts or concentrations of an organic chemical as it is emitted into various defined volumes of media. It consists of a series of four multi-media fugacity models describe the partitioning of chemicals between the media air, water ,soil, sediment and biota at various levels of complexity.*

The method (M-501) can be applied to continuous, intermittent, or changing emissions related to one or more media. Degrading reactions, advective flow, and intermedia transport resistances can be included. The overall environment or region consists of defined volumes of air, water, etc. The media volumes are assumed to be well-mixed and homogeneous.

> ***Method 502: Assessment of Risk to Aquatic Biota***
> *This method is used for screening purpose to calculate relative risks of non-polar organic pesticide applications to aquatic fauna (e.g. forestry or agriculture). It can be extended to degradation products.*

The concentrations of chemicals in the compartments air, water, sediment and biota are estimated by a level-I multi-media fugacity model (M-502), which describes equilibrium partitioning of a fixed amount of chemical among these compartments. A risk factor (RF) is calculated from the expected concentration (EC) in water, the LC^{50} of the chemical, and the expected bio-concentration factor (BCF):

$RF = BCF * EC / LC^{50}$, where RF is expressed relative to a chemical with a known environmental behavior. The output lists risk factors and concentrations in the respective compartments.

> ***Method 503: EEP (Environmental Exposure Potentials)***
> *The EEP system is used in the assessment of a chemical's environmental hazard once the marketing volume per producer or importer reaches one tons/year. It relies on information supplied by the industry to governmental agencies. The primary use of the methodology is to determine the exposure potential from the data available at this stage, taking into account information on use, production volume, physical-chemical properties, degradation and accumulation potential.*

No distinction is made between continuous and intermittent releases, but multiple points or diffuse releases can be assessed using the EEP-methodology (M-503). The system is applicable to all chemicals for which the Mackay[5] compartmentalization calculation is valid (i.e. principally organic substances). Data for sorption and distribution in the environment are needed for dissociating molecules, heavy metal compounds, etc. The criteria used to calculate environmental exposure potential have been assigned scores and weights reflecting their environmental relevance. Quantities released to the environment are based on the annual production rates and use pattern of the chemical in question. Use pattern are classified as "intermediate", "processing" and "end use". The initial partitioning criterion considers the receiving compartments through which these quantities of production volume will enter the environment according to the chemical's use pattern and/or waste disposal. The "quantity in compartment" criterion refers to the tendency of a chemical to move preferentially to one or more environmental media. Abiotic degradation and hydrolysis are not considered. The data needed are usually submitted by notifyers in the case of new chemicals. For existing chemicals data estimation technique can be used.

> ***Method 504: Modified Fugacity Models (levels I and II)***
> *The multi-media model is that proposed by Mackay, modified by the addition of one more compartment: terrestrial plants.*

The method (M-504) is used to predict the environmental distribution and fate of chemicals in defined compartment, so called "unit world". The model is of simple approach and the type of release is single-point, unique emission. The model is used for screening and intermediate analysis of organic chemicals (mainly pesticides) with defined molecular weight and limited dissociation. The terrestrial plants can be introduced as a simple compartment or further divided into subcompartments of roots, stem, and foliage. All the other conditions are the same. The output is expressed in tables showing percentage distribution of the chemical, concentrations in various environmental compartments, and transformation kinetics analysis.

> ***Method 505: ENPART, The environmental partitioning model***
> *It is a steady-state, multi-media partitioning box model to estimate potential chemical distribution among environmental compartments. The model identifies the most important environmental pathways and estimates the substance's environmental persistence and bioconcentration potential.*

The ENPART model (M-505) was developed as first-level screening tool for the review of pre-manufacturing notices and existing chemicals from a minimum set of chemical-specific input data. It can also be use to perform sensitivity analyses. In this box model, the release is assumed to be continuous, and the media compartments are of fixed size and physical characteristics. The model can calculate both equilibrium and dynamic partitioning, the latter taking environmental transformation processes into account. The model is adapted from the fugacity approach to equilibrium partitioning as described by Mackay. In this approach, equilibrium concentration ratios between environmental compartments are calculated from simple physical-chemical data such as molecular weight, solubility and vapor pressure. In ENPART this approach has been modified by the addition of environmental parameters, such as mixing depth, transport and degradation, to yield a second level, of concentration for dynamic partitioning. The output calculates inter compartment concentrations and mass ratios, concentration and mass partitioning factors, and a ranking analysis for both equilibrium and dynamic partitioning. Partitioning estimations are not valid for short-term or high-quantity release scenarios.

> ***Method 506: SIMPLESAL***
> *SIMPLESAL is an application of SIMPLEBOX (Method 604). It is a spreadsheet-based, steady-state or quasi-dynamic multi-media box model of the Mackay type".*

Using Mackay's technology, this box model (M-506) may be classified as "Level III or IV". SIMPLESAL is a diagnostic method for analyzing the environmental fate of micro pollutants (organic compounds as well as heavy metals) in terms of key pathways and key processes. The method can be used to estimate time-dependent or steady-state concentration levels in different environmental compartments from steady or intermittent releases into air, water and soil on the basis of theoretically formulated process rates for inter-compartment transfer and transformation. SIMPLESAL is designed for use as screening tool to predict results of emission control scenarios for existing and new chemicals. The input data required are, among others, the steady emission rates into air, water, and soil; the dimensions and physical characteristics of the compartments; the partition coefficients for inter-compartment equilibrium. The output gives steady-state concentrations in compartment and dynamic response to changes in emission rates and environmental conditions; net rates of advective and dispersive transfer at steady-state; rates of transformation in compartments at steady state.

> ***Method 507: MNSEM-1.3, Multi-phase Non-Steady State Equilibrium Model***
> *MNSEM-1.3 is a simplified kinetic model for rapid evaluation of environmental fate and of typical concentrations of organic chemicals which are continuously loaded into the environment (Japanese' model).*

This model (M-507) comprises four compartments: air, water, soil, and sediment. Each phase is assumed to be well-mixed. It is assumed that the speed rates of inter-phase transfer processes (e.g. sorption, bioaccumulation) are faster than those of interphase transfers (advective outflow, transformation). In addition to diffuse interphase transfers, MNSEM-1.3 considers non-diffuse transfer processes, such as rainfall, surface runoff, settling of suspended solids, and vertical transport in soil (leaching). The model requires the chemical loading rates into environmental compartments, the usual physical-chemical properties, as well as the bioconcentration factors and organic carbon sorption coefficients. The output of the assessment contains a table giving the mass distribution fraction in each phase; the mean residence time; loading rates; clearance time in the environment; and the steady-state mass and concentration in each phase. MNSEM-1.3 is not a site-specific fate model.

> ***Method 508: EXINT, Exposure by Integrating the Fate of Chemicals in the Environmental Media Soil, Water and Air.***
> *EXINT is of simple approach type. The model was developed primarily for the general characterization of the fate of a substance in a multi-media environment.*

The EXINT model (M-508) links the models EXSOL (M-302), for the upper soil zone, EXWAT (M-207), for surface water bodies, and EXAIR (M-404), for the lower tropospheric mixing layer, taking into account *input rates estimated by RLTEC* (M-104), and intermedia exchange by volatilization and deposition. The output consists of quantities ("descriptors") which relate to mobility, accumulation and degradation. MNSEM-1.3 is also part of the system E4CHEM. A standard scenario has been defined for an appropriate linkage of the single-medium models. The output of the EXINT model is a table containing different terms of the mass balance equation for long-term average conditions. Some additional results, such as concentration of the substance in fish, and in plant transpiration water, are also shown.

Method 509: TOX-SCREEN
TOX-SCREEN is a multi-media model to assess the potential for environmental transport and accumulation of neutral organic chemicals released to air, surface water or soil. It is a refined approach type intended for use as a screening device to identify chemicals that are highly unlikely to pose a problem, even under conservative assumptions. Four types of surface water can be considered: lakes, rivers, estuaries, and oceans.

The model (M-509) is capable of addressing single or multiple-point-source or diffuse, time-dependent, stepwise (monthly) releases. TOX-SCREEN is implemented in the graphical Exposure Modeling System (GEMS) developed by the Office of Toxic Substances (US Environmental Protection Agency). An input data management system TOXIN is attached to the model to assist the user in the preparation of input data files. TOX-SCREEN also estimates the bioaccumulation of chemicals in aquatic organisms, terrestrial food animals and plants. TOX-SCREEN provides estimates of the fate of the pollutant once it is released into air, water, and/or soil. The model is overpredictive in nature.

Method 510: SWRRB, Simulator for Water Resources in Rural Basins
The pesticide Runoff Simulator (SWRRB, or PCSWRRB) simulates pesticide runoff and adsorption into the soil of small agricultural basins (in the U.S.).

The SWRRB model (M-510) is based on a hydrologic and sediment yield model (*CREAMS*, Option 1, Smith and Williams, 1980) and requires daily rainfall data.

Method 511: HSPF, Hydrological Simulation Program
The Hydrological Simulation Program (HSPF) is a comprehensive package for simulation of watershed hydrology and water quality for both conventional and toxic organic pollutants. HSPF incorporates watershed-scale models into a basin-scale analysis framework that includes fate and transport in one-dimensional stream channels.

HSPF is the only comprehensive model (M-511) of watershed hydrology and water quality that allows the integrated simulation of land and soil contaminant runoff processes with instream hydraulic and sediment-chemical interactions. HSPF simulates three types of solids (sands, silts, and clays) in addition to a single organic chemical and transformation product of that chemical. The transfer and reactions included are hydrolysis, oxidation, photolysis, biodegradation, volatilization, and sorption (modeled as first-order kinetic process). Data needs for HSPF, which is a general-purpose program, can be extensive. The application is limited to well-mixed rivers and reservoirs. HSPF produces as output a time history of the runoff flow rate, sediment load, and nutrient and pesticide concentrations along with a time history of water quantity and quality at any point in the watershed.

4.2.7. MISCELLANEOUS FATE AND PATH ASSESSMENT METHODS

The methods described below cannot be adequately classified according to environmental media.

- **Methods 601 and 602** are *geochemical models* to assess the fate and pathways of metals in water.
- **Method 603** is a collection of *formulas, tables, and graphs* for preliminary assessment of surface and ground water quality.
- **Method 604** is a *general-purpose computational framework for constructing a box model.* It is based on a spreadsheet approach and allows to construct steady-state or quasi-dynamic box models of the environmental fate of chemicals.

The methods are described in TABLE 4.7.

Method 601: Geochemical Mass Balances
The method proposed allows mass balance accounting of the fluxes of chemicals within a circumscribed system. The primary application of mass balances are to assure data quality and to assess the degree of contamination. The technique consists to establish a mass balance for water and particulate material for the subject system. The water mass balance includes river runoff, atmospheric precipitation, evaporation, etc.

The data needed for particle mass balance are: hydrologic and physical data; suspended particle concentration in boundary and in fresh water inflows and outflows. After the basic water and particle mass balance has been constructed, the construction of a mass balance for a specific constituent or contaminant may begin (M-601).

TABLE 4.7. Miscellaneous methods methods

Miscellaneous Methods					
Method No.:	**601**	**602**	**603**	**604**	
Models / Methods for:					
- Geochemical model	X	X			
- Formulas, Tables, and Graphs			X		
- Computational framework				X	
Hardware / Software requirements					
Manual operation	X		X		
Spreadsheet	X			X	
Not yet computerized	X				
Desk calculator	X		X		
Apple II / Hewlett-Packard					
IBM-PC / compatible (MS-DOS)		X		X	
Workstation					
Mainframe computer (V = VAX-VMS; U = Unix)		V			
Mainframe computer (S = Siemens; C = Cyber)					
Fortran 77/ Basic / Pascal		X		P	
Assessment Effort					
- Training (h / d / w / month)	w	w/m		d	
- Data collection and preparation (h / d / w / month)	d/w/m	d/w	d/w	d/w	
- Assessment (h / d / w / month)	h/d	d/w	d/w	h/d	
Validation / Review					
not applicable	X			X	
limited experience					
validated with model experiments		X			
validated with field experiments / data		X			
validated through reviews/experts		X			
used for regulatory purposes		X			
C = Code/Model ;T = Technique; P = Procedure	No	Yes	P	Yes / P	

Method:

601: mass balances for chemicals in a circumscribed system (water, particles, constituents)

602: MINTQA1 surface water and soil (ground water) model with metal speciation equilibrium

603: surface water/soil (ground water) screening procedure for toxic/conventional pollutants

604: SIMPLEBOX general purpose box-modeling system

The data needed at this stage are: location and variability of the landward and seaward flows; form, rates of decay, and transfer to particulate form; continental runoff; atmospheric deposition; sediment accumulation; offshore boundary exchange. The output of the system is an estimate of the fluxes of environmental chemicals into and out of a system. The method is limited by the quality of the data and the consideration of a finite number of process parameters. The collection of data to construct a mass balance sheet is the most time consuming and expensive part of the process.

> ***Method 602: MINTEQA1, Equilibrium Metal Speciation Model***
> *MINTEQA1 is a versatile, qualitative geochemical model capable of calculating aqueous speciation in equilibrium state, adsorption, has phase partitioning, solid phase saturation states, and precipitation-dissolution of eleven metals (arsenic, cadmium, chromium, copper, lead, mercury, nickel, selenium, silver, thallium, and zinc).*

The MINTEQA1 model (M-602) contains an extensive thermodynamic data base and six different algorithms for calculating adsorption. It is designed to solve a broad range of complex equilibrium problems involving reactions among gases, aqueous solutions, mineral phases, and adsorbed phase species with minimal demands on the user. For waste load allocation problems, MINTEQA1 must be run in conjunction with transport and fate models.

> ***Method 603: Water Quality Assessment***
> *Water quality assessment, a screening procedure for toxic and conventional pollutants in surface and ground waters, is a collection of formulas, tables and graphs that planners can use for preliminary assessment of surface and ground water quality in large river basins.*

Method (M-603) is designed to be used with a minimum of information, consistent with reasonable assumptions. The desktop procedures are appropriate for hand calculators or for implementation as spreadsheet calculations on personal computers. The output of the assessment is typically either a chemical property of direct use in further analysis, or the concentration of a contaminant in a receiving water body or aquifer. The method has been validated with both field data and model experiments.

> ***Method 604: SIMPLEBOX***
> *SIMPLEBOX is a general purpose box modeling system. It is a computational method rather than model. Its spreadsheet-based framework can be used to construct a steady-state or quasi-dynamic box models of the environmental fate of chemicals.*

modeling system Models based on the SIMPLEBOX system (M-604) may be applied to any chemical for which the standard input requirements can be met, particularly with respect to exchange ratios and removal rates. The models may be single or multi-media, and one-, two-, or three-dimensional. Environmental parameters are necessarily constant in time. The quasi-dynamic option is meant to be used only in order to estimate the time scale according to which the system will respond to sudden changes in release rates. The standard input required by SIMPLEBOX consists of: constant emission rates for all boxes; volumes for all boxes; advective flow rates between all boxes; dispersive exchange coefficients between all boxes; pseudo-first-order removal rate constants for all boxes; steady concentration levels outside the system for all boxes; initial (zero-time) concentrations in all boxes, in the case of quasi-dynamic modeling. The output is either steady-state concentrations in all boxes or concentration-time series for all boxes. SIMPLEBOX has been used as a computational framework for the construction of a multi-media distribution model and as an estimation model for the removal of micro-pollutants in waste water treatment plants.

4.3. Models and Computer Based Tools for Integrated Regional Risk Assessment

4.3.1. COMPUTER PROGRAMS FOR RISKS ASSESSMENT OF COMPLEX INDUSTRIAL AREAS AND ENERGY SYSTEMS

Various computer programs for assessing the risk of energy systems and complex industrial areas are made available by different organizations (including those who prepared the codes listed below), that offer programs in this field (TABLE 4.8).

TABLE 4.8. Details of some of the organizations that offer programs in the field of industrial risk analysis

Organizations Active in the Field of Industrial Risk Analysis

- *Arthur D. Little, Acorn Park, Cambridge Ma USA.; Also Cambridge, UK.*
- *Battelle Laboratories, 505 King Avenue, Columbus OH, USA*
- *Ecology and Environment, 195 Sugg Road, Buffalo, New York, USA*
- *Energy Analysts, P.O. Box 1508 Norman OK, USA*
- *ENSR Consulting and Engineering, 696 Virginia Road, Concord MA, USA*
- *JBF Associates Inc. Technology Drive, 1000 Technology Park Center Knoxville TN, USA*
- *NUS Corp. 910 Clopper Road, Gaithersburg - MD USA*
- *Pickard, Lowe and Garrick, 7760 University Drive, Newport Beach, CA, USA*
- *Radian Corporation, P.O. Box 9948, Austin TX, USA*
- *SAFER Corporation, 5700 Corsa Avenue, Westlake Village, CA, USA*
- *SRD, UKAEA, Wigshaw Lane, Culcheth, Warrington, WA3 4NE, UK*
- *Science Applications Int. Corporation, 5150 El Camino Real, Los Altos CA, USA*
- *Technica 355 East Campus View Bld. Suite 110 Columbus OH, USA; also London, UK*
- *TNO Department of Industrial Safety Appeldoorn, Holland, Westinghouse Risk Management, P.O. Box 355, Pittsburgh, PA, USA.*

The TABLES 4.9-4.10, below, give brief details of just some of the computer programs that are suitable for use in the assessment of risks from the types of industry (excluding nuclear risks) to be found in large industrial areas. TABLE 4.9 gives details of programs which tackle individual problems.

TABLE 4.10 gives details of codes which tackle a more comprehensive range of topics, showing the application range, the name of the code and the organization supporting such computer program.

TABLE 4.9. Programs tackling specific problems

Application	**Name**	**Organization**
(a) Release and discharge rates	DEERS	Jaycor Inc. (USA)
	SAFIRE	A.I.Ch.E (USA)
	TRAUMA	SRD (UK)
(b) Flashing and Evaporation	GASP	SRD (UK)
	SPILLS	Shell Development (USA)
(c) Atmospheric Dispersion	CHARM	Radion Corp. (USA)
	DEGADIS	U.S. Coastguard (USA)
	DRIFT	SRD (UK)
	EAHAP	Energy Analysts Inc. (USA)0
	HASTE	ENSR (formerly ERT) (USA)
	SLAB	Lawrence Livermore National Lab (USA)
	TRACE	Safer Corp. (USA)
	WHAZAN	Technica (UK / USA)
(d) Fault Tree Analysis	RIKKE	R. Taylor (Denmark)
	IRRAS-PC	EG & G Services (USA)
	CAFTA-PC	Science Applications Int. Co. (USA)
	TREDRA	J.B.F. Associates (USA)
	GRAFTER	WESTINGHOUSE (USA)
	BRAVO	J.B.F. Associates (USA)
(e) Event Tree Analysis	ETA II	Science Applications (USA)
	RISKMAN	Pickard, Lowe & Cranick (USA)
	SUPER	Westinghouse risk Management (USA)
(f) Human Reliability	SLIM-MAUD	Embrey (UK)
	HEART	Williams (UK)

TABLE 4.10. Programs tackling a more comprehensive range of topics

(a)	WHAZAN	*Technica* (UK/USA) provides a wide range of models and a chemical properties data base. World Bank Manual (produced by *Technica* in 1985) provides the formulas which are implemented in WHAZAN
(b)	SAFETI	*Technica* (UK/USA) provides a complete risk assessment package giving risk contours and F-N curves - this is a fully integrated software package which need some skill to use properly
(c)	EAHAP	*Energy Analysts Inc.* (USA) provides a wide range of consequence calculations.
(d)	INTERCLAIR	(see Chapter 2, Section 2.4)

4.3.2. ENERGY SIMULATION MODELS

Energy Simulation at PSI

The legislation in Switzerland, as defined by the Clean Air Ordinance, specifies sulfur content of the fuels and allowable emissions for stationary sources. Other laws specify emission limits for private automobiles and trucks. Finally, the Clean Air Ordinance foresees a series of measures aiming at reducing emissions back to the level of the 50s, by the year 1995. *PSI (Paul Scherrer Institute, Villigen, Switzerland)* have developed the energy simulation model SMEDE. [6] which was used for the engineering simulation approach, and the IEA process oriented optimization model MARKAL (Market allocation) was used to analyze the energy demand and supply.[7] The energy scenarios developed at Bern by EGES (Expert Group for Energy Scenarios) in 1988 were used. A sectorial demand simulation model has been used (SMEDE) to define the energy demand constraints for use by the MARKAL model. MARKAL is a large scale linear programming dynamic process model driven by exogenous demands for goods and services, and prices of imported energy forms, through the time horizon considered.[8]

Risk-Cost Assessment Methodology

Bolten[9] et al., (1983) have proposed a risk-cost assessment methodology for toxic pollutants from fossil fuel power plants. The methodological framework was developed and applied to the case study of arsenic emissions from a hypothetical coal-fired plant. The models contained in the framework can determine the chronic human health risks associated with different costs of controlling these emissions. In a second report Bolten[10] et al., (1985) discuss (a) alternative models for above mentioned framework; (b) simplified environmental transport models that may be used for initial analysis; and (c) the implication of certain assumptions used in exposure analysis. As a test-case they choosed "selenium emissions from a coal-fired plant".

The general framework of this second report uses a composite multimedia modeling approach for environmental calculations. This approach applies independent models for each medium sequentially to determine the distribution of a pollutant in the air, soil, groundwater, and surface water. The inputs and outputs of the different models are coupled to provide the interactions between environmental media.

Different water models have been reviewed by Onishi[11] et al., (1981), and Anderson[12] (1979). The selection of models encountered in the literature is a difficult task. For surface water transport, hundred of models could be used. These models differ in many ways and can be characterized, among others, by the following properties:

- Whether they are steady-state or time-dependent
- The types of water bodies to which they apply
- The number of kinds of pollutants included
- How transformation and decay processes are treated
- Whether they emphasize sediment processes, water quality, or radionuclides' transport.

Most models have been developed for a particular purpose and may not be suitable for toxic chemical risk calculations.

The models included by Bolten in the framework (computer program *RATE*: *Risk Assessment of Toxic Emissions*) calculate pollutant concentrations and transfer rates for the air, soil, and surface water bodies. The models can run with limited data and is designed to calculate upper and lower bounds to exposure rates and concentrations with extensive sensitivity analysis. The results are of the type "level I analysis". Constant mean exposure rates and stationary, static populations are assumed. Calculation results have shown that assuming a static population may overestimate the health risk, if birth, death, and migration rates are large. Similarly, although some dose-response models are intensive to variations in exposure rates, others will provide significantly different results when used with time-dependent dose information. The health risk may be either over- or underestimated, depending on the nature of the pollutant and its effects.

By performing this screening analysis (level I), the user can isolate the important aspects of a problem and determine where further detailed analysis (level II) and resources are necessary.

4.3.3. REVIEW OF MODELS TO ASSESS HUMAN EXPOSURE TO CHEMICALS

Some special models are briefly described below:

GEOTOX Model

It is a multimedia exposure model (McKone[13], 1986), which calculates chemical partitioning, degrading reactions and interphase transport on both diffusive and non-diffusive types. The concentrations estimated for the different environmental compartments are combined with appropriate human inhalation and ingestion rates and absorption factors to calculate exposure.

Some special models are briefly described below:

The model incorporates the following compartments: air (gas), air (particles), biomass, upper soil, lower soil, groundwater, surface water, and sediments. These media are assumed to be made up of sub-phases of gas, liquid and solid. Chemical partitioning between compartments, interphase transport, reaction and advective losses are described by first order rate constants. The model consists of a set of simultaneous equations in matrix form, which can be solved for constant or time-varying chemical sources.

Contact person: Dr. Tom McKone, Risk Assessment and Environmental Policy, Lawrence Livermore Laboratory, P.O. Box 5507, L-453, Livermore, CA94550, USA (Tel.: (415)-422-7335. Disk available free of charge.

Model B (RAPS)

The RAPS system was prepared for the U.S. Department of Energy for ranking of chemical and radioactive waste disposal sites (Whelan[14] et al., 1987). It is used as a tool for setting priorities and is applied in a comparative mode. The methodology considers four major pathways of contaminant migration: groundwater, surface water, overland and atmospheric. The resulting concentrations in the different compartments are then used to assess exposure to neighboring populations.

An enhanced version of RAPS has been recently developed by Battelle for the U.S. Department of Energy. The model is called MEPAS[15] (Multimedia Environmental Pollutant Assessment System) and ranks environmental problems according to risk to human health and is currently being used by the DOE. It uses dBase III plus and runs on an IBM PC and is written in FORTRAN.

Contact: Dr. James C. Droppo, Jr. (509)376-7652, Pacific Northwest Laboratory, P.O. Box 999, Richland, Washington 99352, USA.

Model C (MNSEM)

Model C is a multi-phase non-steady state equilibrium model . This kinetic model has been designed to predict the environmental fate of organic chemicals under steady-state conditions of continuous loading to the environment (Yoshida[16] et al., 1988).

It calculates the chemical distribution, persistence and concentrations in an evaluative or hypothetical environmental media. It also estimates recovery time after emissions are terminated. The model treat an entire country (Japan). The program is written in MS-DOS basic and is available.

Model D (PERSISTENCE)

This is an aquatic model developed for the National Research Council of Canada (Asher[17] et al.). It is used as a screening tool to produce a systematic method of comparing predicted fates of various organic chemicals, especially pesticides, which may enter the aquatic environment. The model considers four compartments representing suspended solids, invertebrates (or other components of water system excluding fish), sediment and fish.

Default environment for the model are: a standard pond, simulating a small, eutrophic pond, and a standard lake, simulating a deep, oligotrophic lake. Chemicals can only be added to the water compartments, and can occur by finite release, continuous release, or pulses occurring over a period of time. Transfer between water and the three other compartments is characterized by a rate constant. Degradation removal pathways are photodegradation, volatilization, and hydrolysis of water; biodegradation in fish, and microbial degradation in suspended solids and sediments. The dynamic solution, calculated form a set of differential equations, estimates concentrations as a function of time. An overall retentive capacity or persistence is calculated.

Model E: Environmental Exposure Potentials (EEP)
This model is a multimedia exposure model based on fugacity equilibrium compartmentalization used to determine the exposure potential of new organic chemicals (Klein[18] et al., 1988).

It is used by member states of the European Communities This code runs on an IBM-XT with DOS. It is written in Turbo Pascal.

Model F: Environmental Partitioning Model (ENPART)
It is a multi-media compartment model which provides a first-level screening analysis of chemicals released into the environment (USEPA[19], 1982). The model is fugacity-based and calculates the steady-state equilibrium concentration ratios between compartments. It identifies dominant pathways and estimates the chemical's persistence and bioconcentration potential.

The system considers compartments of air, water and soil and has for sub-compartments: sediment, suspended sediments, biota, and ions considered in equilibrium with water. The model is primarily used for chemicals being evaluated for testing. The output is in the form of concentration ratios between compartments rather than absolute values. It runs on VAX or an IBM PC, and is written in FORTRAN. A PC version is available.

Contact: Exposure Evaluation Division, Office of Toxic Substances (TS-798), US Environmental Protection Agency, 401 M Street, SW, Washington, DC 20460 USA, Attn. Russel S. Kinerson.

Model G: TOX-SCREEN
It is a time-dependent multimedia screening model used to determine potential for environmental transport and accumulation of chemicals released to air, water or soil (Hetrick[20] et al., 1984). It is expected to identify chemicals that are unlikely to become a problem.

Atmospheric dispersion downwind a point source is estimated using a modified Gaussian plume dispersion equation. Chemical migration in soil is estimated by means of the seasonal soil compartment model (SESOIL). Aquatic fate and transport are modeled using EXAMS. The code TOXSCREEN estimates bio-accumulation in aquatic organisms, terrestrial food, animals and plants. The code is written in FORTRAN and runs on a VAX-VMS.

Model H: Exposure Commitment Method
This method uses a time-independent approach for assessing human exposure to environmental pollutants (Bennett[21], 1981). Essentially, it calculates the cumulative exposure by various pathways for a constant, continuing environmental release.

The model has been modified for application to organic compounds and metals. Exposure or intake are calculated sequentially via a succession of compartments for each pathway.

Method I: Preliminary Pollutant Limit Value (PPLV)
It is an exposure assessment calculation. This comprehensive assessment method was developed for the U.S. Army to estimate the allowable level of pollutants in environmental media by considering human health effects (Rosenblatt,[22] 1980).

Environmental pollutant pathways considered are through air, water, atmospheric vapor and particles to crops, domestic animals, and fish.

Model J: Multimedia Exposure Fugacity Model
In the last decades a series of models has been developed at the university of Toronto by Mackay, Paterson et co-workers (Mackay[23] et al., 1990). this particular model treats four bulk compartments: air, water, soil and sediments.

The algebraic model solution incorporates expressions for emissions, diffusive, and non-diffusive transfer process, advective flows and degrading reactions. The model calculates the steady-state distribution of a chemical and its persistence in the environment. It runs on an IBM-PC, and is available.

Contact: D Mackay or S. Paterson, Dept. of Chemical Engineering, University of Toronto, Ontario M5S 1A4, Canada; Tel: D. Mackay (4169 978-4019).

Model K: SIMPLESAL
It is a spreadsheet-based, multimedia, fugacity, compartment model, which can be used to estimate steady-state or time-dependent concentrations of organic compounds as well as heavy metals.

The SIMPLESAL model runs on an IBM-PC, using Lotus 1-2-3.

Contact: D. Van de Meent, National Institute of Public Health & Environmental Hygiene (RIVM), P.O. Box 1, 3720 BA Bilthoven, The Netherlands, Tel: 030-743130.

Model L: AERIS
It is a risk assessment multimedia model that runs with an expert system programming environment. AERIS (Aid for Evaluating the Redevelopment of Industrial Sites) uses an environmental and toxicological information to estimate exposures (and therefore the risk) that site users or visitors can receive as a result of contaminants present in the soil.

It is not a comprehensive multi-media since it considers only phases of air, soil and groundwater. AERIS is designed to run on an IBM-PC.

Contact: Brett Ibbotson, SENES Consultant Limited. Tel: (416) 764-9386 or Art Stelzig Industrial Programs Branch, Environment Canada (613) 953-1134.

Model M: Multimedia Compartmental Model (MCM)
This multimedia describes the dynamic distribution of organic pollutants in a six-compartment system, consisting of air, water, soil, sediment, suspended sediment and biota (Cohen[24] *et al., 1985).*

It incorporates diffusive and non-diffusive transfer processes, advective and convective flows, and first-order reactions in various compartments. It runs on a mainframe computer.

Contact: Dr. Y. Cohen, National Center for Intermedia Transport Research, University of California, Los Angeles, CA 90024, USA.

4.4. Estimates of Risks from Continuous Emissions - The Dose-Response Relationship

4.4.1. BACKGROUND

The need to quantify *dose-response curves* is more recent and is tied up directly to quantitative risk assessment. To aid in rational planning and decision making, it is necessary to estimate health risks associated with new developments or review of existing situations. *The method of approach generally adopted consists of hazard identification, followed by parallel steps of exposure assessment and dose-response assessment, which are brought together in risk characterization (NAS*[23]*, 1983).*

This section is concerned with *dose-response assessment*, the evaluation and quantitative characterization of the relationship between level of exposure and health-related response (CCERP[25], 1985). It differs from other parts of risk assessment in that it often is developed independently of the application on which the risk assessment focuses, using data from other sources. A single dose-response assessment may be applicable to many risk assessment applications.

Toxicity is related to dose: a "toxic" agent is observably toxic at some dose level. Below that level it is apparently safe, and may even be beneficial. Depending on the mechanism of effect, there may be a *threshold level* below which the agent is truly safe, or there may be no threshold. In the latter case, with decreasing dose, the level of effect approaches the level of spontaneous occurrence of the same effect so that the effect signal becomes lost in the natural variation of the effect. It is then impossible to detect the presence of an effect in a particular experimental design, even if it exists. In theory, one can detect effects, if they exist, at lower and lower dose levels by going to more powerful experimental designs, but there are practical limits. *There will always be dose levels below which one cannot observe effects directly:* It is statistically impossible to distinguish between a true threshold of effect and a continuing effect which disappears into the background noise of spontaneously occurring disease. Decisions on (a) whether to treat a dose-response relationship as threshold or no-threshold and (b) the form of equation that should be used for the dose-response function should be based on an understanding of the underlying mechanism of disease, not on the data of a single experiment.

In the hazard identification process, *screening experiments* are often used to determine if an effect can be observed. The *standard animal tumor bioassay of the National Toxicology Program* (NTP[26], 1984), for example, is used to determine if a chemical is a carcinogen. This generally consists of exposing animals at three dose levels: the maximum tolerated dose (the highest dose that is not acutely toxic), half the maximum tolerated dose, and a zero-dose control. The question to be answered is basically, "Do the exposed animals have a higher tumor incidence than the control animals?" This is a different question to ask of a data set than, "What is the quantitative relationship between exposure and response?"

In *cancer risk assessment*, the process is generally a two-step process. First, ask if the substance is a carcinogen. If the answer is "yes", then determine the quantitative dose-response relationship for use in risk assessment. Other situations have been handled as a one-step process and the quantitative dose-response relationship is determined directly. Since the determination that the substance is a carcinogen is made at a high-dose level, the confidence limits of the dose-response function may include zero effect at low doses.

Detailed risk assessments are unnecessary if environmental agents are routinely present at levels which are clearly toxic to any substantial group of people; if such an exposure occurred, it could be brought under control by regulatory authority. *Dose-response functions* in this clearly toxic range are useful primarily to assess effects of accidental exposures. Risk assessment of routine, low-level exposures almost always involves dose-response functions in the fuzzy range below the point where effects are clear. *As dose-response functions are extrapolated from high doses (where the data have been collected) to low doses, uncertainty increases.*

Effects model can influence space and time resolution in a given dispersion model. In case of non-stochastic effects, which manifest themselves at high exposure levels, there is certainty that harm will occur sometimes above some threshold level of exposure. *In this case the harm will be a monotonically increasing function of the exposure* (*Figure 4.1*).

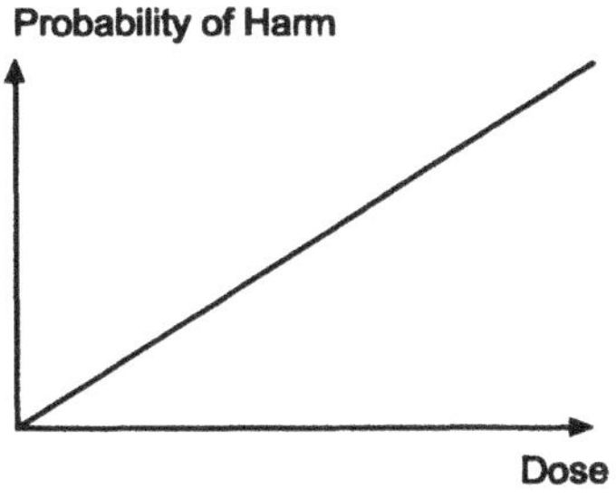

Figure 4.1. Dose-Effect relationship (Case with non-stochastic effects)

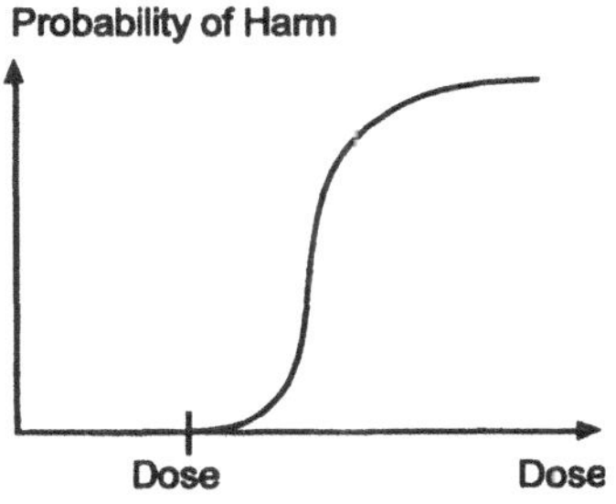

Figure 4.2. Dose-Response relationship (Case with stochastic effects)

Definition of Stochastic effects
Stochastic effects are those in which there is always a probability of harm from any exposure to a contaminant, no matter how small. This situation is given by a dose-response relationship (Figure 4.2).

For the *case of near field of exposure,* peak concentrations are relevant. In the far field one has to define the long-term exposures in order to estimate both the individual and collective exposure risks.

Mathematically, a dose-response function can be extrapolated down to infinitesimally small doses. Practically, limits must be considered. First, at what level does the risk become so low as to be of no practical concern to either society or the individual? This is often referred to as a "*de minimis level*". If the criterion is individual risk, a *de minimis* level can be determined from the dose-response function directly. If the criterion is the population risk, however, the decision goes beyond dose-response assessment and also depends on the size of the population exposed and the level of exposure. Second, at what level does the dose-response relationship become too tenuous to justify its application? As the dose level of concern extends further from the range for which experimental or observational data are available, uncertainty increases. *In general, either a threshold is established or a decision is taken to assume a continuous dose-response function to zero dose.*

The most important current problem in the interpretation of dose-response functions and the risk assessments using them is that of including an appreciation of uncertainty and of individual risk levels in the final risk estimates. Typically, the size of the population at risk and the uncertainty in the dose-response function increase with decreasing exposure levels. It is difficult to judge the importance of estimates of high population risk when large populations are exposed to low doses without a clear understanding of the uncertainty.

4.4.2. KINDS OF EXPOSURES: DOSE-RESPONSE IMPLICATIONS

Exposure and Dose

An understanding of dose-response assessment and its use in risk assessment requires a clear distinction between *exposure* and *dose*:

Definition of Exposure and Dose

*- **Exposure** is the concentration (or amount) of pollutant in the environment to which a person is exposed, (e.g., 25 mg/m^3) in the air one breaths or (e.g., 50 ppb) in the water one drinks.*

*- **Dose** is the amount of pollutant reaching the organ, tissue, or specific cell of interest. Interposed between the two are simple factors such as breathing rate (e.g., an exercising person has a faster breathing rate and thus inhales more pollutant than the sedentary person with the same exposure) as well as complex metabolic and pharmacokinetic processes (NAS[27], 1987).*

The dose of concern may not even be due to the same material which caused the exposure, but may result of a chemically altered metabolite (e.g., the actual cancer-producing agent in an exposure to benzo-[α]-pyrene is its metabolite: 7-[β]-8-[α]- hydroxy-9-[α],10-[α]-epoxy-7,8,9,10- tetrahydro-benzo-[α]-pyrene).

The term *dose-response function* most commonly means exposure-response function. Risk assessments must deal with multi-media exposures including air, water, and food-chains (*Figure 4.3*).

Human Exposure to Hazardous Substances

Step 1: *Define and analyze source and mechanism of toxic release to the environment*

Step 2: *Define and select transport medium:*
Air
Water
Food chain

Step 3: *Define pathway and exposure route in human body:*
Absorption
Resorption
Desorption

Figure 4.3. Human exposure to hazardous substances

The same level of exposure in different environmental media can lead to vastly different doses at the tissue level. The exposure-response function is thus highly dependent on how the person was exposed. For example, while the exposure-response function for lung cancer from benzo-[α]-pyrene in air may be substantial, the exposure-response function for lung cancer from benzo-[α]-pyrene in food may be zero. The true dose-response function for the lung is independent of how the person was exposed or how the pollutant reached the lung. The true dose-response function may be difficult to determine, however, since the actual dose to the target organ is difficult to measure. The target tissue itself may be unknown or in question. Is the target tissue the whole lung or the basal cells of the bronchial epithelium?

Exposure-Response Function. It incorporates the metabolic, pharmacokinetic, and other processes which intervene between the initial point of exposure and the tissue of interest. These processes are often poorly understood, and, from what is known, can be highly non-linear and involve substantial interspecies variation. The current trend is to couple physiologically based pharmacokinetic models, which predict organ or tissue dose, with dose-response models which predict health effects from biologically relevant dose. Unless the relationship between exposure and dose can be quantified, one must exercise care in using a dose-response function based on ingestion in a risk assessment involving an inhalation exposure or even in extrapolating from high doses to low doses.

Averaging Time. Concentrations of pollutants in air vary over time and space. A person standing in one spot is submitted to a continually varying exposure. People moving through their daily activities are exposed to even wider variations in exposure.

> *Averaging times for exposure measurements range from seconds to 24 hours or longer.*

- *For acutely toxic materials,* a single breath exposure may be an appropriate measure, although the more conventional approach is a *30-minute exposure.* The latter defines the level immediately dangerous to life or health or IDLH (NIOSH-OSHA[28], 1981).
- *For some pollutants* (e.g., ozone, nitrogen oxides), peak exposures seem to be important in causing effects, although the role of the time-distribution of the peak or the interval between peaks is not well understood.
- *For other pollutants,* particularly those which accumulate in the body (e.g., lead, cadmium), long term averages are more appropriate measures.

In deriving dose-response functions, the importance of exposure averaging time is often neglected. In many cases, lack of data on effects at the desired averaging time force the analyst to rely on less suitable data. For example, data from short-term occupational or accidental exposures may be used as surrogates for population dose-response functions. Consideration should always be given to the match between the averaging time of the dose-response function (presumably that from the underlying experiment) and that of the exposure data in the risk assessment. When they do not match, assumptions must be made about the relationship between average and peak exposures.

Time-Regimen of Exposure. Uncertainties are introduced in risk assessments by applying dose-response functions based on animal experiments in which the animals were exposed 5-days per week, or from occupational epidemiological studies in which the workers were exposed for 8-hours per day, 5-days per week, to general populations with continual exposures (although varying in magnitude).

> ***Definition:*** *Regimen is the limiting of exposure due to incorporation in order not to exceed an admissible dosis value.*

A correction factor is used to pro-rate the exposure on the assumption that only the long-term average is of importance. But, is there a recovery factor over the weekend? Does this make a difference when carried through to a low-dose extrapolation? Quantitative answers are seldom available to such questions, but some qualitative information, based on knowledge of the action of the chemical, may be available.

Complex Mixtures. No one is exposed to a single pure chemical. *Dose-response functions,* on the other hand, are always expressed in terms of single chemicals, sometimes in a pure state (usually derived from animal experiments), sometimes as an *index of a mixture* (usually derived from epidemiological studies). The *index-based dose-response function* presumably would overestimate the risk in this situation. In general, the state-of-the-art precludes more than simple assumptions of independence of action in most cases, but further information is available in Calabrese[29] (1990), Gray et al.[30] (1987), NAS[31] (1988).

Measurement Techniques

Technical considerations in *measuring exposure, dose, or effect* are often ignored. A dose-response function is developed from a study in which both exposure and response were measured. It is then applied in a risk assessment in which only exposure is measured (or estimated with models). This is particularly true for composite indices such as "fine particles" or "total organics". Although this is not the principal source of error in risk assessments, one must give some consideration to the compatibility of the exposure measures in the risk assessment and the dose-response function. *From the standpoint of dose-response, there is, thus, an obligation to provide adequate documentation on the basis of the exposure.*

Environmental Measurements. They are made for many different purposes and in many different ways. Those made in conjunction with dose-response studies must be designed to provide estimates of exposure to people. *Ambient concentrations may not be the same as exposure.* Surprisingly, only in the past decade has it become commonly considered in air-pollution epidemiology and risk assessment *that people spend the majority of their time indoors, and that outdoor measurements may be poor indicators of exposure (Morgan[2] et al., 1977).* Personal monitoring; monitoring of micro-environments, including indoor monitoring and monitoring in vehicles; and personal activity pattern analyses are now almost a sine qua non in air pollution health effects studies. *Recent studies are following the concept of "total exposure" (Wallace[32], 1987).*

4.4.3. KINDS OF EFFECTS AND THEIR IMPLICATIONS FOR DOSE-RESPONSE

Health End-Points. There are many health end-points for which dose-response functions are available or might be desirable for risk assessment. Commonly used end-points depict diseases which are of greatest concern, e.g., cancer, heart disease, or reproductive effects. They also include injury, mortality, and effects on future generations. These end-points have obvious social significance that can be understood easily by the decision maker. For example, a risk assessment which concludes that a given action will produce ten additional cancers per year fits easily into a decision process. *If desirable, such results can readily be translated into monetary terms to allow calculation of benefit-cost ratios in like units.*

> ***Remarks Concerning Health End-points***
> *There is some feeling that such end-points may over-simplify the risk, reducing it to "body counts" that are often artificial. This is particularly pertinent in the case of low-level effects spread over a large population. The numbers may be a useful index to the analyst but have quite a different meaning to the public.*

Two recent advances in the study of health effects are a focus on sensitive subgroups in the population and on early biological and biochemical markers of disease:

■ *Sensitive Populations:* Important effects may be missed in large population studies if they occur only in *sensitive subgroups* that are not identified and examined separately or in sufficient numbers.

■ *Early Markers of Disease:* Parallel to the increasing use of *biochemical markers of dose,* research efforts have recently begun to focus on early *biochemical indicators of disease*, rather than the disease itself. These may or may not have direct significance themselves and so are more difficult to incorporate in a risk assessment.

■ *Morbidity:* Morbidity is the recognized presence of disease. It is generally expressed as *incidence* (number of new cases developing annual per 1000 people in a population) or *prevalence* (number of cases extant per 1000 people in a population at a given time). The former is more appropriate for dose-response relationships. Dose-response relationships usually relate exposure to a specific pollutant with a specific disease. For example, sulfur dioxide with respiratory disease, lead with neurotoxic diseases, cadmium with kidney disease. Special subcategories of morbidity are reproductive and developmental effects. The former is associated with preconceptual exposure to the mother or father while the latter is associated with exposures (through the mother) to the fetus. Risk assessment applications of these effects are discussed in NAS[33] (1986).

■ *Mortality:* Total mortality is frequently used as an *end-point in dose-response functions.* An important consideration for risk assessment in mortality studies has been the length, quality, or value of life lost. Expressing the response in terms of years of life lost would help solve the problem regarding the length of life lost. The *mortality rate equation* developed on a linear multiple regression for the case of an industrialized urban area (Seoul) when CO and SO_2 pollutants are taken into consideration, is of the following form (Kwi-Gon, 1991):

$$\text{Mortality} = -0.29065 * 10^{-5} * \text{CO}\,[\text{kg}] - 0.39953 * 10^{-5} * \text{SO}_2[\text{kg}] + 4.96441$$

Since such equations are very site and case-specific, they cannot be directly transposed to other places. For these reasons new equations based on local data available have to be worked out.

Remark Concerning the Mortality Rate
The results of linear multiple regression analysis are not consistent with the hypothesis that air pollution increases the mortality rate. Mortality rates presumably reflect the influence of technology (e.g. medical improvements) a great many environmental and time factors. However, this analysis does not account for factors other than pollution.

4.4.4. DATA SOURCES AND THEIR IMPLICATIONS

Data for *dose-response functions* come from three basic sources:

- *Studies of human populations,* both epidemiological and clinical
- *Toxicological studies* on whole animals, generally mammals, and
- *Laboratory studies* on human or animal cells or tissues, or on lower life forms such as bacteria.

Any study which is to provide the basis of a dose-response function must include information on both exposure (or dose) and the health response. Before attempting to derive a dose-response relationship from basic data, however, the risk analyst should seek relationships already developed and available in the literature. Suitable quantitative relationships are available for many pollutants along with considerable background analysis and discussion. For other pollutants, qualitative or semi-quantitative reviews may be available.

Integrated Risk Information System
The *EPA (US Environmental Protection Agency)* has developed numerous quantitative dose-response functions, especially for carcinogens. These are available in computerized form, along with other dose-response information, in *EPA's Integrated Risk Information System (IRIS).* IRIS includes some on-line documentation, but cites appropriate source reports.

Other Sources
, Other useful sources include *WHO's series on Environmental Health Criteria* and the publications of the *International Agency for Research on Cancer.* Numerous commercial publications compile available dose-response functions or supplementary information. These include Calabrese and Kenyon[34] (1990), Cothern[35] et al., (1988),

Howard[36] (1989), Lappenbusch[37] (1989), Lioy and Daisey[38] (1987), Sittig[39] (1985), and Weiss[40] (1986).

4.4.5. DERIVING THE DOSE-RESPONSE RELATIONSHIP

A dose-response relationship specifies a quantitative increase in a specific health effect associated with an increase in exposure to a pollutant. The effect may be in absolute terms (number of increased cases per 1000 people per unit of exposure) or in relative terms (percentage increase in background rate per unit of exposure). Dose-response relationships are usually derived through the application of a mathematical model to data from epidemiological, toxicological, or clinical studies. Mathematical models simplify the underlying biological mechanisms and often include assumptions that are not experimentally verifiable.

> ***Deriving quantitative dose-response relationships***
> *Mathematical and statistical methods of deriving quantitative dose-response relationships from a toxicological or epidemiological data set are well known. Almost never is a single study sufficient to determine the form of the dose-response function. In fact, for extrapolations to the low-dose region, there is seldom sufficient information on which to select a functional form; it must be assumed. Ideally, the form of the function (i.e., the shape of the dose-response curve), be it linear, quadratic, exponential, threshold or no-threshold, is based on the entire body of knowledge available on the mechanisms involved in producing the observed effects from the kind of agent of concern.*

A *dose-response function* can take several forms. *Qualitatively, as dose increases, different effects of increasing severity occur within an individual.* Carbon monoxide, for example, at low levels of exposure causes measurable (though not noticeable) visual impairment and decreased manual dexterity; at increasingly higher exposure levels the progression of symptoms includes headache, dizziness, nausea, vomiting, collapse, coma and ultimately death. Because of differences in individual susceptibility, the threshold for each level of effect will differ among individuals. *Quantitatively, the distribution of these thresholds describes a population dose-response function for that individual effect.* That is, at increasing levels of dose, the particular effect will occur in an increasing number of people. This is called a *statistical or tolerance distribution model.* This is an appropriate form for threshold phenomena, and is the basis of classical toxicology in which dose-response functions are often represented as probit curves.

Deriving a Dose Response Function for a Population

In some cases, detailed dose-response function equations are themselves incorporated into risk assessments. More frequently, however, results of such dose-response modeling are reduced to a single coefficient, the slope of a linear portion of the dose-response curve. This coefficient is then simply multiplied by the exposure to yield the effect on the population. While the state of knowledge may not warrant a more complicated approach, it is important to be aware of the assumptions behind this. It assumes a linear dose-response function, at least within the range of exposures in the population of interest. Equally important, it assumes a dose-response function in one variable, excluding any effect of concurrent exposures or population-based factors such as age or susceptibility.

Let us consider a dose-response function for the exposure of interest (F_i) is unknown, but there is a potency estimate (P_i). For a similar exposure, both the dose-response function (F_s) and the potency estimate (P_s) are known. In the most simple form, the desired dose-response function is estimated as: $F_i = F_s\,(P_i/P_s)$.

This is an example of combining different kinds of health effects information to produce a dose-response function for which insufficient information was available from a single kind of information.

It is generally recognized that drawing on all the information available is more likely to provide a better dose-response function than one based on a single study. While information from other studies is often used to support the dose-response function based on a single study, there is no generally recognized analytical method to mathematically combine data from several sources to form a combined dose-response function, although such integration of results is obviously done subjectively.

4.4.6. DOSE-RESPONSE MODELS FOR TIME-DEPENDENT DOSING

Making carcinogenic risk estimates from animal or epidemiologic dose-response data requires that a mathematical model be fit to the experimental data and then used to extrapolate to the dose range of interest. In animal experiments, the dose variable often corresponds to a *constant lifetime dosing rate*, and the dose response model chosen must only relate the constant dose to the fractional response rate of an exposed population. A relatively standard set of dose-response models are used to make risk estimates. These include probit (Finney[41], 1971), multistage (Armitage and Doll[42], 1954), and related power series model (Guess and Crump[43], 1976), Weibull (Carlborg[44], 1981), the one-hit (Iverson and Arley[45], 1950), multi-hit (Cornfield and Mantel[46], 1977; Rai and Van Ryzin[47], 1981). However, these models are not generally formulated to account for time-dependent dosing.

In some animal experiments and epidemiologic situations, the dose rate is not constant and the dose response model must take time-dependence into explicit account. Furthermore, the ability to take dose-time dependence into account must also be given when one uses a fully parameterized model to make prospective risk estimates, since these must often be made for arbitrary dosing scenarios (e.g., accidental release of toxic bursts). The usual approach is then to use the *constant dose form* of above mentioned models, taking the time-averaged dose as the constant dose. Such a procedure is not generally valid for all models. The results of risk estimates can vary, depending on whether time-averaged carcinogen concentrations or time-dependent concentrations were used. A related problem involves migration and/or birth-death of the exposed population. In the latter case a population dynamic model must be used (Bolten[9] et al., 1983; Bolten[10] et al., 1985).

In conclusion, it is extremely difficult to incorporate time-dependent phenomena into the dose-response calculations. Whether or not the details of a time-varying concentration of toxicants need be taken into account in estimating toxic response depends on the dose-response model chosen. Failure to include full time dependence may lead in some cases to an overestimation or underestimation of risk. The multi-stage and the Weibull models require the most careful treatment of time-dependent phenomena. Order of magnitude errors may occur if short, intense exposures (incidental burst) are treated as time-averaged quantities, particularly when these models have parameter values that strongly prevent their re-expression in terms of time-averaged concentrations. These authors have shown that use of a time-averaged concentration in the one-hit and multi-hit models provides a rigorous risk estimate, whereas in the case of the Weibull model, its use leads to an error that can be estimated from the exposure time / response time ratio and two of the Weibull parameters.

4.4.7. LEVELS OF AGGREGATION IN CONSIDERING A POPULATION AT RISK

Each individual responds uniquely to an environmental exposure, but it is impracticable to make environmental risk assessments at the individual level. *Instead, the population must be divided into groups with similar characteristics.* The degree of detail in grouping depends primarily on information available for exposure and dose-response. Thus, the more details available in the dose-response function, the more flexibility for grouping is available in the risk assessment. *It is always possible for the risk assessment to be conducted at a more aggregated level than the dose-response function, but seldom possible to meaningfully work at a more detailed level.*

There are basically three classes of grouping:

- *by demographic factors,* (e.g., age, sex, and race)
- *by constitutional factors,* (e.g., genetic pre-dispositions, pre-existing disease, constitutional susceptibilities resulting from earlier disease, and susceptibilities or sensitivities resulting from previous exposures)
- *by exposures,* , especially the exposure level of the particular agent of interest but also others, including smoking, diet, and concurrent occupational or environmental exposures.

Some of these factors can easily be included in a dose-response function and used in a risk assessment. These include the demographic factors and many exposure factors. Others are more difficult. Information on genetic susceptibility, for example, may be impossible to obtain. Often, surrogate factors such as socio-economic level are used as indicators, too.

The effects of a migrating and proliferating population on estimates of added risk have been estimated by Bolten[10] et al., (1985). Failure to account for exposure (residence) time distributions, as caused by migration or birth-death of a population, may lead to a significant overestimation of risk. Such an overestimation may occur for both the multi-hit and Weibull models provided the exponent parameter on exposure time and the migration or death-rates are sufficiently large.

4.4.8. UNCERTAINTY AREAS IN QUANTITATIVE RISK ASSESSMENT

Dose-response functions, particularly for low-level exposures, are inherently subject to considerable uncertainty. If this uncertainty is not explicitly incorporated in the dose-response function, results derived from using the dose-response function can be misleading. Several reports provide useful information on the characterization of uncertainty in dose-response functions and their application in risk assessments (Morgan et al.,[2] 1985; Griffiths,[48] 1985; Niehaus et al.,[49] 1985; Cothern et al.,[35] 1986). Crump[50] (1984) lists *five areas in quantitative risk assessment* which have important uncertainties:

■ *High- to low-dose extrapolation:* This can apply to dose-response functions derived from occupational as well as animal studies.

■ *Animal to man extrapolation.*

■ *Extrapolation from long-term to short-term exposures:* In the case of carcinogens, the studies from which dose-response functions are drawn must be long-term; the application of these dose-response functions to short-term exposures introduces uncertainties. In other situations, dose-response functions developed from short-term studies might be extrapolated to long-term exposures, also introducing uncertainty.

■ *Subject's age.* It is the subject's age at the time of exposure

■ *Extrapolation from one route of exposure to another.*

In any situation in which a model is used, two basic sources of uncertainty must be considered:

a) uncertainty in the appropriateness of the functional form of the dose-response model;
b) uncertainty in specifying the parameters of that model, which has to do with the validity of data and the stochastic nature of events.

Use of Confidence Levels

Because use of 95% confidence levels are so ingrained in science, and because risk assessment draws heavily on science, *95% confidence levels* are often used in risk assessments. *The degree of uncertainty risk assessment, and especially in dose-response assessment, is usually so great that estimating 95% confidence bounds requires assumptions about the shape of probability distributions that are unwarranted.* Much better is to present 67% or 80% confidence bounds in which one has confidence than 95% confidence bounds that may be misleading.

Confidence Levels of Dose-response Functions
The ideal, and generally practicable solution, in the case of dose-response functions, is to explicitly include sufficient uncertainty information so that decision-makers (who may be the public), who are in a better position to judge the appropriate level of confidence for a particular analysis, can draw their own conclusions.

4.4.9. GUIDANCE NOTE

Dose-response functions, in general, represent biological relationships which are common world-wide. The same dose-response function can thus be used in risk assessments in different settings and cultures. In any case, however, even in situations seemingly similar to that in which the dose-response function was derived, the specific applicability of the dose-response function should be investigated. Areas of particular importance to assure compatibility include characteristics of exposure and of the population at risk. Are the exposure measurement techniques, concurrent exposures to other materials, and the relationship between the overall complex mixture to the index compound in the risk assessment compatible with those from which the dose-response function was derived? Is the effect associated with particularly sensitive subgroups of the population or does it depend on a particular distribution of sensitivity in the population? Ideally, dose-response functions should be desegregated in such a way that dependencies on characteristics of the population at risk are explicit.

Uncertainty of Dose-response Functions
All quantitative dose-response functions involve considerable uncertainty at low-dose levels. If this uncertainty can be adequately characterized and expressed, however, quantitative dose-response functions can be usefully applied in health risk assessments to guide planning and policy.

4.4.10. FOOD-CHAIN MODELS

A large number of conceptual models and computer codes were developed recently for assessing human exposures to radionuclides in food. Most of these models could also be adapted relatively easily to assess exposures to heavy metals, but not as easily to assess organic compounds or other pollutants metabolized or transformed in biological systems. Some care is required in adapting them. Many are specifically designed to model special circumstance, such as accumulation of strontium-90 in milk, and do not contain the necessary structure to model other circumstances.

Determination of Food-Chain Pathways
Food chain pathways that should be evaluated can be determined from examination of diets, local sources of food, and the likely pathways for contamination of these foods. One special "food" that should be considered is ingestion of soil, which is common in children two through six years of age.

Although food-chain models are usually categorized as terrestrial or aquatic, they do not differ in concept. All are based on an assumption of equilibrium transfer rates among "compartments" representing different parts of an ecosystem. They differ only in their relative complexity -- the number of compartments included, and the number of variables influencing each compartment, including multiple interconnections with other compartments.

They range from *simple transfer coefficients or bioaccumulation factors* expressing the proportion of contamination deposited to water or ground that is ingested by humans (single-compartment) to *complex ecosystem models with multiple transfers among many ecosystem compartments* (generally represented by Box-type models). Bioconcentration factors can be derived from field measurements or they can be generalized from results of more complex modeling of the contaminant through the food chain under representative conditions. The use of dynamic models for biological processes is restricted to research, particularly in the context of large-scale accidental releases.

These models were developed primarily for assessing *long-term releases.* They can be applied to *short-term accidental releases,* but at considerable increases in uncertainty associated with the values of model parameters. Their scale is necessarily medium-range; small-scale contamination is easily prevented from reaching the food-chain and long-range dilution usually reduces exposures to insignificance. *In general, the more complex the models, the more site-specific data they require. The most complex ones are so site-specific that the effort required to adapt them to other sites is generally not justifiable.*

Among the generic data required by all compartment-type models are the *concentration and bioaccumulation factors* for different substances under various conditions. Values for such coefficients are found in extensive tabulations available in the literature.[51, 52]

Hoffman[53] et al., (1981) compared predictions resulting from six terrestrial food-chain models; they concluded that differences in the results are due more to model assumptions than to uncertainties in data. The present state of knowledge concerning transport and distribution of radionuclides in terrestrial and aquatic ecosystems has been reported by Coughtrey[54] et al., (1983).

4.5. Estimates of Risks from Accidental Gas Releases

The estimation of the source term for the release of chemically toxic or flammable vapors is not simple. It may not even be possible to predict with certainty the possible size of a hole which may be formed by failure. This has proven to be the case especially for rail and road tankers accidents (real or simulated), but it is also the case for refineries and chemical installations.

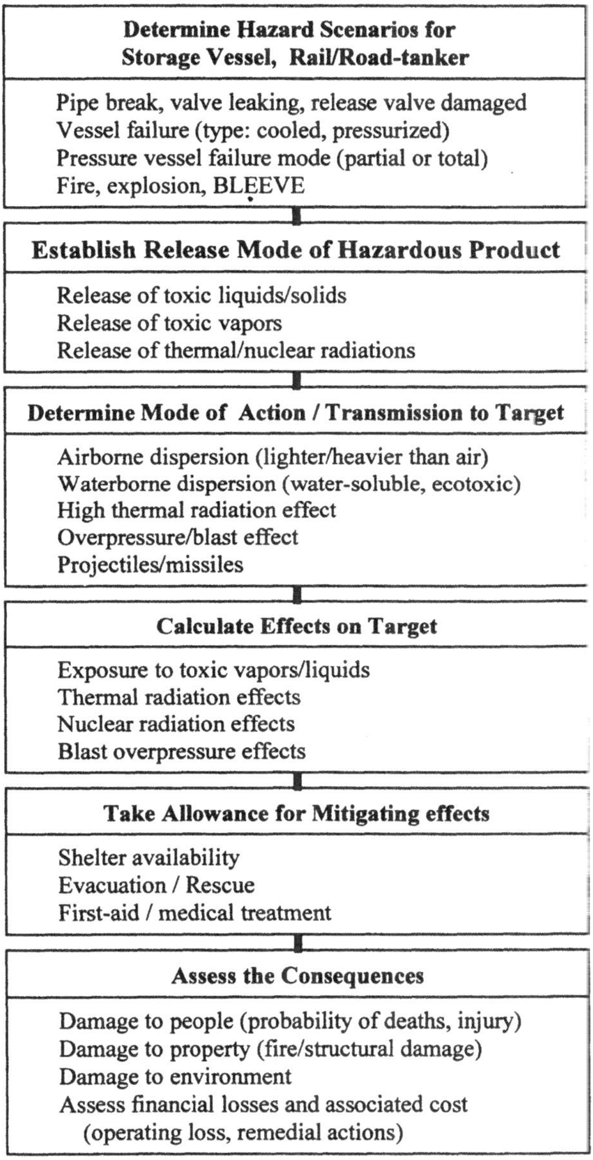

Figure 2.1. Overall approach to consequence estimations

A superficial approach would be to postulate the worst possible case by assuming the total failure of a storage vessel and the immediate discharge of the entire contents into the surrounding atmosphere, but the likelihood may be so small that the risk from such an event becomes insignificant when compared to other risks, if the likelihood is more significant, the combined frequency of the event chain needed to complete the disaster may become so low as to be negligible.

The assessment of the consequences of an accidental release of a hazardous gas involves the sequences shown in *Figure 2.1.* The matter will be treated with more details in the 2nd volume on Integrated Regional Risk Assessment by the same authors (Nicolet-Monnier and Gheorghe[55], 1995).

In order to perform some calculations and assess the consequences of accidental releases on men and property, considerations should be given to the choice of appropriate models and to the effects of mitigation measures:

- physical models
- effect models
- consideration of mitigating effects.

4.5.1. THE MODEL ARCHIE

The ARCHIE[56] model was designed by the US-FEMA and the DOT (Department of Transportation) to provide planning personnel with *integrated methods for use in assessing the vapor dispersion, fire, and explosion impacts related to discharges of hazardous materials into the terrestrial environment.* By doing so, the program not only enhances understanding of the nature and sequence of events and hazards associated with potential accidents, but also provides a basis for emergency planning.

> ***The ARCHIE Model***
> *The program is based on the US "Handbook of Chemical Hazard Analysis Procedures"*
> *Users of ARCHIE may contact agencies of the Federal Government. One method to communicate with these agencies is via use of the Hazardous Materials Information Exchange (HMIX) computerized bulletin board system, which is operated by FEMA and the DOT. The system includes a conference section that is dedicated to ARCHIE. HMIX can be accessed by modem at telephone number 1-312-972-3275. A users manual for the system and/or technical assistance* can *be obtained By calling 1-800-PLAN-FOR (1-800-367-9592 in Illinois). Users unable to access HMIX may submit questions or comments regarding the computer program in writing to: ARCHIE Support (DHM-51 / Room 8104), U.S. DOT Office of Hazardous Materials Transportation, Research and Special Programs Administration, 400 7th Street, S.W. 2 Washington, D.C. 20590.*

The computer code includes models corresponding to the sequence of procedures suggested for assessing the hazards of:

a) *Discharges of gas or liquid from a tank or pipeline leading to flammable, or toxic vapor dispersion, or fires, or explosions involving such materials*

   - Computation of discharge rate and duration for vapor/gas discharge direct to atmosphere
   - Computation of pool area for liquid spills
   - Computation of pool evaporation rate
   - Evaluation of downwind toxic gas or vapor dispersion hazards

- Evaluation of pool fire hazards
- Evaluation of vapor cloud fire hazard
- Evaluation of vapor cloud explosions

b) *Fireball and blast effects resulting from the explosion of an inadequately vented tank engulfed in a fire, i.e., BLEVE (Boiling liquid expanding vapor explosion)*

- Evaluation of tank overpressurization explosion hazard
- Evaluation of fireball radiation hazard upon tank rupture

Fire/explosion models of ARCHIE only apply to flammable or combustible materials upon their ignition.

c) *Detonation of a solid or liquid*

- Evaluation of true explosive force of solid or liquid explosive.

Explosion models of ARCHIE do not consider hazard of airborne fragments.

A great deal of effort was put into the program to ensure that users do not apply the evaluation procedures in an inconsistent and inappropriate fashion. Nevertheless, the complexity of the processes being evaluated did not permit the development of a fully foolproof program, so it is necessary for users to apply common sense at each stage of an analysis to ensure that input provided to the program and the scenario being considered are consistent and reasonable.

A Word of Caution
First, it must realized that the procedures provided with ARCHIE are simplified versions of more sophisticated (and more difficult to use) methods available to experts in the field. Thus, be advised that ARCHIE is intended to provide approximate answers sufficient for general planning. It will usually (when used as instructed) produce results that overestimate hazards but occasional exceptions are possible and likely Application of safety factors by users is encouraged. A second issue involves the fact that the procedures in ARCHIE are designed for spills of relatively pure substances. Mixtures can only be considered in special cases by the knowledgeable user by provision of appropriate input data to the program. The program is not restricted to gases heavier than air.

4.5.2. SIMPLIFIED METHOD FOR ESTIMATING GAS DISPERSION

We present below a simple method for quickly estimating the dispersion, which is based on a scaling law relating the dispersion range to the mass released. Every escape or release can be considered equivalent to a discrete mass, and the dispersion of such mass can be related to the next appropriate dependent variable in the chain of events which has to be worked through. The independent variable that has been found most appropriate is the down-wind range to a given *gas concentration*. This in turn may relate to a *lethality criterion* in the case of a toxic gas, or to a *flammability criterion* in the case of a gas which may ignite.

For that purpose, the release of hazardous liquids can be classified broadly into two classes:

- near instantaneous release from a vessel which has suffered catastrophic failure (conveniently described by a mass release)
- slower release rate from a partial failure of the vessel, or from a pipe, or other device connected to the vessel system (described by a mass flow).

Since the development of a standard release pattern may rest upon an estimation of the sizes and frequencies of the larger possible events, *it is helpful to transform mass flows into equivalent mass release according to the scaling law which relates the downwind range to the mass released (method of Marshall[57]), when constructing and comparing release patterns with generic and plant-specific data.*

Case 1: Instantaneous Release from a Pressure Vessel

We assume that most if not all of the vessel contents are released. The thermodynamic equation for standard release is:

$$\varnothing = 1 - \exp\left[-\frac{C_p}{\Delta H_v}(\theta_1 - \theta_2)\right] \tag{2.1}$$

where

$\varnothing$ = mass fraction vaporized
C_p = specific heat of the liquid
ΔH_v = latent heat of the liquid
θ_1 = storage temperature
θ_2 = boiling point of stored liquid

In the case of escape from a catastrophically ruptured container, however, turbulence caused by rapid boil will add spray to the flash fraction given by calculation from the formula. This may result in an ejection of the total content of the vessel as a spray of liquid droplets. The resulting turbulence will promote a cloud formation with air entrainment up to ten times the original mass.

Case 2: Slow Release from a Pressure Vessel

Where the release is not instantaneous, three kinds of release may be considered:

- non-flashing flow
- flashing liquid flow
- gaseous discharge

Case 2a: Non-flashing Flow

The following equation applies to the first case:

$$\dot{m} = Ka\sqrt{2\rho\,(P_1 - P_2)} \tag{4.2}$$

where

$\dot{m}$ = mass discharge rate (kg s^{-1})
K = discharge coefficient (-)

a = area of discharge (m^2)
ρ = liquid density ($kg\ m^{-3}$)
P_1 = upstream pressure (Pa)
P_2 = downstream pressure (Pa).

Case 2a: Flashing Flow

The determination of flashing flow is problematic, but a simple approximate method has been described in the second Canvey report[58] and empirical methods have been described elsewhere.[59] Whilst empirical methods may not be applicable in all cases, it is often found in multiphase flow that the reduction factor between each of the three phases is 4. As a result this approximation gives for multiphase flow an addition of 25 per cent to the non-flashing flow for the case of flashing liquid flow, and 6 per cent for sonic gas release.

To transpose the mass release rates into equivalent cloud masses. the method of Marshall is followed.[57] The following empirical equation gives the equivalent mass of material in a cloud between flammable limits:

For Atmospheric Dispersion

$$Q_{FL} = \left(\frac{0.321}{D^{0.59}}\right)\left(\frac{\dot{m}_o^{1.59}}{u^{-1.59}}\right)\left(\frac{1}{\chi_L^{0.59}} - \frac{1}{\chi_U^{0.59}}\right) \tag{4.3}$$

For Jet Dispersion

$$Q_{FL} = \left(3.40\rho_a^{1.5}\right)\left(\frac{m_o^{1.5}}{w_o^{1.5}}\right)\left(\frac{1}{\chi_L^2} - \frac{1}{\chi_U^2}\right) \tag{4.4}$$

where

Q_{Fl} = quantity in the cloud (kg)
D = a constant
$\dot{m}$ = mass flow rate ($kg\ s^{-1}$)
u = wind velocity ($m\ s^{-1}$)
χ_L = concentration at lower flammability limit ($kg\ m^{-3}$)
χ_U = concentration at upper flammability limit ($kg\ m^{-3}$)
ρ_a = density of air ($kg\ m^{-3}$)
w_0 = jet velocity ($m\ s^{-1}$)

An example, which refers to a hypothetical hydrocarbon release with lower and upper flammability limits of 0.039 and 0.176 respectively, is given below to illustrate this relationship.

Equivalent Cloud Masses for a Hypothetical Hydrocarbon Release
Assuming the lower and upper flammability limits to be 3.9 and 17.6 vol.-% respectively, and

- for a wind velocity, u = 10 m/s, and with neutral weather conditions (Pasquill category D), we can approximate the mass in the cloud (kg) according to the formula:

$$QFl \cong 0.65\ \dot{m}^{1.59}$$

- for a wind velocity, u = 1 m/s, and stable weather conditions (Pasquill category F), we find the relation:

$$QFl \cong 60\ \dot{m}^{1.59}$$

where, $\dot{m}$ = *mass flow rate (kg* s^{-1}*)*

In the second Canvey report[58] published by HSC an empirical equation for the radius of a vapor cloud has been given:

$$R = 30\, M^{1/3} \tag{4.5}$$

where

R = Radius of the gas cloud (kilometers)
M = Mass of the gas cloud (tons)

It is based on experimental data and makes use of the assumption that the ratio of the cloud to its height is 5 : 1. For neutral-density dispersion the equation of Pasquill suggests that for a continuous release, the downwind concentration is directly proportional to the mass. For a puff-release the result will lie somewhere in-between these two. For an *average release* we obtain the following formula:

$$R = (1/C)^{0.76} \tag{4.6}$$

whilst for low toxic gas concentration (*C* ranging from 1 - 5 %) the next formula applies:

$$R = k\,(Mass)^{n} \tag{4.7}$$

where

n = 0.40 - 0.43.
k = factor depending on the physical properties of the gas and on the weather conditions

The values of the k-factor for different gases are given in TABLE 4.11. These values must be used with equation 4.7, taking a value for n equal to 0.42.

But the dependency of the concentration range upon density is likely to be a complex matter. Two important aspects of the weather which affects dispersion are speed and direction of the wind, and the stability of the atmosphere. Light wind and heavy gases allow the cloud to drift downwind without undue dispersion. Initial increase of wind speed or decrease of density increases the range, but as the wind becomes stronger and the gas gets lighter the dispersion rate increases so that the range decreases again.

TABLE 4.11. Values of the *k*-coefficient for different gases

	Values of the *k*-parameter as a function of weather Pasquill categories (C-F)				
Toxic Gas	**C5**	**D5**	**D2**	**E2**	**F2**
Ammonia	0.10	0.14	0.17	0.20	0.23
Chlorine	0.20	0.28	0.26	0.31	0.36
Propane	0.10	0.15	0.20	0.24	0.29
LNG	0.07	0.09	0.11	0.13	0.15
Butane	0.09	0.12	0.11	0.13	0.15
Hydrogen Fluoride	0.10	0.13	0.15	0.18	0.21

TABLE 4.12. Lethal concentration range (LC_{50}) for ammonia and chlorine

Toxic Gas	**Concentration**	**Range (R, in km)**	
		at D5	**at F2**
Ammonia (100-t release)			
LC_{50} (at 30 min.)	10,000 ppm	1.03	1.70
LC_{50} (at 10 min.)	17,000 ppm	0.69	1.10
Chlorine (50-t release)			
LC_{50} (at 30 min.)	500 ppm	1.45	1.86
LC_{50} (at 10 min.)	866 ppm	1.10	1.41

Influence Exerted by the Pasquill Categories

The individual risk due to drifting vapor clouds increases at night due to the weather conditions represented by the Pasquill categories E and F. The resulting dispersion is quite symmetric and in such a case the risk contours can be represented by a set of concentric circles. The risk contours for the daytime conditions cover a much smaller area and are conveniently represented by ellipses with the long axis oriented in the wind direction. For most practical purposes the combined average day and night contours may be represented by circles.

The hazard analysis for any particular gas container must consider all possible conditions of the ambient atmospheric turbulence, and the possibility of release occurring at any time of the day or night during any season of the year.[60]

When assessing the risk of toxic gas dispersion it is important to establish for a given region a table of weather probabilities distributed on the Pasquill categories A-F and also considering the period of the day (night/day time). The all weathers' probability is for a given place is equal to one and represents the sum of the weather probabilities of the different Pasquill categories for that location.

Using the scaling law some downwind ranges for chlorine and ammonia have been calculated for two different weather conditions, i.e., at D5 and F2. They correspond to a lethal concentrations (LC_{50}).

Caution is required when extrapolating to large catastrophic releases, using the above mentioned formulas. It must be emphasized that the formulas provide a simple basis for calculating the downwind range.

4.6. Environmental Guidelines and Their Value to Risk Management

Voluntary guidelines and legally enforceable standards for contaminants in air and water are needed by analysts attempting to determine the hazards presented by environmental contamination, and the benefit of applying different pollution control strategies.

> ***Environmental Guidelines***
> *Guidelines and standards for various substances are frequently determined by environmental agencies or by ministries of environment. In the absence of such sources, voluntary guidelines published by international agencies such as the World Health Organization (WHO) should be consulted.*
>
> *World-wide limits for toxic and hazardous chemicals in air, water and soil have been published by Marshall Sittig[61] (1994).*
> *Another handbook by the same author is concerned with toxic and hazardous chemicals and carcinogens (Marshall Sittig[39], 1985).*
>
> *A compilation of air pollutant emission factors (Vol. II: Mobile Sources) has been published by the U.S. Environmental Protection Agency[62] (1985).*

The environmental guideline values for specified contaminants in air and water, which were published by WHO (WHO[63], 1993; WHO[64], 1987), are presented. In developing these guidelines, a consistent process of assessment was used. The primary aim of these guidelines is to provide a basis for protecting public health from the adverse effects of air and water pollution and for eliminating or reducing those contaminants that are known or likely to be hazardous to human health and welfare. The guideline values should not be considered standards in themselves. Standards, which have to be determined by scientists and administrators making risk management decisions, should be consistent with these guideline values, taking into account also other factors such as specific environmental, social, economical conditions.

WHO clearly indicates that *numerical values are to be regarded as indications;* they are proposed in order to help avoid major discrepancies in reaching the goal of effective protection against recognized hazards. *The guideline values should be used and interpreted in conjunction with the scientific information that are at their basis.*

4.6.1. GUIDELINES FOR AIR QUALITY

A summary of the Guidelines on Air Quality for Europe is reproduced here.[64] For some pollutants no guideline values are recommended, but risk estimates are indicated instead. TABLE 4.13 summarizes the different endpoints on which guideline values and carcinogenic risk estimates have been based for organic and inorganic substances, showing that all relevant biological effects (endpoints) were evaluated and sometimes more than one endpoint was considered for guideline recommendations.

> ***A Word of Caution***
> *The numerical guideline values and the risk estimates for carcinogens should be regarded as the shortest possible summary of a complex scientific evaluation process. Scientific results are an abstraction of real life situations, and this is even more true for numerical values and estimates based on such results. Numerical guideline values, therefore are not to be regarded as separating the acceptable from the unacceptable, but rather as indications. They are proposed in order to help avoid major discrepancies in reaching the goal of effective protection against recognized hazards. Moreover, numerical guidelines for different substances are not directly comparable. Variations in the quality and the extent of the scientific information and in the nature of critical effects result in guideline values which are only comparable between pollutants to a limited extent.*

TABLE 4.16-TABLE 4.19 show *air quality guideline values or carcinogenic risk estimates for organic and inorganic substances recommended by WHO for Europe.*[64] The emphasis in the guidelines is based on *exposure.* Since this is the element that can be controlled to lessen dose and hence lessen response.

Guideline Values Based on Effects Other than Cancer

The starting point for the derivation of guideline values based on effects other than cancer or odor was to define the lowest concentration at which adverse effects are observed (TABLE 4.14).

It is on the basis of the body of scientific evidence and judgments of protection (safety) factors, that the guidelines (WHO[63, 64], 1987) were established. For some of the substances, a direct relationship between concentrations in air and possible toxic effects is very difficult to establish, because ingestion, more than inhalation, could highly contribute to the body burden (e.g., Cr, Pb, and Cd). For instance, available data show that the food chain is, for most people, the critical route of non-occupational exposure to lead and cadmium. On the other hand airborne lead and cadmium may contribute significantly to the contamination of food by these metals. Complications of this kind were taken into consideration and WHO has made an attempt to develop guidelines which would also prevent those toxic effects of air pollutants that resulted from uptake through both ingestion and inhalation. The averaging times of exposure that have been chosen for the guidelines are based on the characteristic effects of the substance.

TABLE 4.13. Established guideline values and risk estimates

Substance	IARC Group Classification	Risk Estimate based on carcinogenic endpoint	Guideline value(s) based on		
			Toxicological Endpoint	Sensory Effects or Annoyance Reaction	Ecological Effects
Organic substances					
Acrylonitrile	2A	x			
Benzene	1	x			
Carbon disulfide	-		x	x	
1,2-Dichloroethane	-		x		
Dichloromethane			x		
Formaldehyde	2B		x		
Polynuclear aromatic hydrocarbons (Benzo-α-pyrene)	-	x			
Styrene	3		x	x	
Tetrachlorethylene	3		x	x	
Toluene	-		x	x	
Trichlorethylene	3		x		
Vinyl chloride	1	x			

Note: IARC (International Agency for Research on Cancer)

TABLE 4.13. (continued)

Substance	**IARC Group Classification** [a]	**Risk Estimate based on carcinogenic endpoint**	**Guideline value(s) based on**		
			Toxicological Endpoint	**Sensory Effects or Annoyance Reaction**	**Ecological Effects**
Inorganic Substances					
Arsenic	1		x		
Asbestos	1		x		
Cadmium	2B		x		
Carbon monoxide	-		x		
Chromium (VI)	1	x			
Hydrogen sulfide	-		x	x	
Lead	3		x		
Manganese	-		x		
Mercury	-		x		
Nickel	2A [b]				
Nitrogen dioxide	-		x		x
Ozone / photochemical oxidants	-		x		x
Radon	-	x			
Sulfur dioxide and particulate matter	-		x		x
Vanadium	-		x		

[a] IARC Monographs on the Evaluation of the Carcinogenic Risk of Chemicals to Humans, Vol. 32.
[b] Exposures from nickel refineries are classified in Group 1.
Note: The definition of the IARC Group classification is given in Section 4.6.2. of the present volume.

TABLE 4.14. WHO Air quality guidelines values for individual substances based on effects other than cancer, or odor / annoyance

Substance	Time-weighted average	Averaging Time
Cadmium	1-5 ng/m^3 10-20 ng/m^3	1 year (rural areas) 1 year (urban areas)
Carbon disulfide	100 μg/m^3	24 hours
Carbon monoxide	100 mg/m^3 § 60 mg/m^3 § 30 mg/m^3 § 10 mg/m^3	15 minutes 30 minutes 1 hour 8 hours
1-2, Dichloroethane	0.7 mg/m^3	24 hours
Dichloromethane	3.0 mg/m^3	24 hours
Formaldehyde	100 μg/m^3	30 minutes
Hydrogen sulfide	150 μg/m^3	24 hours
Lead	0.5-1.0 μg/m^3	1 year
Manganese	1 μg/m^3	1 year
Mercury	1 μg/m^3 §§	1 year
Nitrogen dioxide	400 μg/m^3 150 μg/m^3	1 hour 24 hours
Ozone	150-200 μg/m^3 100-120 μg/m^3	1 hour 8 hours
Styrene	800 μg/m^3	24 hours
Sulfur dioxide	500 μg/m^3 350 μg/m^3	10 minutes 1 hour
Tetrachloroethylene	5 mg/m^3	24 hours
Toluene	8 mg/m^3	24 hours
Trichloroethylene	1 mg/m^3	24 hours
Vanadium	1 μg/m^3	24 hours

§ Exposure at these concentrations should be for no longer than the indicated times and should not be repeated within 8 hours.

§§ The value is given only for indoor pollution.

For certain compounds, such as *organic solvents*, the proposed health-related guidelines are order of magnitude higher than current ambient levels. The fact that existing environmental levels, for some substances, are much lower than the guideline levels by no means implies that pollutant burdens may be increased up to the guideline values. Any level of air pollution is a matter of concern, and the existence of guideline values never means a license to pollute the environment.

For substances with *malodorous properties*, at concentration below those where toxic effects occur, the guideline values were established to protect the public from odor nuisance.

TABLE 4.15 shows the rationale and guideline values based on sensory effects or annoyance reactions, using an averaging time of 30 minutes (WHO[64], 1987).

The aspects and respective levels considered by WHO in the *evaluation of sensory effects* are:

- the *intensity* (where the detection threshold level is defined as the lower limit of the perceived intensity range)
- the *quality* (recognition threshold level); and
- the *acceptability and annoyance*, where the nuisance threshold level is defined as the concentration at which less than 5% of the population experience annoyance for less than 2% of the time.

TABLE 4.15. Rationale and guideline values based on sensory effects or annoyance reactions, using an averaging time of 30 minutes

Substance	**Detection Threshold**	**Recognition Threshold**	**Guideline Value**
Carbon disulfide in viscose emissions	-	-	20 $\mu g/m^3$
Hydrogen sulfide	0.2 - 2.0 $\mu g/m^3$	0.6 - 6.0 $\mu g/m^3$	7 $\mu g/m^3$
Styrene	70 $\mu g/m^3$	210 - 280 $\mu g/m^3$	70 $\mu g/m^3$
Tetrachlorethylene	8 mg/m^3	24 - 32 $\mu g/m^3$	8 mg/m^3
Toluene	1 mg/m^3	10 mg/m^3	1 mg/m^3

Guideline Values Based on Carcinogenic Effects

In establishing criteria upon which guidelines could be based, it became apparent that carcinogens and noncarcinogens would require different approaches. Carcinogenic risk estimates were made by WHO for substances which, according to the International Agency for Research on Cancer (IARC), are considered proven human carcinogens or probable human carcinogens with at least limited evidence of carcinogenicity in humans and sufficient evidence for carcinogenicity In animals. In these guidelines the risk associated with lifetime exposure to a certain concentration of a carcinogen in the air has generally been estimated by linear extrapolation, assuming *no-threshold dose* (i.e., there is no threshold for effects, which means that there is no safe level). Therefore, risk managers are faced with two decisions: either to prohibit a chemical or to regulate it at levels that result in an acceptable degree of risk. Indicative figures for risk and exposure assist the risk manager to reach the latter decision (TABLE 4.16).

The *carcinogenic potency* is expressed as the incremental unit risk estimate, defined as *"the additional lifetime cancer risk occurring in a hypothetical population in which all individuals are exposed continuously from birth throughout their lifetimes to a concentration of 1 $\mu g/m^3$ of the agent in the air they breathe"* (WHO[64], 1987; US EPA[65], 1985).

Air Quality Guideline Values or Carcinogenic Risk Estimates for Organic and Inorganic Substances
Compliance with proposed guideline values does not guarantee the absolute exclusion of effects at lower levels owing to the existence of highly sensitive groups, especially those impaired by concurrent diseases or other physiological limitations.

Separate consideration is given to risk estimates for *asbestos* (TABLE 4.17) and *radon* daughters (TABLE 4.18).

Formaldehyde represents a chemical for which cancer bioassays in rats have resulted in nonlinear exposure response curves. In view of the nonlinearity of the tumor incidence with exposure concentration, Starr and Buck[66] (1984) introduced the concept of *"delivered dose"* (i.e., amount of formaldehyde covalently bound to respiratory mucosal DNA). The latter represents a measure of the exposure into several low-dose extrapolation models. Results showed considerable differences in the ratio between risk estimates based on administered dose and those based on the delivered dose, with a great variance of ratios between models.

It is important to note that quantitative risk estimates may give an impression of accuracy which in fact they do not have. An excess of cancer in a population is a biologic effect and not a mathematical function, and uncertainties of risk estimation are caused not only by inadequate exposure data but also, for instance, by the fact that specific metabolic properties of agents are not reflected in the models.

TABLE 4.16. Carcinogenic risk estimates based on human studies

Substance	Unit Risk[a]
Acrylonitrile	2×10^{-5}
Arsenic	4×10^{-3}
Benzene [a]	4×10^{-6}
Chromium (VI)	4×10^{-2}
Nickel	4×10^{-4}
Polynuclear Aromatic Hydrocarbons (PAH) [b]	9×10^{-2}
Vinyl Chloride	1×10^{-6}

a) *Cancer risk estimates for lifetime exposure to a concentration of 1 μg/m³.*
b) *Expressed as benzo-[α]-pyrene.*

TABLE 4.17. Risk estimates for asbestos

Concentration	Range of Lifetime Risk Estimates
500 Fibres/m³ (0.0005 Fibres/ml) *Fibres are measured by optical method*	10^{-6} - 10^{-5} (Lung cancer in a population where 30% are smokers) 10^{-5} - 10^{-4} (Mesothelioma)

TABLE 4.18. Risk estimate and recommended action level for radon daughters

Exposure	Lung cancer excess lifetime risk estimate	Recommended level for remedial action in buildings
1 Bq/m^3 EER	(0.7×10^{-4}) - (2.1×10^{-4})	≥ 1000 Bq/m^3 EER (annual average)

Guidelines Based on Ecological Effects on Terrestrial Vegetation

TABLE 4.19 shows WHO guideline values for individual substances based on effects on terrestrial vegetation which occur at concentrations below those known to be harmful to humans. It is worth mentioning that the pollutants selected in WHO guidelines (SO_x, NO_x, and ozone/photochemical oxidants) are only a few of the pollutants that are harmful to the ecosystem and that only the effects to the vegetation are considered here.

TABLE 4.19. Guideline values for individual substances based on effects on terrestrial vegetation

Substance	Guideline Value	Averaging Time	Remarks
Nitrogen Dioxide	95 $\mu g/m^3$ 30 $\mu g/m^3$	4 hr 1 year	In the presence of SO_2 and SO_3 levels which are not higher than 30 μm^3 (arithmetic annual average) and 60 μm^3 (average during growing season), respectively
Total Nitrogen Deposition	3 g/m^3	1 year	Sensitive ecosystems are endangered above this level
Sulfur Dioxide	30 $\mu g/m^3$ 100 $\mu g/m^3$		Insufficient protection in the case of extreme climatic and topographic conditions
Ozone	200 $\mu g/m^3$ 65 $\mu g/m^3$ 60 $\mu g/m^3$	1 hr 24 hr averaged over growing season	
Peroxy-acetyl-nitrate	300 $\mu g/m^3$ 80 $\mu g/m^3$	1 hr 8 hr	

WHO guidelines are for individual chemicals (except that for sulfur dioxide and suspended particulates). Pollutant mixtures can yield differing toxicities, but data for synergetic effects are at present insufficient for establishing guidelines.

4.6.2. GUIDELINES FOR WATER QUALITY

TABLE 4.20 shows *WHO guideline values for various substances or contaminants in drinking water* (WHO[63], 1984). WHO states that, if properly implemented, the guidelines will ensure the safety of drinking-water supplies.

> ***Remarks on the Use of Guideline Values***
> *- A guideline value represents the concentrations of a constituent that does not result in any significant risk to the health of the consumer over a lifetime of consumption.*
> *- Although the guideline values describe a quality of water that is acceptable for lifelong consumption, the establishment of these guidelines should not be regarded as implying that the quality of drinking-water may be degraded to that recommended level. In this context, the specified guideline values have been derived to safeguard health on the basis of lifetime exposure. Short term exposure to higher contaminant levels that might occur following an accident may be tolerated, but should be examined on a case-by-case basis.*
> *- When a guideline value is exceeded, this should be a signal: (i) to investigate the cause with a view to taking remedial action; (ii) to consult with, and seek advice from the authority responsible for public health.*
> *- In the case of radioactive substances, screening values for gross alpha and gross beta activity are given, based on a reference level dose.*

In developing the guideline values, WHO took into consideration the total intake of each contaminant from air, food and water. For the majority of the substance evaluated, the toxic effect in man is predicted from studies with animals. Furthermore, because of the uncertainties in applying animal data to humans, and because of the doubts about the reliability of extrapolating from high doses to low doses, arbitrary safety factors ranging from 100 to 1000 were applied.

Microbial Aspects
The use of water contaminated by human and animal excreta, particularly faeces, can transmit infectious diseases. These can be caused by pathogenic bacteria, viruses, and protozoa, or by parasites, which are the most common and widespread health risk associated with drinking water.

The human pathogens that can be transmitted orally by drinking-water and that present a serious risk of disease include bacteria, viruses, pathogen, and helminths:

- *Bacteria*: Salmonella spp.; Shigella spp.; pathogenic Escherichia coli; Vibrio cholerae; Yersinia enterocolitica; Campylobacter jejuni; and Campylobacter coli.
- *Viruses*: Adenoviruses; Enteroviruses; Hepatitis A; Enterically transmitted non-A, non-B hepatitis viruses, and hepatitis E; Norwalk virus; Rotavirus; Small round viruses, and
- *Parasites*: Giardia spp; Cryptosporidium spp; Entamoeba hystolica, and Dracunculus medinensis.

Most of these pathogens are distributed worldwide. Besides these pathogens, there are a number of diverse organisms that have no public health significance, but which are undesirable because they produce turbidity, taste and odor, or because they appear as visible animal life in water.

Chemical Aspects

The contribution of drinking-water to daily exposure includes direct ingestion as well as some indirect routes, such as inhalation of volatile substances and dermal contact during bathing or showering.

There are two principal sources of information on health effects resulting from exposure to chemicals that can be used in deriving guideline values. The first is studies on human populations, and the second is toxicity studies using laboratory animals. Furthermore there is a need to extrapolate the result of laboratory animal studies (made at relatively high doses) to the low doses to which human populations are usually exposed. It is then necessary to select the most suitable experimental animal study on which to base the extrapolation. Data from well-conducted studies, where a clear dose-response relationship has been demonstrated, are preferred.

The tolerable daily (TDI) intake is an estimate of the amount of a substance in food or drinking water, expressed on a body weight basis (mg/kg or μg/kg of body weight), that can be ingested daily over a lifetime without appreciable health risk.

The chemical contaminants of drinking-water supplies can be broadly classified into the following categories:

Inorganic Constituents. These include aluminum, ammonia, antimony, arsenic, asbestos, barium, beryllium, boron, cadmium, chlorides, chromium, copper, cyanides, fluorides, hydrogen sulfide, iron, lead, manganese, mercury, molybdenum, nickel, nitrate and nitrite, selenium, silver, sodium, sulfate, inorganic tin, uranium, zinc.

Further the *total dissolved solids* (TDS) comprise inorganic salts (principally calcium, magnesium, potassium, sodium, bicarbonates, chlorides and sulfates) and small amounts of organic matter that are dissolved in water.

Hardness in water is caused by dissolved calcium and, to lesser extent, magnesium. It is usually expressed as the equivalent quantity of calcium carbonate. Depending on pH and alkalinity, hardness of above 200 mg/liter can result in scale deposition, particularly on heating.

Organic Constituents. These include *chlorinated alkanes* (carbon tetrachloride, dichloromethane, dichloroethane, trichloroethane, etc.); *chlorinated ethenes* (vinyl chloride, dichloroethene, trichloroethene, tetrachloroethene); *aromatic hydrocarbons* (benzene, toluene, xylenes, ethylbenzene, styrene, polynuclear aromatic hydrocarbons); *chlorinated benzenes*; miscellaneous *organic constituents*; *pesticides*; *disinfectants* and by-products (chlorine, chloramines, etc.).

TABLE 4.20. WHO Drinking-water quality guidelines for inorganic and organic contaminants of health significance

Contaminant	Guideline Value
Aldrin and Dieldrin	0.03 μg/l
Arsenic	0.05 mg/l
Benzene	10 μg/l [a]
Benzo[a]pyrene	0.01 μg/l [a]
Cadmium	0.005 mg/l
Carbon Tetrachloride	3 μg/l [a]
Chlordane	0.3 μg/l
Chloroform	30 μg/l [a]
Chromium	0.05 mg/l
Cyanide	0.1 mg/l
2,4-D	100 μg/l [b]
DDT	1 μg/l
1,2-Dichloroethane	10 μg/l [a]
1.1-Dichloroethene	0.3 μg/l [a]
Fluoride [c]	1.5 mg/l
Gamma-HCH (lindane)	3 μg/l
Gross alpha activity	0.1 Bq/l
Gross beta activity	1 Bq/l
Heptachlor & heptachlor epoxide	0.1 μg/l
Hexachlorobenzene	0.01 μg/l [a]
Lead	0.05 mg/l
Mercury	0.001 mg/l
Methoxychlor	30 μg/l
Nitrate	10 mg/l (N)
Pentachlorophenol	10 μg/l
Selenium	0.01 mg/l
Tetrachoroethene	10 μg/l [a]
Trichloroethene	30 μg/l [a]
2,4,6-Trichlorophenol	10 μg/l [a, b]

a) *These guideline values were computed from a conservative hypothetical mathematical model which cannot be experimentally verified and values should therefore be interpreted differently. Uncertainties involved may amount to two orders of magnitude (i.e., from 0.1 to 10 times the number).*

b) *May be detectable by taste and odor at lower concentrations.*

c) *Local or climatic conditions may necessitate adaptation.*

Carcinogenic Aspects

The evaluation of the potential carcinogenicity of chemical substances is usually based on long-term animal studies. Sometimes data are available on carcinogenicity on humans, mostly from occupational exposure. A classification of substances according to their potential carcinogenic risk has been made by IARC. The following groups are defined as follows:

Group 1 : the agent is carcinogen to humans
Group 2A : the agent is probably carcinogen to humans
Group 3 : the agent is not classifiable as to its carcinogenicity to humans

Group 4 : the agent is probably not carcinogen to humans.

The actual methods of extrapolating data from animal to man deal with exposures to single substances; therefore, effects from exposure to mixtures are not considered. Guideline values are also proposed for carcinogenic substances, taking into account appropriately conservative risk factors.

Radiological Risk Assessment for Drinking water

Environmental radiation originates from a number of naturally occurring and man-made sources. It has been estimated that exposure to natural sources contributes to more than 98% of the radiation dose to the population (excluding medical exposure). there is only a very small contribution from nuclear power production and nuclear power weapons testing. The global average human exposure from natural sources is 2.4 mSv/year. there are large local variations in this exposure depending on a number of factors, such as height above sea level, amount and type of radionuclides in the soil, and the amount taken into the body in air, food, and water. The contribution of drinking-water to the total exposure is very small, and is due largely to naturally occurring radionuclides in the uranium and thorium decay series.

> ***Guide Line Values for Radioactive Substances***
> *In the case of radioactive substances, the term guideline value is used in the sense of "reference level" as defined by ICRP. The values shown in TABLE 4.21 for gross alpha and gross beta activity are to be referred to all sources of radioactivity, natural and man-made. These values were calculated so that the associated dose corresponds, according to ICRP data, to a total risk in the range 10^{-7}-10^{-6} fatalities per year (WHO, 1984).*

For practical purposes, the recommended guideline activity concentrations are 0.1 Bq/liter for gross alpha and 1 Bq/liter for gross beta activity. The *"reference level" of dose* needs to be expressed as an activity concentration of radionuclides in drinking water. Metabolic and dosimetric considerations have been included in the development of dose conversion factors, expressed as Sieverts per Becquerel, which relate a dose expressed in Siverts to the quantity (in Becquerels) of radioactive material ingested.

Examples of the doses arising from 1 year's consumption of drinking water is presented in TABLE 4.21. Examples of radionuclide concentrations (reference concentrations) corresponding to the reference level of dose, 0.1 mSv/year, are given in TABLE 4.22.

The method proposed for assessing the *radiological risk of drinking-water* is shown in *Figure 4.1*. If either the gross alpha activity concentration of 0.1 Bq/liter or the gross beta activity concentration of 1 Bq/liter is exceeded, then the specific radionuclides should be identified and their individual activity concentrations measured. From these data, a dose estimate for each radionuclide should be made and the sum of these doses determined. Where the following formula is satisfied, no further action is required:

$$\sum_i \frac{C_i}{RC_i} \leq 1 \qquad (4.8)$$

where C_i is the measured activity concentration of radionuclide i and RC_i is the reference activity concentration of radionuclide i that, at an intake of 2 liters per day for 1 year, will result in a committed effective dose of 0.1 mSv (TABLE 4.22).

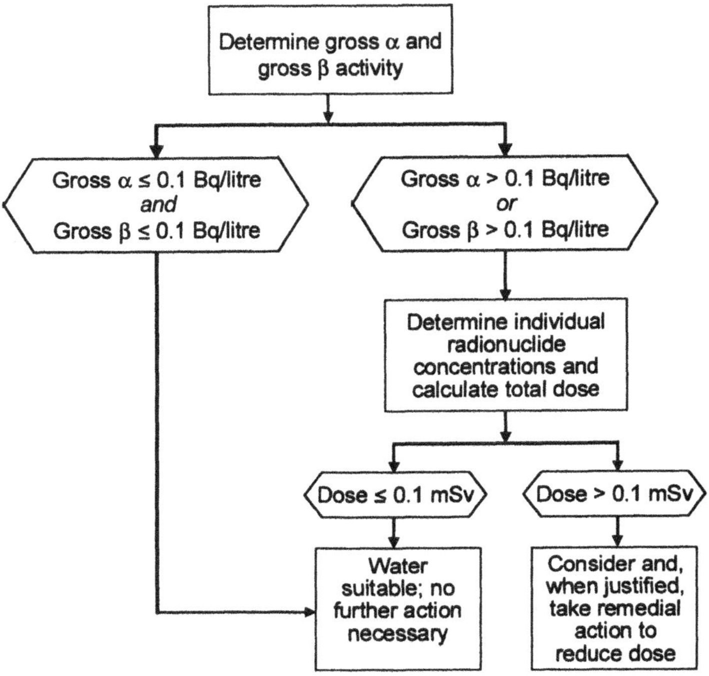

Figure 4.4. Application of recommendations on radionuclides in drinking-water based on an annual reference level of dose of 0.1 mSv (Source: WHO[63], 1993).

If alpha-emitting radionuclides with high dose conversion factors are suspected, this additive formula may also be invoked when the gross alpha and gross beta activity screening values of 0.1 Bq/liter are approached. Where the sum exceeds unity for a single sample, the reference level of dose of 0.1 mSv would be exceeded only if the exposure to the same measured concentrations were to continue for a full year. Such a sample dose result does not mean that the water is unsuitable for consumption and should be regarded only as a level at which further investigation (including additional sampling) is needed.

There are difficulties in applying the reference level of dose to derive activity concentrations of radon (Rn^{222}) in drinking-water, because radon is easily released from water during handling. The global average dose from inhalation of radon from all sources (soil, water, air) is about 1 mSv/year, which is roughly half of the total natural radiation exposure. In comparison the global dose from ingestion of radon in drinking-water is relatively low.

TABLE 4.21. Examples of the doses arising from 1 year's consumption of drinking water

Radionuclide	Dose (mSv)
Alpha Emitters (0.1 Bq/liter)	
Po^{210}	0.045
Ra^{224}	0.006
Ra^{226}	0.016
Th^{232}	0.130
U^{234}	0.003
U^{238}	0.003
Pu^{239}	0.040
Beta Emitters (1 Bq/liter)	
Co^{60}	0.005
Sr^{89}	0.003
Sr^{90}	0.020
I^{129}	0.080
I^{131}	0.016
Cs^{134}	0.014
Cs^{137}	0.009
Pb^{210}	0.950
Ra^{228}	0.200

Note (1): It is assumed that the water is containing any of the given alpha-emitting radionuclides at an activity concentration of 0.1 Bq/liter, or of the given beta-emitting radionuclides at an activity concentration of 1 Bq/liter.

Note (2): Appropriate dose conversion factors were taken from National Radiological Protection Board, Committed equivalent organ doses and committed effective doses from intakes of radionuclides, (Chilton, Didcot[67], 1991).

TABLE 4.22. Activity concentration of various radionuclides in drinking water corresponding to a dose of 0.1 mSv from 1 year's intake

Radionuclide[a]	**Dose Conversion Factor** (Sv/Bq)[b]	**Accumulated Dose per Year** Calculated rounded value (Bq/liter)
H^3	$1.8x10^{-11}$	7800
C^{14}	$5.6x10^{-10}$	250
Co^{60}	$7.2x10^{-9}$	20
Sr^{89}	$3.8x10^{-9}$	37
Sr^{90}	$2.8x10^{-8}$	5
I^{129}	$1.1x10^{-7}$	1
I^{131}	$2.2x10^{-8}$	6
Cs^{134}	$1.9x10^{-8}$	7
Cs^{137}	$1.3x10^{-8}$	10
Pb^{210}	$1.3x10^{-6}$	0.1
Po^{210}	$6.2x10^{-7}$	0.2
Ra^{224}	$8.0x10^{-8}$	2
Ra^{226}	$2.2x10^{-7}$	1
Ra^{228}	$2.7x10^{-7}$	1
Th^{232}	$1.8x10^{-6}$	0.1
U^{234}	$3.9x10^{-8}$	4
U^{238}	$3.6x10^{-8}$	4
Pu^{239}	$5.6x10^{-7}$	0.3

[a] *It should be not necessarily assumed that the reference level of dose has been exceeded simply because the gross beta activity concentration approaches or exceeds 1 Bq/liter. This may results due to the presence of a naturally occurring radionuclide* K^{40}.

4.6.3. HEALTH EFFECTS OF VARIOUS POLLUTANTS

It is useful to include short information on *the health effects of exposure to various concentrations in air of some pollutants.* Guideline values for combined exposure to sulfur dioxide and particulate matter are indicated in UNEP[68] (1987).

TABLE 4.23 - TABLE 4.28 concern health effects of sulfur dioxide and sulfate, sulfur dioxide and particulate matter, nitrogen oxides, carbon monoxide and ozone.

TABLE 4.29 - TABLE 4.30 - (OECD[69], 1991) show the major health effects of selected toxic trace air pollutants, (i.e., organic compounds and metals), without specifying the dose absorbed and the exposure time.

Risk estimate figures should not be applied in isolation when regulatory decisions are being made, but they should be combined with data on exposure levels and individuals exposed. In this way they may be a useful contribution to risk assessment. The decision on the acceptability of a certain health risk should anyway be taken by the national authorities in the context of a broader risk management process. Risk assessment can be used together with technological, economic and other considerations in the risk management process.

TABLE 4.23. Guidelines values for combined exposure to sulfur dioxide and particulate matter [a]
Source: Air Quality Guidelines for Europe, WHO, European Series No. 23.

				Gravimetric Assessment	
	Averaging time	Sulfur dioxide (μg/m^3)	Reflectance assessment black smoke (μg/m^3)[b]	Total suspended particulates (TSP) (μg/m^3)[c]	Thoracic particles (TP) (μg/m^3)[d]
Short Term	24 hours	125	125	120	70
Long Term	1 year	50	50	-	-

[a] *No direct comparison can be made between values for particulate matter in the right and left hand sections of this table, since both the health indicators and the measurement methods differ*
[b] *Nominal μg/m^3 units, assessed by reflectance. Application of the black smoke value is recommended only in areas where coal smoke from domestic fires is the dominant component of the particulates. It does not necessary apply where diesel smoke is an important contributor*
[c] *TSP: measurement by high volume sampler, without any size selection*
[d] *TP: equivalent values as for a sampler with ISO-TP characteristics (having 50% cut-off point at 10μm); estimated from TSP values using site-specific TSP/ISO-TP ratios.*
[e] *Values to be regarded as tentative at this stage, being based on single study*

TABLE 4.24. Health effects of sulfur dioxide and sulfate

Concentration SO_2 (μg/m^3)	**Concentration Sulfates** (μg/m^3)	**Averaging Time**	**Health Effect**
300-400	NA	24 hours	Increased mortality
365	8-10	24 hours	Aggravation of symptoms in elderly
180-250	6-10	24 hours	Aggravation of asthma
220	11	Annual mean	Decreased lung function in children
90-100	9	Annual mean	Increased acute lower respiratory disease in families
95	14	Annual mean	Increased prevalence of chronic bronchitis
106	15	Annual mean	Increased acute respiratory disease in families
NA	13	Annual mean	Increased respiratory disease-related illness, absence in female workers

Source: U.S. Environmental Protection Agency, 1974.

TABLE 4.25. Synergetic effects of sulfur and suspended particulate matter

SO_2-Concentration (μg/m^3)	**Concentration SPM** (μg/m^3)	**Averaging Time**	**Adverse Effect**
500	500	daily average	Excess mortality and hospital admissions
500-250	250	daily average	Deterioration of patients with pulmonary disease
100	100	mean	Annual arithmetic Respiratory symptoms
80	80	annual geometric mean	Visibility and/or human annoyance effects

Source: "Health Hazards of the Human Environment", WHO, Geneva, 1973.

TABLE 4.26. Health effects of carbon monoxide

CO-Concentration (μg/m^3)	**Health Effects**
0 - 4,250	No symptoms
4,250 - 12,750	Slight headache, tightness across forehead, shortness of breath with vigorous exertion, dilation of cutaneous blood vessels
12,750 - 21,250	Slight to moderate headache and throbbing in temples, shortness of breath with moderate exertion
21,250 - 34,000	Decided to severe headache, weakness, dizziness, dimness of vision, nausea, vomiting and collapse, irritable judgment disturbed
34,00 - 51,000	Same as previous item with more possibility of collapse and syncope, especially with exertion and increased respiration and pulse; slight confusion
51,000 - 76,500	Fainting, increased respiration and pulse, coma with intermittent convulsions and irregular respiration
76,500 - 119,000	Coma with intermittent convulsions, depressed hearing action and respiration and possible death
119,000 - 204,000	Weak pulse and slow respiration, respiratory failure and death
204,000	Rapidly fatal
204,000	Immediately fatal

Source: 'Medical Aspects of Air Pollution'. Society of Automotive Engineers Inc. Michigan, 1971.

TABLE 4.27. Health effects of nitrogen oxides

Average NO_2 Concentration ($\mu g/m^3$)	**Health Effects**	**Population Studied**
300 - 1130	Decrease in several measures of pulmonary function as compared to controls	Japanese railroad workers
150 - 280	Borderline decrease in lung function test	Chattanooga school-children, aged 7-8
100 vs. 80	No difference in various measures of pulmonary function	Central City vs. suburban policemen in Boston
96 vs. 43	No difference in various measures of pulmonary function	Seventh Day Adventists in Los Angeles San Diego
20 - 70	Twofold excess in acute respiratory disease compared to unexposed group	Czechoslovakian children aged 7-12
10	Excess in acute respiratory disease ranging from 11 to 27%	USSR adolescents in chemical and fertilizer plants
580 - 1120	Forty-four per cent increase in physician visits for respiratory, visual, nervous system and skin problems	Individuals living within 1 km of a USSR chemical plant
150 - 280	Excess in acute respiratory disease 1-17% in children, 9-33% in adults	Families in Chattanooga, Tennessee
150-280	Infants exhibited 10-58% excess of acute bronchitis; children 6-9 years showed 39-71% excess of acute bronchitis	Infants and children 6-9 in Chattanooga

Source: U.S. Environmental Protection Agency, 1976

TABLE 4.28. Relationship of ozone and photochemical oxidant exposure to human health effects and recommended alert and warning system levels

Recommended Episode Levels	Duration of Exposure	Health Effects
Emergency	2.0hr.	Soreness of upper respiratory tract; tendency to cough while taking deep breaths; significant increase in breathing difficulty. These conditions were made worse by 15 minute of light exercise.
	2.75 hr.	Measurable biochemical changes in blood sera enzyme levels and in red blood cell membrane integrity; some subjects become physically ill and unable to perform normal jobs for several hours
	2.0 hr.	Impairment of pulmonary function in young adults probably due to a decreased lung elastic recoil, increased airway resistance and small airway obstruction
	2.75 hr.	Significant biochemical changes in blood sera enzyme levels and in red blood cell membrane integrity, but less severe than at 0.50 ppm; some subjects become physically ill and unable to perform normal jobs for several hours.
Red alert	-	Precipitous increase in rates of cough and chest discomfort in young adults.
	-	Greater number of asthma attacks in patients on days when daily maxima equaled or exceeded 0.25 ppm during a 14-week period.
	2.75hr.	Biochemical changes in blood sera enzyme levels.
Yellow alert	1 hr.	Breathing impaired
	-	Tokyo elementary school children had significantly reduced respiratory function associated with ozone levels less than 0.1 ppm during a long-term epidemiological study. Beginning of headache without fever in young adults; median age 18.6 years.
Watch	2 hr (average)	Impairment of performances of student athletes during running competition
	15-30 min.	Threshold of respiratory irritation
	-	Decreased electrical activity of the brain.

Source: E.J. Calabrese. 'Pollutants and High-Risk Groups'. Wiley-Interscience, New York, 1978.

TABLE 4.29. Health effects of selected toxic trace air pollutants (organic compounds)

Pollutant	Major Health Effects
Acryronitrile (CH_2-CH-C-N)	Dermatitis; hematological changes; headaches; irritation of eyes, nose and throat; lung cancer
Benzene (C_6H_6)	Leukemia; neurotoxic symptoms; bone marrow injury incl. anemia, chromosome aberrations
Carbon disulfide (CS_2)	Neurologic and psychiatric symptoms, incl. irritability and anger; gastro-intestinal troubles; sexual interferences
1,2 Dichloroethane ($C_2H_2Cl_2$)	Damage to lungs, liver and kidneys; heart rhythm disturbances; effects on central nervous systems, incl. dizziness; animal mutagen and carcinogen
Formaldehyde (HCHO)	Chromosome aberrations; irritation of eyes, nose and throat; dermatitis; respiratory tract infections in children
Methylane chloride (CH_2Cl_2)	Nervous system disturbances
Polychlorinated bi-phenyls (PCB) (coplanar)	Spontaneous abortions; congenital birth defects; bioaccumulation in food chains
Polychlorinated dibenzo-dioxins and furans	Birth defects; skin disorders; liver damage; suppression of the immune system
Polycyclic Organic Matter [incl.benzo-(α) -pyrene (BaP)]	Respiratory tract and lung cancers; skin cancers
Styrene (C_6H_5-CH-CH_2)	Central nervous system depression; respiratory tract irritations; chromosome aberrations; cancers in the lymphatic and hematopoietic tissues
Tetrachloroethylene (C_2Cl_4)	Kidney and genital cancers; lymphosarcoma; lung, cervical and skin cancers; liver dysfunction; effects on central nervous system
Toluene (C_6H_5-CH_3)	Dysfunction of the central nervous system; eye irritation
Trichloroethylene (C_2HCl_3)	Impairment of psychomotoric functions; skin and eye irritation; injury to liver and kidneys; urinary tract tumors and lymphomas
Vinyl chloride (CH_2-CHCl)	Painful vasospastic disorders of the hands; dizziness and loss of consciousness; increased risk of malformations, particularly of the central nervous systems; severe liver disease; liver cancer; cancers of the brain and central nervous system; malignancies of the lymphatic and hematopoietic system

Source: OECD

TABLE 4.30. Health effects of selected toxic trace air pollutants (metals)

Pollutant	Major Health Effect
Arsenic (As)	Lung cancer; dermatological disorders, incl. ulcerative dermatitis; hematological effects, incl. anemia
Beryllium (Be)	Dermatitis; ulcers; inflammation of mucous membranes
Cadmium (Cd)	Acute and chronic respiratory disease; renal dysfunction; animal carcinogen
Chromium (Cr)	Lung cancer; gastro-intestinal cancers; dermatitis
Lead (Pb)	Interference with blood forming processes; liver and kidney damage; neurological effects
Mercury (Hg)	Effects on nervous system, incl. deficits in short-term memory, disturbance of sensory and coordination functions; kidney failure
Nickel (Ni)	Respiratory illnesses, incl. asthma; impairment of respiratory defense system; birth defects and malformations; nasal and lung cancers
Thallium (TI)	Bioaccumulation; toxic to plants and animals
Vanadium (V)	Respiratory irritation; asthma; nervous disturbances; changes in the blood formula

Source: OECD[70] (1990)

4.6.4. CHARACTERIZING UNCERTAINTY IN INTEGRATED ENVIRONMENTAL MODELS.

The use of integrated computer models for environmental research, management, policy, analysis and regulatory purpose is becoming widespread.[70] The characterization of uncertainty in model predictions also has become an important dimension of model applications. The development of integrated environmental models typically arises from a need for policy guidance and/or regulatory decisions. In such applications, relationships are sought between the magnitude of a potential environmental hazard or risk and the levels of pollutant emissions subject to regulation or control.

Morgan[71] *et al.,* (1989) present an *excellent review of the literature on uncertainty.* Uncertainties in model inputs often exist in the empirical quantities of a given model, and about the actual model form or structure. A common quantitative representation of uncertainty is to specify a range of values for a particular quantity (e.g., low, nominal, high value). For empirical parameters, a probability distribution also may be specified. This could either be a discrete distribution over alternative values (e.g., an x percent chance of value v_1 and a y percent chance of value v_2, etc.), or a continuous distribution specified either partially (e.g., by a mean and a standard deviation, or by a confidence interval), or completely (e.g., as a uniform, normal, or other standard parametric distribution, or as points on an arbitrary density function). Non-probabilistic representation of uncertainty, such as fuzzy set membership function over alternative values, also may be specified, though this method is not commonly employed in environmental modeling.

The *uncertainties in model outputs*, where quantitative measures are specified, may be obtained in a number of ways, using for instance scenario analysis (e.g., by choosing single values for each uncertainty when running the scenario model). Due to a combinatorial explosion, a subset of interesting values are usually chosen. Another approach is based on first order (Gaussian) error propagation in which the derivative of an output with respect to each uncertain input is used, together with the uncertainty variance for each input, to evaluate the total output uncertainty and the contribution from each input. This approximation is only to be used if the output is roughly linearly dependent on the input over the range of uncertainty.

Two other methods for propagating the uncertainty employ probabilistic methods:

- *Decision tree:* here, a combinatorial scenario analysis is performed to compute the outcome value for all combinations of decisions and uncertain variables, and
- *Stochastic or probabilistic simulation:* here, the uncertainty induced in each model output is estimated from either continuous or discrete probability distribution specified for model inputs, using a Monte Carlo simulation. A large number of scenarios is generated by sampling from the probability distribution for each model output. Latin Hypercube sampling can be employed to minimize computational efforts.

Uncertainties in model results can be expressed by either a probability density function (PDF) or cumulative distribution function (CDF) for a quantity (e.g., pollutant) in a given year. Alternatively, the probability distributions for a particular scenario can be used to plot trend lines and specified confidence intervals. Another useful way of displaying the consequences of a given scenario is to examine the net change in some environmental effect over the simulation period. A distribution function also can be developed for the net difference between two scenarios in a given future year, when all model parameters except emissions are sampled probabilistically in an identical manner.

Integrated Models of Acid Deposition. Integrated models for assessing acid rain impacts have been proposed in the literature:

ACIDRAIN[72]: Cambridge, Decision Analysis (UK). Emissions and UK cost control. Atmospheric deposition; damage functions for lake, fish, forests, crops, buildings, monuments, human health; unit damage prices. Uncertainty Methods: scenario analysis; probabilistic simulation; alternate model forms. Level I integrated model.

ADAM: Carnegie Mellon University (USA). Atmospheric deposition assessment model. Emissions; control costs; atmospheric deposition; precipitation; pH,; lake acidity; fish viability; plus user-specified models for other effects or damage valuation. Uncertainty Methods: Scenario analysis; probabilistic simulation; alternative models. Level II integrated model.

ADEPT: Decision Focus Inc., (USA). Acid deposition tree. Emissions; control costs; atmospheric deposition; precipitation; pH; lake impacts; forests impacts; other impacts;

lake liming; effects valuation; research costs. Uncertainty Methods: Scenario analysis; decision trees; alternate model forms. Level I integrated model.

BICRAM: Beijer Institute/University of York (Sweden/UK). Resource assessment and management. Energy use; emissions; control costs; atmospheric deposition; receptor sensitivity classes. Uncertainty Methods: scenario analysis. Level II integrated model.

RAINS: International Inst. for Applied Systems (Austria). Regional acidification information and simulation. Emissions; control costs; atmospheric deposition; pH; lake acidity; groundwater sensitivity; direct forest impacts. Uncertainty Methods: scenario analysis. Level II integrated model.

A major U.S. effort also is in progress to link some 20 to 30 different environmental models for a major assessment of acid rain deposition in 1990 (NAPAP[73], 1989).

The concepts of critical loads or target loads have emerged as a basis for designing emissions reduction strategies which achieve deposition levels judged to be protective or adverse environmental effects in a given region. From the point of view of integrated modeling, this approach simplifies the problem since effects and damage evaluation models are not explicitly required. The principal objective of integrated modeling is thus to identify efficient cost-effective strategies for achieving deposition targets.

Air pollution information system tools for desk top air pollution management on a local scale could be associated with a structure for data acquisition and handling form decentralized databases, as given in TABLE 4.31.

TABLE 4.31. Structure of a possible air pollution information system

Scenario Menu	**Dispersion Menu**	**Report Menu**	**Mutation Menu**
Scenarios Definition	Dispersion Calculation	Report Generation	Data Management
Scenario Type	Model Type	Report Type	- Data Type
Substances	Substances	Substances	- Background - Meteorological information
Emissions - type - area - factors	*Sources* - type - area Apply factors *Calculate* - concentrations - doses	*Emissions* - type - area Add up emissions	- Data to add - Data to remove - Data to change - Update Files - Preparation for batch processing
Compare scenarios	Produce standard tables	Produce tables isopleths	Show mutations

Different types of reports must be predefined, varying according to the type of data and type of aggregation. Algorithms should be included to calculate the deposition/washout as product of concentration and deposition velocity, both depending on atmospheric conditions, and a source depletion type model based on losses during transportation. Different predefined types of scenarios are allowed in which different properties are used to group the emissions, or the same properties play different roles, that is, appear at different levels in the hierarchy. The system might be coupled with dBase-III and CLIPPER (a dBase-like language), and for instance with a geographical information system GIS.

Some Additional Comments

This chapter is presenting different methods and ecological models used by government, industry and research institution in OECD Member countries to assess:

a) the sources of release of chemicals to the environment

b) the fate and pathways of chemicals in the environment.

In the past, contaminant assessment generally focused on only one environmental pathway, such as groundwater, surface water, or atmosphere. However, contaminants released into one medium eventually migrate through and interact with several media. Thus to more accurately portray the complexities of the interactions of contaminants and the environment, modern assessment methodologies must consider integrated multimedia transport and contaminant fate.

The models have been classified into :

a) Source assessment methods
b) Fate and Pathways assessment methods
c) Surface water methods
d) Soil (ground water methods)
e) Air methods
f) Multi-media methods
g) Miscellaneous methods

Some of these methods are used to assess the sources of release, or fate and pathways, of a large number or a broad class of chemicals. Other are designed to be more chemical-specific. Most of the methods presented here have been developed and/or used in the areas of chemicals control, air and water pollution control, or waste management.

Using energy simulation models and risk-cost assessment methodology for toxic pollutants, allows risk managers to compare the cost of implementing new technologies or taking alternative remedial or preventive measures to protect the environment, and to propose legal or technical changes to the competent authorities, that are based on financial cost estimates.

Multimedia exposure models are also presented, which are based on compartment or fugacity models. Some of the proposed methods are used for screening purpose and for preliminary exposure assessment calculation.

A further section is concerned with dose-response assessment, the evaluation and quantitative characterization of the relationship between level of exposure and health-related response. Further considerations are given to time-dependent dose-response relationships.

Finally guideline values for specified contaminants in air and water, which were recently published by WHO, are also presented. They can serve as a basis for comparing monitored pollution values with international standards and defining areas of activities where effort should be focused on.

Summary (Chapter 4)

This chapter includes models and assessment methods for environmental exposure and health impacts with regard to integrated regional risk assessment. Key issues in an ecological risk assessment program are summarized, and assessment methods are indicated and reviewed (e.g. matrix approach, thresholds, functions, simulation models, probabilistic models, expert judgment). Special attention is given to environmental exposure assessment methods for chemicals; source assessment methods, fate and pathways assessment methods, assessment methods for surface water and for soil (groundwater), assessment methods for air and for multi-media. Special attention is given to models and computer based tools for integrated regional risk assessment, complex industrial areas and energy systems, energy simulation models, models to assess human exposure to chemicals. The chapter includes methodology for i) estimating the risks from continuous emissions by using the dose-response relationship, and ii) estimating of risks from accidental gas releases with associated uncertainty analysis.

References (Chapter 4)

[1] Barnthouse, L.W., et al., Methodology for Environmental Risk Analysis, Oak Ridge National Laboratory, Oak Ridge, TN, USA, Report ORNL/TM-8167 (1982).

[2] Morgan, M.G.; Moris, S.C., Needed: A National R&D effort to develop individual air pollution monitor instrumentation,J. Air Pollution Control Association, 27 (1977) 670-673.

[3] OECD Environment Monographs, No. 27, Compendium of Environmental Exposure Assessment methods for Chemicals, OECD (Organisation for Economic Co-operation and Development), (July 1989).

[4] Hydroscience, Inc., Simplified mathematical modeling ofwater quality, Westwood, New Jersey, March (1971).

[5] Mackay, D.; Paterson, S., Fugacity revisited, Environmental Science and Technology, 16(2) (1982), pp. 654A-660A.

[6] Kypreos, S., Technical options for a stringent clean air concept in Switzerland, in J. Fenhann, H. Larsen, G.A. Mackensie and B. Rasmussen (eds.), Environmental Models: - Emissions and Consequences, Developments in Environmental Modelling 15, Riso International Conference 22-25 May, 1989, Elsevier, Amsterdam, (1990), pp. 203.

[7] Compilation of air pollutant emission factors. U.S. Environmental Protection Agency, Vol. II: Mobile Sources, Ann Arbor (1985).

[8] Berger, C.; Haurie, A.; Loulou, R., Modeling long range energy/technology choices: the MARKAL approach. GERAD working paper, Ecole des HEC, Montreal (1987).

[9] Bolten, J.G.; Morrison, P. F.; Solomon, K., Risk-Cost Assessment Methodology for Toxic Pollutants from Fossil Fuel Power Plants (R-2993-EPRI, June 1983).

[10] Bolten, J.G.; Morrison, P. F.; Solomon, K; Wolf, K., Alternative Models for Risk Assessment of Toxic Emissions, RAND / N-2261-EPRI, April 1985.

[11] Onishi, Y.; Serne, R.J.; Arnold, E.M., et al., Critical Review: Radionuclide Transport, Sediment Transport, and Water Quality Mathematical Modeling; and Radionuclide Adsorption/Desorption Mechanisms, Battelle Pacific Northwest Laboratories, January 1981.

[12] Anderson, M.P., Using models to simulate the movement of contaminants through groundwater flow systems, CRC Critical review in Environment Control, 9(2), (1979), pp. 97-156.

[13] McKone, T.E.; Layton, D.W., Screening the potential risks of toxic substances using a multimedia compartment model: "Estimation of human exposure", Regulatory Toxicol. Pharmacol., 6 (1986) 359-380.

[14] Whelan, G.; Strange, D.L.; Steelman, B.L.; Buck, J.W., The remedial action priority system (RAPS): "mathematical formulation", prepared for the U.S. Department of Energy, Office of Environment, Safety, Health, (1987).

[15] Buck, J. W.; Strenge, D. L.,; Droppo, J. G., Analysis of risk indicators and issues associated with applications of a screening model for hazardous and radioactive waste sites, Risk Assessment, National R+D Conference on the Control of Hazard. Materials, Feb. 1991, Anaheim, California, Hazardous Materials Control Institute, (1991) pp. 98-102.

[16] Yoshida, K.; Shigeoka, T.; Yamauchi, F., Estimation of environmental fate of industrial chemicals, Toxicol. Environ. Chem., 17 (1988) 69-85.

[17] Asher, S.C.; Lloyd, K.M.; et al., A critical examination of environmental modelling - modelling the fate of chlorobenzenes using the persistence and fugacity models. NRCC's Associate Committee on Scientific Criteria for Environmental Quality. Report NRCC 23990.

[18] Klein, W.; Kördel, et al., Systematic approach for environmental hazard ranking of new chemicals, Chemosphere 7 (1988) 1445-1462.

[19] Environmental Partitioning Model Draft Documentation, Office of Toxic Substances, US EPA, (1982).

[20] Hetrick, D.M.; Mc Dowell-Boyer, L.M., User's Manual for TOX-SCREEN: A Multi-media Screening-level Program for Assessing the Potential Fate of Chemicals Released to the Environment, Oak Ridge National Laboratory, (1984), Report ORNL-6041; and USEPA, Washington DC, Report EPA-560/5-83-024.

[21] Bennett, B.G., The exposure commitment method in environmental pollutant assessment, Environ. Monit. and Assessment, 1 (1981) 21-36.

[22] Rosenblatt, D.H.; Dacre, J.C.; Cogley, D.R., An environmental fate model leading to preliminary pollutant limit values for human health effects, in R.J. Convay (ed.), Environmental Risk Analysis for Chemicals, Van Nostrand Reinhold Co., New York, (1980).

[23] Mackay, D.; Paterson, S., Evaluating the regional multimedia fate of organic chemicals: a level III fugacity model, Environ. Sci. Technology. In Press, (1990).

[24] Cohen, Y.; Ryan, P.A., Multimedia modelling of environmental transport: trichlorethylene test case, Environ. Sci. Technol. 19 (1985) 412-417.

[25] CCERP (1985). Risk Assessment and Risk Management of Toxic Substances. Committee to Coordinate Environmental and Related Programs, US Department of Health and Human Services, Washington DC, (1985).

[26] NTP (1984). Report of the NTP Ad Hoc Panel on Chemical Carcinogenesis Testing and Evaluation, Board of Scientific Counselors, National Toxicology Program, US Department of Health and Human Services, (1984).

[27] NAS (1987). Pharmacokinetics in Risk Assessment, Drinking Water and Health, Vol. 8 (1987). Subcommittee on Pharmacokinetics in Risk Assessment, National Research Council, National Academy Press, Washington DC.

[28] NIOSH-OSHA (1981). Pocket Guide to Chemical hazards, National Institute of Occupational Safety and Health, US Department of Health and Human Services, Washington DC, (1981).

[29] Calabrese, E.J., Multiple Chemical Interactions, Lewis Publishers, Inc., Chelsea, Michigan, USA, (1990).

[30] Gray, R.H.; Chess, E.K.; Mellinger, P.J.; Riley, R.G.; Springer, D.L. (eds). Health and Environmental Research on Complex Organic Mixtures (CONF-851027), Pacific Northwest Laboratory, Richland, Washington, USA, (1987).

[31] NAS (1988). Complex Mixture, Methods for In Vivo Toxicity Testing. Committee on Methods for the In Vivo Testing of Complex Mixtures, National research Council, Natioanl Academy Press, Washington DC, (1988).

[32] Wallace, L.A., The Total Exposure Assessment Methodology (TEAM) Study: Summary and Analysis, Vol. 1, US Environmental Protection Agency, Washington DC. Report EPA/600/6-87/002a.

[33] NAS (1986). Drinking water and health, Vol. 6, Safe Drinking Water Committee, National Research Council, National Academy Press, Washington DC, (1986).

[34] Calabrese and Kenyon (1990), Reference is missing!

[35] Cothern, C.R.; Coniglio, W.A.; Marcus, W.L., Development of Quantitative Estimates of Uncertainty in Environmental Risk Assessment when the Scientific Data Base is Inadequate, Environment International, 12 (1986) 643-647.

[36] Howard, P.H., Handbook of Environmental Fate and Exposure data for Organic Chemicals, Lewis Publishers, Inc., Chelsea, Michigan. (1989).

[37] Lappenbusch, W.L., Contaminated Waste Sites Property and your Health. Lappenbusch Environmental health, Inc., Alexandria, Virginia, (1989).

[38] Lioy, P.J.; Daisy, J.M., Toxic Air Pollution - Non Criteria Pollutants, Lewis Publishers, Inc., Chelsea, Michigan, (1987).

[39] Sittig, M., Handbook of Toxic and Hazardous Chemicals and Carcinogens, Noyes Publications, Parc Ridge, New Jersey, (1985).

[40] Weiss, G., Hazardous Chemicals Data Book, Second edition, Noyes Publications, Parc Ridge, New Jersey, (1986).

[41] Finney, D.J.; Probit analysis, 3rd Edition, Cambridge University Press, New York, (1971).

[42] Armitage, P.;Doll, R., The age distribution of cancer and a multistage theory of carcinogenesis, British Journal of Cancer, 8 (1954) 1.

[43] Guess, H.A.; Crump, K.S., Low-dose-rate extrapolation of data from animal carcinogenicity experiments-Analysis of a new statistical technique, Mathematical Biosciences, 32 (1976) 15.

[44] Carlborg, F.W., Dose-response functions in carcinogenesis and the Weibull model, Food and Cosmetic Toxicology, 19 (1981) 255.

[45] Iverson, S.; Arley, N., On the mechanism of experimental carcinogenesis, Acta Pathology and Microbiology Scandinavia, 27 (1950) 773.

[46] Cornfiel, J.; Mantel, N., Discussion Appendix to H.O. Hartley and R.L. Sielken, Estimation of safe doses in carcinogenic experiments, Biometrics, 33 (1977) 1.

[47] Rai, K.; Van Ryzin, J., A generalyzed multi-hit dose-response model for low-dose extrapolation, Biometrics, 37 (1981) 341.
[48] Griffitths, R.F., Chemical Plant Risk Assessment:Uncertainties and development Needs, Environmental International, 10 (1985) 523-530.
[49] Niehaus, F.; Novegno, A.; Vuori, S., Areas of Uncertainty in Comparing Risks of Energy Systems, Environment International, 10 (1985) 531-539.
[50] Crump, K., An overview of factors that influence the quantitative risk assessment of carcinogens, in Workshop on Problem Areas Associated with Developing Carcinogen Guidelines (BNL 51779), Brookhaven National Laboratory, Upton, N.Y., (1984), pp. IV-16 - IV-19
[51] Blaylock, B.G., Radionuclide data bases available for bioaccumulation factors for freshwater biota, Nuclear Safety, 23 (1982) 427.
[52] Colsher, C.S.; Thompson, S.E., Soil-to-Plant Concentration Factors for Radiological Assessments, Reports NUREG/CR-2975, and (UCID-19463); and Transfer Coefficients for Assessing the Dose from Radionuclides in Meat and Eggs, Report NUREG/CR-2976 (UCID-19464), NRC (Nov. 1982).
[53] Hoffman, F.O.; Bergstrom, U; Gylander, C.; Wilkens, A.B., Comparison of predictions from international recognized assessment models for the transfer of selected radionuclides through terrestrial food chains, Nuclear Safety, 22 (1981) 766.
[54] Coughtrey, P.J.; Thorne, M.C., Radionuclide distribution and transport in terrestrial and aquatic eco-systems, Vol. 1, 2, and 3, Publisher: A.A. Balkema, Rotterdam, (1983).
[55] Nicolet-Monnier, M.; Gheorghe, A.V., Integrated Regional Risk Assessment, Vo. II, - Consequence Assessment of Accidental Releases, Kluwer Academic Publishers B.V., Dordrecht, The Netherlands, (1995).
[56] ARCHIE, Handbook of Chemical Hazard Analysis Procedures, Federal Emergency Management Agency (FEMA), U.S. Department of Transportation, U.S. Environmental Protection Agency, (1987).
[57] Marshall, J.G., I. Chem. E. Symposium Series 58 (1980) 11.
[58] Canvey Second Report, London: HMSO, (1981).
[59] Wallis, J.B., Int. J. Multiphase Flow, 6 (1980) 97; Akker, H.E. et al., Discharge of pressurized liquefied gases through apertures and pipes, I. Chem. E. Symposium Series 80, 1, E23, (1983)
[60] Chatwin, P.C., The influence of basic physical processes on the statistical properties of dispersing heavy gas clouds, Proc. 7th Biennial Symposium on Turbulence, Rolla, MO, (1981).
[61] Sittig, Marshall, World-Wide Limits for Toxic and Hazardous Chemicals in Air, Water and Soil, Noyes Publications, New Jersey, USA (1994).
[62] U.S. Environmental Protection Agency, Compilation of air pollutant emission factors, Vol. II: Mobile Sources, Ann Arbor (1985).
[63] WHO (1993). Guidelines for Drinking-water Quality, WHO Geneva, (1993).
[64] Air Quality Guidelines for Europe, WHO Regional Publications, European Series No. 23, World Health Organization, Regional Office for Europe, Copenhagen, (1987).
[65] U.S. EPA (1985). Health Assessment Document for Nickel, EPA Research Triangle Park, N.C., Report EPA-600/8-83-012F, (1985).
[66] Starr, T.B.; Buck, R.D., The importance of delivered dose in estimating low-dose cancer risk from inhalation exposure to formaldehyde, Fundamental and Applied Toxicology, 4 (1984) 740-753.
[67] National Radiological Protection Board , Committed Equivalent Organ Doses and Committed Effective Doses From Intakes of Radionuclides, (Chilton, Didcot, 1991). Report NRPB-R245.
[68] UNEP (1985). The Environmental Impacts of the Production and Use of Energy Part IV, United Nations Environment programme, Paris, (1985).
[69] OECD 1991. The State of the Environment, OECD Paris, (1991).
[70] OECD (1990). Environmental models: Emissions and consequences, in J. Fenhann; H. Larsen; G.A. Mackensie, and B. Rasmussen (Eds.), Developments in Environmental Modeling 15, Riso International Conference, 22 - 25 May 1989, Elsevier, Amsterdam, (1990).
[71] Henrion, M.; Wishbow, N., 1st Demos user's manual: Version 3, Dept. Eng. and Public Policy, Carnegie Mellon University, Pittsburgh, P.A. (1987).
[72] Watson, W.S.; Hope, C.W., ACIDRAIN: A model for analyzing the costs and benefits of pollution control, Final Report, Cambridge Decision Analysts, Cambridge, England.

[73] NAPAP (1989). Plan and Schedule for NAPAP Assessment Reports, National Acid Precipitation Assessment Program, Washington DC.

CHAPTER 5

TOWARDS INTEGRATED RISK ASSESSMENT AND SAFETY MANAGEMENT AT REGIONAL LEVEL

5.1. The Potential Use of Comparative Risk Assessment

Methodological key issues (Gheorghe[1], 1994) to ensure a credible comparison to be made among different technologies or technological systems in a given region can be summarized as follows:

a) The delineation of consistent and comparable boundaries for different technologies or chains of technologies. The definition of principles and delineation of "reference technological chain" for risk comparison purposes is a priority issue in this regard.

b) Methods and dose-effect relationship for the risk assessment of the health effects of different technological systems (e.g., chemical plants, electricity generation systems, biotechnological installations) including, in particular, the deviation of consistent indicators of health risks for comparative assessment.

c) Methods and relationships for the assessment of impacts of different technologies (chemical works, electricity generation plants, biotechnological installations, etc.) on the natural environment and associated ecosystems, and particularly the derivation and application of indicators of environmental risks for comparative assessment.

d) How to deal, in the comparative risk assessment process, with the time and technological dependencies of risk estimation, given variations in technologies and in the state of technological development (over time). Can the principles of "risk discounting" be used to compare existing risks by postulating and accounting for possible future technological development. How can future technological development be accounted for in the comparative assessment process.

e) How to deal in the comparative risk assessment process with the issue of "uncertainties". Such uncertainties exist in particular in the type, nature and extent of various impacts and with the variable nature of some impacts over time. Uncertainties also exist in the estimation (both qualitatively and quantitatively) of the various risks, both in terms of consequences and probabilities.

f) Comparative risk assessment for severe accidents in different technological systems. The main issues here relate to the tools and criteria for comparison, the methods for presenting the results and the comprehensiveness of the comparative results, particularly in relation to environmental impacts from severe accidents.

g) The integration of the different elements of risks, including the question as to whether one overall integrated indicator of risk is possible or feasible; the applicability of comparative risk assessment to regional safety management studies; the weighting of the various dimensions of different risks in the comparative risk assessment process.

h) Methods and tools for the presentation of the results of comparative risk assessment.

5.1.1. SETTING BOUNDARIES FOR COMPARATIVE RISK ASSESSMENT OF DIFFERENT TECHNOLOGIES FOR REGIONAL SAFETY MANAGEMENT

The definition of consistent boundaries for the different technological chains or technologies which might operate in a given area is important for credible comparative risk assessment results. When dealing with comparative risk assessment of various technologies which might coexist in a given region, a consistent boundary definition has to be chosen.[2] When comparing technological or activity risks in a given area there is obviously a need to agree on some boundaries.

The principles for setting system boundaries are as follows:

- The setting of boundaries has to be adapted to the purpose of the assessment
- The boundaries should be set taking into account:
  - time (horizon)
  - space (workers and public exposure, environment)
  - phases of the systems under investigation
  - impacts to be considered
  - material balance consideration in terms of risk
  - per net production (e.g., energy) output
  - acceptable threshold criteria.
- At a regional level, the setting of boundaries will greatly benefit from structured public scoping/participation process, so that all concerned could be involved in boundary setting. This process would greatly enhance the credibility and acceptance of the study process.

5.1.2. ASSESSMENT OF ENVIRONMENTAL RISKS (IMPACTS)

Usually in a region where complex risk assessment studies are made, one has to take in consideration that a large number of compounds are released into the atmosphere and water and disposed on land with associated environmental impacts (risks). These impacts may be immediate or long term. Implicit "environmental impacts" are often a direct or indirect "inference" of "health impacts". TABLE 5.1 and TABLE 5.2 show a generalized environmental transfer model outlining the various essential components of health and environmental risk estimations in large industrial complexes /regions.

TABLE 5.1. Categories of risks usually adopted to assess and compare the health impacts of different pollutant sources

Health Risk			
Source	**People at Risk**	**Exposure**	**Effects**
Routine or accidents	Workers and the public	Short, or medium and long term	Fatal and non-fatal Immediate/delayed respectively Long-term/delayed

TABLE 5.2. Categories of risks usually adopted to assess and compare the environmental impacts of different pollutant sources

Environmental Risk		
Source	**Effects**	
	Duration	**Extent**
Routine or Accidents	Short or medium and long term	Local, regional and global

A linear pollutant pathway model[3] indicating the amount reaching the receptor (target at risk) as a function of the amount emitted, altered by dilution and removal and enhanced by environmental accumulation factors is shown in *Figure 5.1*.

As shown, distance (space) and rate of movement (time) are critical parameters to an environmental impact assessment study. The complex interdependencies between time, space and feedback mechanisms (the degree of resilience of a given environment to external factors) are not fully known; very often it is difficult to normalize them on a common scale for comparison.

In comparative risk assessment of different technologies one has to "compare risks", which may be different in a subjective way from impacts, effects, emissions, etc. Two major limitations have to be considered when dealing with the assessment of environmental impacts of different technologies or technological chains, namely:

- the effects are not always susceptible to quantification
- there is no general agreement on what should be quantified.

Methods of relevance for making comparisons of environmental impacts in integrated regional risk assessment and safety management studies are:

- ranked matrix environmental assessments
- emission values and ambient quality indices
- critical loads and critical levels.

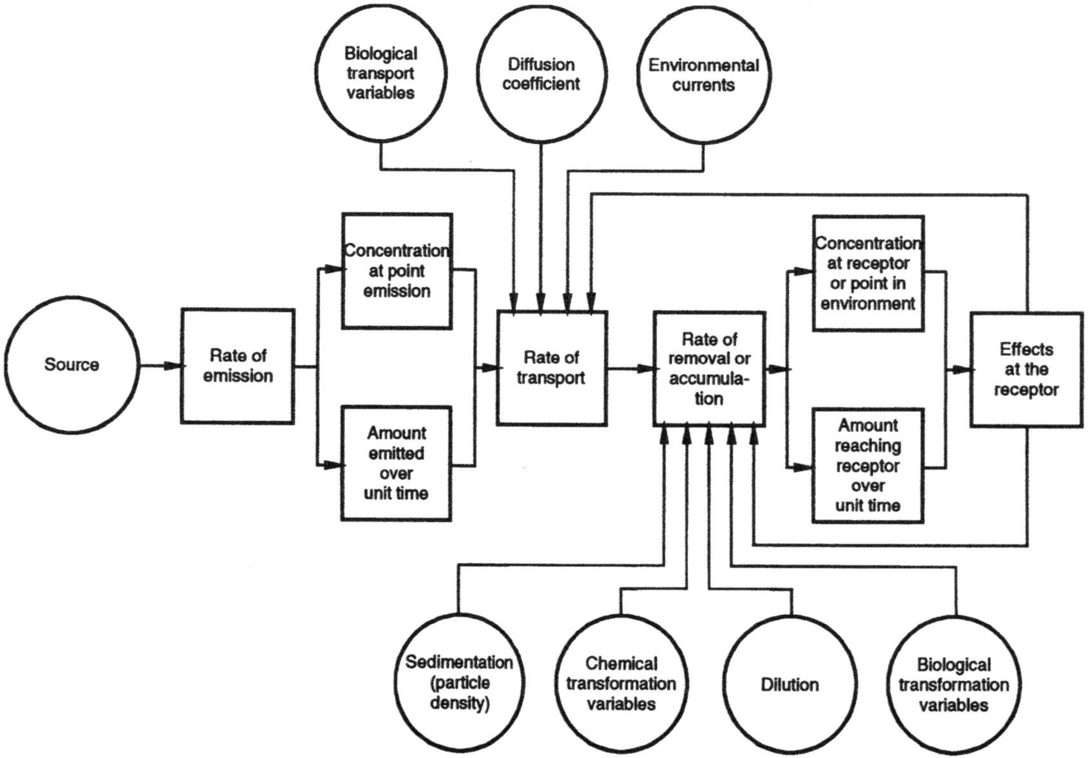

Figure 5.1. Pollutant pathway model used for assessing environmental impacts from emissions

- *A* ***matrix approach*** *is a ranking method to quantification of environmental impacts. It attempts to relate dose to response.*
- *The* ***emission values method*** *is a dose oriented approach.*
- ***Critical load model*** *whilst being dose based, is in fact essentially of a response type, the response being the outset of damage systems.*
- *The* ***Critical Loads concept*** *provides an adequate scientific basis to set limits to environmental impacts above which damage would occur.*

Critical Levels. It expresses the concentration of pollutants in the atmosphere above which direct adverse effects on receptors such as the ecosystem, may occur, according to present knowledge.

Critical Loads. It is a quantitative estimate of an exposure to one or more pollutants below which significant harmful effects on specified sensitive elements of the environment which do not occur, according to present knowledge.

The Critical Loads approach is a procedure for developing optimized abatement strategies by which leveled emission reductions are obtained on the basis of scientifically derived critical values.

Additional concepts associated with the critical loads and levels are:

- target levels and loads
- actual loads and levels.

The critical loads approach for regional environmental impact assessment requires the following types of information:

- inventories of current emissions and projections of future emissions rates
- estimates of the potential emissions reductions
- long range transport models
- maps of critical loads and target loads
- integrated assessment modeling.

In general it is considered that, for example, large scale electricity production systems which might exist at the regional level have an exponentially increased value for the environmental impacts due to the "effect of scale". There are economic and technological limitations to the elimination of all emissions or waste products and their associated impacts. Some emissions could be "imported" from long distances (e.g., transboundary effects of heavy metals, or acid rain); this has to be stated in the risk assessment studies for regional areas in view of proper results regarding comparative risk assessment of various technologies.

Indicators of environmental risks: Qualitative and quantitative indicators could be used for environmental impact indicators of different technologies. Such indicators are: site specific, society specific, value judgment dependent. Synergistic effects could be taken into consideration when dealing with comparative environmental impact assessment.

Environmental performance measurements already accepted by international organizations and practice and which can be used in comparative analysis are:

- environmental performance indicators (e.g., river quality, air quality, soil quality etc.)
- environmental goals (critical loads, sustainability index)
- environmental emissions (SO_x, NO_x, CO_2, Trade in Forest Products, etc.).

5.1.3. COMPARATIVE HEALTH RISK ASSESSMENT

The assessment of health risk is usually based on:

i) data from epidemiological studies for particular situations
ii) estimations based on dose-effect relationships for different levels of exposure.

Mathematical models enable the determination of ambient concentrations of pollutants in different media, and the prediction of health effects through dose-effect relationships. The state of knowledge concerning exposure paths, dose-effect relationships and the health risks for different technologies varies amongst different systems.

Both approaches to estimate health effects from various technological systems which can coexist in a given region, only rarely deliver quantitative risk coefficients with high confidence intervals. Dose-effect relationships for many chemical pollutants are not available.

There is a particular lack of data concerning the delayed (chronic) health effects of many chemical pollutants. The health effects at the low spectrum of continuous exposure need particular attention in this regard. Another issue of research need relates to the indirect health effects of pollutants where relatively little information is available.

A short list of parameters, being used as indicators of health risks for comparative assessment of various technological systems, includes:

- number of fatalities (acute)
- number of fatalities (delayed)
- incidence of injury and disease
- projected incidence of genetic effects
- years of lost life (YOLL)
- relative utility loss (RUL).

The following comments are offered in relation to the use of the above indicators:

■ Although compound indices such as YOLL and RUL offer the prospect of improved reliability in assessing and comparing health detriments, simple indicators such as numbers of fatalities and serious injuries should continue to be used in parallel, at least for the time being.

■ Some classes of detriment should not be aggregated (e.g., prompt and delayed fatalities).

■ Health risks to different groups and of different types should be distinguished, although it may in addition be permissible to aggregate in some cases.

Time Dependence in Comparative Risk Analysis

In comparative risk assessment (CRA), appropriate time frames must be defined for different stages of technological systems (e.g., impacts of thousands of years). In this respect one can consider micro-time (e.g., fraction of seconds) and macro-time (e.g., decades, centuries) as possible concepts to be used in comparative risk assessment.

5.1.4. UNCERTAINTIES IN COMPARATIVE RISK ASSESSMENT

When dealing with comparative risk assessment for activities which take place in large (regional) areas, uncertainties occur in the prediction of scenarios, models and data.

Uncertainties in scenarios:

- erroneous probabilities
- factors not considered
- factors screened out.

Uncertainties in models:

- imperfect conceptual model
- imperfect mathematical model
- imperfect computer model.

Uncertainty in data:

- general vs. site specific data
- measurement errors
- data reduction.

Expert judgment is inherent in the evaluation of uncertainties. Uncertainties must be delineated and exposed whenever appropriate and attempts made to deal with them. Uncertainties take many forms and it is essential that a coherent and clearly visible approach is adopted, both in the computational process and in the interpretation of results.

Within the complex study process of integrated regional risk assessment and safety management, the uncertainties arise from a number of sources:

Data. All data are subject to sampling errors. Statistical uncertainty requires the reporting of range or confidence levels.

Models. Estimates of risks through scientifically based models (lack of direct observations) brings additional uncertainty in a comparative risk assessment. When models are the only tool to be used for risk assessment, uncertainty analysis of the results should be exposed and taken into consideration.

Knowledge. Lack of knowledge or uncertainty in the foundation of basic physical phenomena can lead to erroneous results and faulty predictions. At the present time, this uncertainty in results due to knowledge is solved by using simulation procedures or inter-comparative computer exercises.

There are some techniques already in use which can help to evaluate uncertainties; among them are sensitivity analysis, inferential statistics (Bayesian methods), complex statistical and probabilistic analyses, Monte Carlo simulation. In case of lack of knowledge, bounding calculation and conservative assumptions can be used. As a rule of thumb, one can consider that risk assessment should be realistic and, if this is not possible, then the results should be considered as conservative.

There are different mathematical approaches to dealing with uncertainty: probability theory, Markow models, possibility theory (fuzzy sets), event trees, influence diagrams, heuristic models, etc.

The fact that uncertainties exist does not invalidate the use of comparative risk assessment as a useful tool for overall assessment of the impacts of various technologies in a given area, as long as such uncertainties are clearly understood, exposed and acknowledged in the presentation of the results of the study.

5.1.5. UNCERTAINTY ESTIMATION IN ENVIRONMENTAL REGIONAL RISK ASSESSMENT

In risk estimate, uncertainty pervades each element of environmental risk assessments, as an examination of estimations of the risks to humans from the release of a harmful substance into the environment.

The mathematics in a quantitative risk estimate may enter the following elements:

*Risk = Probability * Quantity * (1/Dispersion) * Population* Uptake* (Dose vs. Response)*

It is clear that an increase in any of these six elements leads to higher estimates of risk levels. There is always some uncertainty about each of these elements in the risk estimation process. In the current discussion, the terms "extreme certainty" and "extreme uncertainty" denote opposite ends of the spectrum of how well a particular phenomenon is understood. Qualitative modifiers "high", "fair", and "moderate" depict gradations of this understanding . These gradations, though arbitrary, are appropriate to the range of uncertainties found in environmental risk assessment.

1) *Uncertainties on the probability of a release of a harmful substance:*

   - May be unnecessary for many environmental risk assessment
   - Estimating the probability or frequency of such accidental releases is the first step in many environmental risk assessment studies
   - There is sizable uncertainty about estimates of the probability of accidental releases of harmful substances (N.B., historical data offer some guide to this

probability for frequently occurring accidental releases, but are less information on some kinds of accidents involving certain substances)
- Mathematical models produce fairly uncertain estimates of the probability of accidental releases of radioactive substances, it would appear that most estimates of the probability of chemical releases would be no better than fairly uncertain.

2) *Uncertainty on the quantity of a harmful substance released:*

- Such estimates improves with the quantity of the information gathered on the harmful substance released into the environment.

3) *Dispersion of a harmful substance and uncertainties on the resulting concentrations of that substance in the environment:*

- In practical situations estimates of concentrations at particular points in the dispersion pathway range from fairly to highly uncertain
- When measurements of concentrations are available, risk analyst can omit estimation of the probability of release of a given substance
- Actual measurements of concentrations of particular substances in the environment reveal that concentrations are highly variable ambient concentration of some pollutants are moderately invariable to moderately variable from day to day in a particular season. These concentrations may vary by a factor of 5 to 10 between warm and cold seasons and among different years. Human exposure to some chemicals in the workplace may be extremely variable among individual work situations.

4) *Uncertainty on the population exposed to release of a harmful substance:*

- In the absence of site-specific data, any estimate of the population potentially exposed to the release of a harmful substance from such sites would be extremely uncertain.

5) *Uncertainties on the uptake of harmful substances by humans and other organisms:*

- Pharmaco-kinetic models reflect how the physiology of humans differs from that of test animals with respect to uptake, metabolism, and excretion of particular chemicals. Pharmaco-kinetic models may modify risk estimates in significant ways, but their structure and the data they contain make these modifications moderately to extremely uncertain.

6) *Uncertainty on the relationship between dose of a harmful substance and adverse toxicological response:*

- The toxicity of a particular dose of a substance not only varies across species, but among individuals of the same species.
- Most acute toxicity estimates are moderately uncertain.
- Cancer potency estimates are fairly to extremely uncertain.

7) *Measurement error:*

- Measurement error increases the uncertainty associated with each of the above elements in an environmental and health risk assessment.
- Among all the elements of risk assessment, measurement error generally introduces the least uncertainty.

8) *Overall uncertainty:*

- Uncertainty in each of the elements of a quantitative risk assessment combines to produce an overall uncertainty in the estimate of a particular risk.
- Only a few studies have carefully integrated the uncertainty associated with individual elements of a risk estimation into an overall estimate of a risk and its overall uncertainty.
- Estimates of human cancer risks due to exposure to chemicals have been shown to be moderately to extremely uncertain.

9) *Credibility of Risk Analysis; Uncertainty on Policy Implications:*

- Uncertainties in environmental and health risk assessment are present in all elements of the assessment process.

There are several approaches to deal with uncertainty:

i) Risk analysts should present the particular level of uncertainty in the risks they are estimating.
ii) Risk managers should reduce uncertainty when the reductions would be cost-effective and would affect the choice of risk management option.
iii) When conservative assumptions must be applied, this must be done in the context of the particular uncertainties of the risk being addressed.

Reducing the uncertainty in any one or more of the elements of a risk assessment will reduce the overall uncertainty of a risk estimate; not all such reductions are equally cost-effective. Risk managers can also use the economic concept of the efficient frontier to determine which reductions in uncertainty are likely to reduce overall uncertainty the most and at the lowest cost:

- with regards to making conservative assumptions in the context of uncertainty, risk mangers should consider when and how to apply these assumptions
- some critics assert that conservative assumptions have resulted in grossly inflated risk estimates and, thus, in regulations that are too stringent; others assert that these assumptions are needed in order to compensate for various uncertainties.

Some guidelines to deal with uncertainty

■ <u>Action 1</u>. Explicitly acknowledge and approximate the uncertainties in each element of the risk estimation process:

■ <u>Action 2</u>. Carry the uncertainty through the calculation process and calculate not only the median or mean risk, but its overall uncertainty as well:

■ Action 3. Develop and apply decision criteria that explicitly address acceptable risk levels for individuals and the overall population.

■ Action 4. Offer to the decision makers a description of the nature and the magnitude of risks, including the uncertainty of these risks.

TABLE 5.3 presents some verbal description pertaining to the range of uncertainty.

TABLE 5.3. Range of uncertainty and variability in environmental risk assessments

Range of uncertainty / Variability	**GSD Descriptive Interpretative term phrases**	
More than 10.00	Extremely uncertain	Extremely variable
10.00	Highly uncertain	Highly variable
5.00	Fairly uncertain	Fairly variable
3.00	Moderately uncertain	Moderately variable
2.00	Moderately certain	Moderately invariable
1.50	Fairly certain	Fairly invariable
1.25	Highly certain	Highly invariable
1.05	Extremely certain	Extremely invariable
1.00	There is absolute certainty	

5.1.6. INTEGRATED ENVIRONMENTAL AND HEALTH RISK ASSESSMENT IN LARGE INDUSTRIAL AREAS

Uncertainty in Risk Analysis and Results Representation in Data Bases
There is a need to represent the uncertainty bounds of such data and knowledge in comprehensive studies and in specialized data bases. The purpose of this section is:

i) to help in identifying (direct) uncertainty measures of data and knowledge specific to health and environmental risk assessment studies in large industrial regions and
ii) their representation in documented and specialized data bases.

No measured or calculated value related to environmental and health impact (risk) assessment has any meaning unless it is accompanied by a declaration of quality, in form of an uncertain statement. The uncertainty declaration(s) can be regarded as the quality assurance tool for that process.

Uncertainties associated with model predictions in risk assessment
Uncertainties in models may result from transformation of the given model into a mathematical model. One has to distinguish between parameter, model and conceptual uncertainties, and treat them separately. Often results from modeling activities are stored in a data base. Models for assessment of environmental, health and safety impacts are important elements in integrated and comparative risk assessment for different technologies which might coexist in a given region.

A formal procedure of parameter uncertainty analysis is given next:

1) List all the parameters that are potentially important contributors to uncertainty in the final model prediction.
2) For each parameter listed, specify the maximum conceivable range of possibly applicable alternative values:
   - Specify the degree of belief (in percentage) that the appropriate parameter value is not larger than specific values selected from the range established above and select a probability distribution that best fit the quoted degrees of belief.
   - Account for dependencies among model parameters by introducing suitable restrictions, by quoting appropriate degree of belief, or by specifying suitable measure of the degree of association.
3) Set up a subjective *probability density function (pdf)* for the combined range of parameter values. This will subsequently be referred to as a joint pdf. Propagate this joint pdf through the model to generate a subjective probability distribution of predicted values:
   - Derive quantitative statements about the effect of parameter uncertainty on the model prediction:
   - Rank the parameters with respect to their contribution to the uncertainty in model prediction:
   - Present and interpret the results of the analysis.

Expert judgment elicitation is preferred when one of the following situations exists:

- no other means are available for quantifying an important issue.
- the information available is characterized by high variability.
- some expert question the applicability of the available data.
- the existing results from code calculations need top be supplemented.
- analysts need to determine the current state of knowledge.

Uncertainty declarations in the databases
There are reasons for including uncertainty indicators in the databases which store information relevant to integrated risk assessment or comparative risk assessment of various technological systems. Uncertainty declarations are quality declarations and a good description of uncertainties is decisive for good practice and scientific acceptance.
Uncertainty criteria may be used as acceptance criteria in quality assurance for choosing data to be accepted and included in the data base. A good scientific practice is a strategy for accepting material in the data base.

Validation. The advanced assumptions (theories) should be subject to a validation procedure.
Criticizable. The assumption in risk assessment should be formulated in a well defined manner so that it can be criticized.
Acceptability. The assumptions for risk estimation should be discussed when necessary in a forum (e.g., steering committee).

5.1.7. INTEGRATION OF THE DIFFERENT ELEMENTS OF RISK AND THE ROLE OF COMPARATIVE RISK ASSESSMENT FOR SAFETY MANAGEMENT

Lack of a single integrated risk indicator; the elements of risks from various technological systems can be broadly categorized as indicated in TABLE 5.3. In reviewing risk assessment and management results for their use in area safety planning, one should take in consideration the followings:

■ The various elements and dimensions of environmental and health risks cannot be integrated into one overall indicator of total risk. No comparison on the basis of a single indicator is possible. The comparative risk assessment process must specify on which basis (indicator) the comparison is being made.

■ It is necessary to expose all the dimensions and elements of health and environmental risks in the comparative process. Differences between regions and societies make the development of one overall indicator meaningless. There is no "global" risk value. The results should not be transferred from one study to another without appropriate investigations of differences between regions and countries.

■ One approach to develop an overall indicator is the estimation of "external costs" of impacts in terms of monetary value.

■ The integrated approach to health and environmental risk means all risks should be identified, assessed and considered in the comparative risk and management process.

More Comments on Continuous Emissions

The most serious risks (impacts/effects) from continuous emissions from different industrial activities (e.g., power generation, chemical plants, heating, etc.) may be expected to be those related to human health and of environmental effects. The following impacts were identified by at large for making priority risk analysis due to continuous emissions:

- effects of atmospheric pollution on:
    - human health
    - materials
    - crops
    - forests
    - freshwater fisheries
    - unmanaged ecosystems
- effects of altered land use on unmanaged ecosystems
- assessment of global warming potential of greenhouse gas emissions
- effects of discharges to water and abstraction on health and water supplies
- effects of discharges to water and abstraction on aquatic ecosystems
- effects of noise on human health
- effects of noise on amenity.

The above list is not exhaustive and the process of prioritization is subjective.

A damage function approach can be used to assess and value the impacts resulting from activities which are taking place in a large industrial area. This type of approach allows calculation of marginal damages; many impacts are very often side and technological dependent. A pathway is chosen to show the progression of impact from the industrial activity and corresponding emission through effects on plants in natural ecosystems to economic valuation.

In order to quantify different risks from a specific industrial site it is necessary to identify the distribution of receptors that may be affected by operation of the plant in some detail. The type of data necessary to make this calculations are of the following form:

- Distribution of people in different risk categories
- Distribution of various crop species
- Distribution of various forest species
- Distribution of various species in different risk categories.

It is important that health and environmental damages are assessed over as much of the area as possible. One has to be aware that the artificial truncation of the reference environment for risk assessment will lead to underestimation of impacts.

Impact estimation can be performed on a marginal basis using at least three types of procedures

- Statistical relationships (proportion of the population who will develop respiratory symptoms due to low level pollution in the analyzed region)
- Dose-response functions (linking the rate of corrosion of steel to pollutant levels, known as critical level concept)
- Mathematical models containing a series of cause-effect relationship (e.g., *magic model* used to predict the rate and extent of ecosystem acidification and chemical changes for different pollutant deposition rates).

At the present time, international team of experts were convened (e.g., in EEC) to discuss, agree and recommend the most applicable dose response functions and models for risk assessment. Large degree of uncertainty is associated to the estimation of acidification impacts on forestry and the calculation of global warming damages. Valuation can be performed most easily when the commodity is directly traded (agricultural produce, timber, building repairs, etc.). When this is not possible, it is necessary to assess values indirectly in terms of *Willingness to Pay (WTP)* for improved environmental quality, or *Willingness to Accept (WTA)* increased environmental damage.

External Costs. They are defined to be the costs which fall on one group of people due to the social or economic activities of another group, and where the latter group does not take these costs into account. In the case of electricity generation, this is equivalent to costs which are not reflected in the market price of the electricity generated. There are often difficulties in defining to what extent damage to health or the environment is an external cost.

TABLE 5.4. Indicators representing damage categories

Public health:	
- PM_{10} mortality	- PM_{10} respiratory hospital admission
- SO_2 mortality	- SO_2 cough episodes for children
- PM_{10} bronchitis in children	- SO_2adult chest discomfort days
- PM_{10} symptoms days	- SO_x-eye irritation
- PM_{10} asthma attack	- NO_x-eye irritation
- PM_{10} restricted activity day	- Transport accidents
- PM_{10} emergency room visits	- Construction accidents
Occupational health:	
- Mining (respiratory diseases; mortality and morbidity)	- Transport (death, major/minor injuries)
- Mining (death, major/minor injuries)	- On road accidents (death, major/minor injuries).
Agriculture:	**Aquatic impacts:**
- SO_2 -direct effects on wheat, barley, rye, oats, peas and beans, or other crops	- Mining (groundwater and surface water)
- Acidic deposition -soil liming	- Acid deposition -fisheries
- Effects of ozone	**Forests:** **Biodiversity:**
Materials:	
- Stone refacing	- Galvanized steel replacem0ent
- Mortar repointing	- Aluminum replacement
- Concrete replacement	- Cultural impacts
- Repainting	- Building damage from subsidence
- Steel replacement	
Noise:	**Global warming impacts:**
- Plant operation	- Aquatic life
- Road traffic	- Air quality (ozone)
- Rail traffic	- Glacier regression
	- Agricultural/Forest growth
	- Desertification

In order to be comprehensive, consistent and transparent it is necessary to consider the environmental impacts and their monetarized values over all space and time (the life cycle analysis). Some risks due to technological systems (e.g., electricity generation, production of different chemicals, etc.) due extend far beyond the operating lifetime, for example forest decline and fisheries or the global warming effect.

All part of the analysis are subject to uncertainty, some of which is very considerable. In many cases it is not even possible to describe the uncertainty in terms of distribution of values. The valuation of uncertain events (risk evaluation) poses considerable challenges. Discounting of the monetary values of environmental and health impacts is controversial.

Impact Categories
A list with guiding priority impact categories is given next:

- atmospheric pollution/human health
- accidents (continuous emissions) occupational and public
- atmospheric pollution/materials
- atmospheric pollution/crops
- atmospheric pollution/forests
- atmospheric pollution/freshwater fisheries
- land use/natural ecosystems
- global warming
- mining/ground and surface waters
- emissions to water/drinking water
- emissions to water/ecosystem
- noise/human health
- noise/amenity.

Damage Categories Indicators
Appropriate indicators for damage categories related to the above impact categories' list are given in TABLE 5.4. They have to be considered as indicative, which can guide the analyst in his effort to quantify consequences in the risk assessment stage.

5.2. Some Methodological Issues in CRA for Severe Accidents

The assessment and hence the comparative risk assessment (CRA) of different technologies or technological systems as regard major accidents, focused almost entirely on estimating acute fatalities to people from historical records or using techniques of Probabilistic Safety Assessment (PSA) as a predictive tool.

Methodologies for the estimations of the late health effects to people and of the environmental impacts from major accidents are limited, or in need of significant development and applications (with the exception of nuclear technologies).

A definition for severe accidents should encompass all the elements of health, environmental risks and damage to plant, equipment, buildings and should be expressed in terms of both the potential as well as the actual damage and risk.

5.2.1. METHODS AND TOOLS OF COMPARATIVE RISK ASSESSMENT FOR SEVERE ACCIDENTS

The comparison cannot be made on the basis of the consequence of such accidents in isolation. The likelihood (or probability) of occurrence should also be taken into account. Hence, estimation of the frequency of such accidents is relevant. Such estimations necessitate reliable information on the past records of such accidents and their effects and/or the application of probabilistic methods that predict the likelihood of their future occurrence.

It is difficult to assess and compare the frequency and the health and environmental damages caused by severe accidents because such data are not systematically collected by a single national agency.

There are no data concerning in particular the delayed effects on health from severe accidents for non-nuclear energy systems. All health effects in such cases are reported in terms of immediate fatalities, with immediate injuries reported in a few cases. This makes complete comparison difficult, since the total impact may be underestimated.

The ultimate long term environmental effects, particularly from severe accidents, are difficult to establish; it may be difficult to establish whether the effect is irreversible or whether a recoverable effect is possible.

Presentation of Results of Comparative Risk Assessment for Severe Accidents
In regional risk assessment and safety management studies, in presenting the comparative risk assessment results for severe accidents, the following principles should be adopted:

a) Risks from severe accidents should be presented and compared separately from the risks resulting from routine operations.
b) Data based on historical (actual) occurrences should not be compared directly with data on probabilistic predictions of likely future events.
c) Comparative results should not be presented in terms of aggregate risk in isolation.
d) All dimensions and elements of environmental and health risks should be exposed and presented.

The presentation of results in the form of F-N curves (frequency distribution) is a useful presentation technique. However, the values usually presented without additional information may not by themselves be sufficient since other dimensions of risks may not be exposed. Another method of presenting the results in tabular form should ensure the completeness of all relevant dimensions of risk rather than relying on isolated information (e.g., on fatalities).

The above principles are to be considered for accidental situations, which are mainly investigated in the second volume of this monograph.

5.3. Basic Aspects in Decision Aiding Techniques

It is already acknowledged that the process of initiating, promoting risk analysis and implementing safety management studies for regional areas involves complex decisions as well as the participation of a large number of actors (e.g., public, experts, systems analysts, environmentalists, safety engineers, administrators, politicians etc.). The process of *integrating various aspects of risk* such as environment, health, hazardous installations, safety culture, management, involves decision aiding techniques which are close to the management science field known as decision analysis.[4]

There are, in general, positive and negative aspects associated with decisions. Indeed, many decisions are made intuitively by experts and do not use structured processes or techniques. For many decision problems related to risk assessment and safety management in large and complex industrial areas, the solutions and then advantages and disadvantages may not be immediately apparent because of the complexity of the issues involved. There is a need for systematic processes to be followed that help structure thinking and analysis, and allow different viewpoints to be taken into consideration. Structuring helps avoid inappropriate *ad hoc* decisions and allows the process of reaching a decision to be more open and the decision itself to be more readily defensible (decisions which are taken today very often have long term effects). In the end, the use of various decision aiding techniques and the overall process and technology of decision analysis allows the integration of various risk at regional and area level. The integration of various risks into the decision making process is the appropriate mechanism which allows displaying various risks and choosing the most appropriate safety improvement strategy.

There are many inputs, influences and constraints that a decision maker will consider when deciding which actions to initiate regarding risk reduction or safety management to a particular plant or for the region under analysis. Decision aiding techniques (DAT) are tools for decision makers; they are decision aiding techniques, but not decision making techniques. In the content of regional safety management such techniques could i) assist structuring the complexity of issues derived from hazard identification and prioritization, ii) assist in improving safety of individual installations, iii) siting of new installations to comply with numerous criteria and performance risk indicators, etc. There are complex decisions that need to be made in the field of regional safety management which may involve conflicting positions. A large number of tools are available to assist in solving and structuring decisions of such complexity.

The main purpose of this section is:

- to emphasize and strengthen the use of DAT for integrated regional risk assessment and safety management
- to provide guidance as to where such processes may or may not be appropriate (to avoid the application of sophisticated processes to trivial problems).
- to recognize that DATs are tools, not panaceas.

The main stages of DAT in relation to the integrative process of various risks which could coexist in a given region, are:

- <u>Step 1</u>: Definition and description of the problem.
- <u>Step 2</u>: Consideration and definition of appropriate quality assurance requirements.
- <u>Step 3</u>: Formalizing the descriptive model of the problem.
- <u>Step 4</u>: Obtaining the necessary information for modeling.
- <u>Step 5</u>: Analyze in order to determine the set of alternatives and criteria.
- <u>Step 6</u>: Ensure the selection of the proper method to make the decision regarding the proper integration of various risks in the region and their minimization

■ Step 7: Establish a clear record of the process and any decisions taken as a result of the integration process of various types of risk and appropriate safety measurements.

To make a decision means to select a way of action out of a set of possible alternatives. The decision process is complex and sometimes iterative; the set of alternatives or criteria may vary from an interaction to iteration.

Decision making involves three major elements:

a) alternatives among which the "best" one will be chosen
b) criteria for judgment
c) methods for selecting one alternative from the whole set.

Decision aiding process (DAP) is needed because it helps to generate a degree of shared understanding among interest parties who are concerned with issues on risk assessment and safety management in regions with complex and large industrial installations. There are a few issues to be highlighted: in this domain one has to involve complexity, uncertainty or even fuzziness, multiple criteria and group interests with the objectives in conflict. DAP can provide models that integrate these features into an adequate methodology. DAP provide a framework within which informed discussions of key issues (e.g., environmental or health risks) can be conducted, and facilitate generation of consensus and integration of such issues. The relative costs and benefits to the user of applying a decision aiding process should be compared to the costs and benefits of not using one of these decision aiding processes. The cost of not using a decision aiding process can be high, especially in terms of inefficient use of human resources and in the end the incapacity of integrating various types of risks and selective safety improvement measures in the overall decision process.

In the risk management phase for studies concerning complex industrial areas, all possible options (e.g., technological, organizational, managerial etc.), have to be listed exhaustively as well as all relevant choice factor from which the appropriate comparison of the set of options will be made. After having identified the sets of options and factors, each option has to be assessed and weighted according to each factor. These assessments involve generally uncertainties, especially when probabilistic events are concerned for which occurrence probabilities and possible associated consequences cannot be easily evaluated.

When the best option has been selected, one has to check the robustness of this choice in relation with the variety of risk impact indicators, safety improvement measures and a sound cost effectiveness strategy. Various uncertainties or value judgments have to be tested in a sensitivity analysis phase. At the end of this selecting process, results have to be presented as simple and as clear as possible.

The following steps have to be considering when involving DAP for integrated regional risk assessment and safety management:

- Choice of options and relevant technical, environmental, health and economical factors.
- Assessment of options according to each relevant factor and the relative importance of each factor in the political or technical decision environment.

- Choice of the most appropriate and relevant decision aiding techniques.
- Estimation of weighting factors.
- Use of sensitivity analysis.
- Presentation and interpretation of results.

The wide practice of the use of decision analysis shows that no decision aiding technique is appropriate for all circumstances. A variety of techniques which could be used in risk management in complex industrial areas are presented in this chapter. Examples of the types of problems in the field of risk assessment and safety management where DAT can be helpful are outlined below:

■ Siting of new industries, land fills, or other hazardous installations, taking into account factors such as alternative land uses, environmental effects, and population density.

■ Development of strategies for emergency planning (e.g., sheltering, evacuation, relocation) taking into account the social implications of moving people in all types of weather, as well as financial and logistical considerations.

■ Deriving levels of pollution contamination in water and ground at which emergency countermeasures may be introduced following an accident.

■ Determining requirements for additional safety equipment.

■ Determining the appropriate degree of redundancy of safety devices and plant equipment, taking into account possible safety effects (e.g., domino effect) on other plant equipment.

■ Optimizing the protection and other aspects of maintenance and repair operations, choosing between major alternatives such as replacement or repair of large components.

■ Selection of optimum plant equipment and procedures for normal and abnormal circumstances.

Decisions related to risk management in large (regional) areas involve negotiations between people with different viewpoints or value judgments, such as representatives of the cantons, health officials, the environment departments, representatives of the workers and the management of a given facility, and the public. One objective of the negotiations is to determine the "choice factors" which are representative of the different viewpoints.

The range of possible decisions in any investigated area is generally limited by certain constraints which may be imposed, for example, by a regulatory authority at the federal or cantonal level, by operating policies, or by generally accepted good practice.

Other constraints may be less tangible, such as acceptance by the public or by the industry practice, limited perceived threats to environmental quality and health risks, and political conditions. New techniques are now available (decision conferencing) which enable the integration of various aspects into a comprehensive decision policy.

5.3.1. OVERVIEW OF DECISION AIDING TECHNIQUES

Before carrying out any decision aiding process for integrated risk assessment and safety management at regional level, it is essential to identify all of the available options and relevant factors that would influence the outcome. "Brainstorming" can present a constructive approach. The use of multi-disciplinary scientific and engineering expertise groups should be considered for dealing with the more complex assessment problems, particularly those involving the need to choose between availability of alternative or competing action pathways.

Generally, one option is devoted to a given objective function (e.g., reduction of the probability of occurrence of an undesired event, reduction of various releases, etc.). One can thus structure a brainstorming session as follows:

- What are the involved functions?
- Is each function necessary?
- It is possible to ensure a given function by different ways?
- For each way, is it possible to ensure the function differently?

The selection of the relevant factors can be aided by the application of analytical hierarchy process or value trees which enable the complexity of a given problem to be broken down into smaller constituent parts. As an example, TABLE 5.1 and TABLE 5.2 illustrate the criteria hierarchy for the overall health risks' As shown, distance (space) and rate of movement (time) are critical parameters to an environmental impact assessment study. The complex interdependencies between time, space and feedback mechanisms (the degree of resilience of a given environment to external factors) are not fully known; very often it is difficult to normalize them on a common scale for comparison.

In comparative risk assessment of different technologies one has to "compare risks", which may be different in a subjective way from impacts, effects, emissions, etc. Two major limitations have to be considered when dealing with the assessment of environmental impacts of different technologies or technological chains, namely:

- the effects are not always susceptible to quantification
- there is no general agreement on what should be quantified.

Methods of relevance for making comparisons of environmental impacts in integrated regional risk assessment and safety management studies are:

- ranked matrix environmental assessments
- emission values and ambient quality indices
- critical loads and critical levels.

objective (e.g., routine, accidental, public, occupational etc.).

When different groups are directly involved in the decision making process, it is important to take into account the different objectives of each group In this context it is necessary to recognize the fundamental objectives and the boundary conditions, together with any particular preferences with respect to these criteria. For instance, for the objective "minimize hazardous potential" the associated criterion is *"hazardous potential"* and the preference is for *minimization*.

When a particular option cannot be uniquely prescribed by a single figure within acceptable confidence limits, it may be possible to use alternative values or distribution of values, or distribution functions, or even possibility distributions or fuzzy sets evaluations.

5.3.2. DECISION AIDING TECHNIQUES IN USE FOR INTEGRATED RISK ASSESSMENT AND SAFETY MANAGEMENT FOR LARGE AND COMPLEX INDUSTRIAL AREAS

A number of decision-aiding techniques are currently available. For the purpose of this book, the application of Cost Effectiveness Analysis (CEA), Cost Benefit Analysis (CBA), Multi-Attribute Utility Technique (MAUT), and Multi-Criteria Outranking Technique (MCOT) are considered in further details.

Cost Effectiveness Analysis (CEA)

CEA technique can be applied when two related factors are considered, one of which being a cost. The relative interest of each option is assessed through a "cost effectiveness ratio" illustrating the amount of effort (cost) devoted to a decreasing of one unit of risk. This technique is relatively easy to apply and can be recommended when only two relevant factors are envisaged because in this case the sensitivity analysis is very easy to make.

Cost Benefit Analysis (CBA)

CBA. originates from the economic theory of welfare; it compares the benefits and harm associated with different options The principal characteristic of this technique is that all relevant factors have to be expressed in monetary terms. When all these factors are expressed in the same unit it is very easy to aggregate all these different costs in a total cost. The best option is then the option presenting the minimum total cost.

Multi-attribute Utility Technique (MAUT)

The essence of this technique is to define a scoring scheme (or a multi-attribute utility function), measured on a scale between 0 and 1, with the property that if the score (or utility) is the same for two options there is no preference for one or the other. If however the utility for *option i* exceeds that for *option j*, then *option i* is preferred to *option j*. Such utility functions are established for each relevant factor (or attribute) and then aggregate in a total utility function representing the global interest for each option. The MAUT is an interesting decision analysis model for high level strategy problems involving different factors (quantitatively or qualitatively). Recent experiments show that this technique can be implemented in practice when dealing with different actors involved in a complex decision making environment. Fuzzy sets theory can also be adopted to this technique in order to accommodate for imprecision in evaluating of the factors.

Multi-Criteria Outranking Technique (MCOT)
-The aggregation techniques described in the previous sections combine all the evaluations according to the different relevant factors into a single figure of merit expressing the global interest of each option. Instead of expressing the performances of each option in terms of a single overall figure of merit, outranking techniques compare each *option i* to every *option j* in order to evaluate whether *option i* outranks (or is preferable to) *option j.*

5.3.3. SENSITIVITY ANALYSIS

Any decision aiding procedure involves the use of data, models, techniques, value judgments which contain uncertainties of various types such as:

- Uncertainties associated with an imperfect knowledge of the performances of the options under different circumstances (scenarios) or of the parameters and data used in the assessments.
- Uncertainties associated with the use of models.
- Uncertainties due to imperfect knowledge (e.g., about the future of each option).
- Intrinsic uncertainties resulting from the statistical treatment of the variable (this is very important when assessing the expected outcomes from low probability events).
- Value judgments.

The sensitivity analysis highlights the aspects which have the greatest influence on the results, where it may be desirable to attempt to reduce the uncertainties, if possible.

Types of sensitivity analysis involve the following process:

- Modification of a given parameter.
- Assign probability distributions to the important variables (probability encoding).

5.3.4. PRESENTATION OF THE RESULTS FROM APPLYING A DECISION AIDING TECHNIQUE

Aspects to consider when presenting the results of integrated decision for risk assessment and safety management at regional level are as follows:

- List of the assumptions, hypothesis and initial conditions considered within the case study
- Comment on the weighting factors
- Comment and reference on the various models and techniques used as well as their relevance and integration within the overall study
- Indication of the main uncertainties, quality of data
- Presentation of the sensitivity analysis results.

From these different information the final decision will be made by the decision maker or the group involved in preparing recommendations and policy strategy at area level.

5.3.5. ADVANCED TECHNIQUES FOR DAP

Artificial intelligence models have been lately implemented for computer aided design for risk assessment and management, as well as for expert systems development and implementation. A few comments is worth considering:

i) Expert systems have been developed in assisting the probabilistic safety analysis for complex nuclear and other industrial facilities and processes; they are relevant in integrating various aspects concerning risk in a given region.

ii) Decision support systems by means of expert systems, operation research, multi-criteria decision models with objectives in conflict are already in use in making decisions concerning risk reduction and cost optimization for marginal safety measures implementation.

iii) Connectionist expert systems (neural networks)[5] which exhibit characteristics and capabilities not provided by any other technology; this a complementary technique designed to assist solving ill-designed risk assessment and management problems (e.g., risk to special targets at risk - people with asthma and bronchitis - exposed to low level pollution and a complex meteorological environment and possible changes as well as their integration).

iv) Integrated knowledge based decision support systems by using systems analysis techniques and procedures and by integrating in an efficient way expert systems and neural network technology into an advanced methodology of systems analysis and information technology.

v) Influence diagrams which enables capturing numerous elements which could lead to an unsafe operational environment.

Decision Conferencing

Decision Conferencing is an efficient tool for integrated regional health and environmental risk assessment and safety management. *Decision Conferencing* is a relatively new decision aiding tool, which has been developed from the practice and need of integrating various aspects of risk, which has grown up in a number of organizations or projects, when complex situations and difficult issues arise. In a recent report ("International Chernobyl Project - input from the Commission of the European Communities to the Evaluation of the Relocation Policy adopted by the former Soviet Union[6]) it is highlighted the followings:

> *"Rather than discuss the matter amid the hurly-burly of their day to day activities, the decision makers take time to go away from their regular working place in order to concentrate fully and solely on the issues before them. Sometimes straightforward discussion can lead to a clear decision and a view of*

a way forward, but the complexity of the issues and the uncertainties involved may be too great for simple discussion to resolve. In such cases, it has become the practice in some organizations for the team responsible for the decision to meet together for two days or more away from their normal working environment to discuss and explorer the issues".

Decision conferencing uses the services of a moderator, who is skilled in the process of group decision. The moderator is seldom a person with experience in the context of the issues at hand or have expertise in the discussion. The moderator has a very definite role, namely to smooth the team's work, to help the process and make the team productive and more creative. Within a decision conference event, the content of the discussion comes from the team themselves. A decision conference is a two-day event in which all the 'owners' of the problem (e.g., integrating health and environmental risks in a given region and making appropriate decisions for safety improvements) gather together to agree upon a strategy. The decision makers, or advisors (the problem owners) are supported by a moderator and a decision analyst. Two important comments are made:

i) - the moderator has the role to lead the conference (he acts as a disinterested chairperson guiding the discussion forward constructively, building and interpreting decision models): this would help the decision makers or their advisors to appreciate various facets of the problem before them

ii) - the analyst deals with the details of the model building, by using information technology (models and computers).

The technology of decision conferencing brings together knowledge, skills and techniques arising from the following fields:

- *Decision analysis* which provides a structure and language (e.g., multi-attribute value theory) in which decision makers can think and talk about their problem.

- *Information technology* assists in "real time" the decision making process; the results of decision analysis can be presented to the decision makers in simple, easily comprehended ways.

- *Group dynamics* an awareness of which helps the moderator and analyst ensure that the group of decision makers or advisors interact constructively.

- *Within a decision conference* the team can discover the importance of differences in opinion between the team members and the sensitivity of their conclusions to these differences and to those judgments of which they are most unsure. Controversial discussions are unimportant when discussion is focusing on the critical issues without being deflected. The moderator is expert in group dynamics and rational decision theory, as well as a skilled communicator.

In practice, each decision conference is different. It evolves according to the needs of the team and not according to some fixed agenda. The topic of integrated regional risk assessment and safety management and its complexity is an ideal topic for decision conferencing. No model should be taken as definitive: revised models are build and much sensitivity analysis undertaken until no further insights are obtained and agreed conclusions are reached. Decision conferences are creative events; they create an environment where participants create, evaluate, modify and re-evaluate options, building a strategy which they all support.

In accordance with the practice in this field: "Surprisingly, the analysis in decision conferences needs much less hard data than one would, at first, think. Strategies have to be priced: that is clear. But the cost estimation need only be rough. It is a broad brush picture that the event seeks to create. Detail can be added at a later date".

The lack of a single risk indicator to deal with risk assessment, health and environmental impacts at the regional level, should be compensated by the use of this specialized model - *decision conferencing* - able to integrate objective and subjective decision aspects when dealing with complex issues and costly options and decisions.

Summary (Chapter 5)

Integrating various types of risks, and their sources, existing in a region is not an easy or trivial task. Various types of techniques, methods and tools could be employed in helping the integration process. One potential / relevant instrument for the integration process of various types of risks in a region is that of Comparative Risk Assessment. The chapter investigates the process of setting boundaries for comparative risk assessment of different technologies, the assessment of environmental risks / impacts, comparative health risk assessment, uncertainties in comparative risk assessment. Methods and tools of comparative risk assessment for severe accidents are also discussed.

Recently, Decision Aiding Techniques play a significant role in the integration process of various types of risks and their management at the regional level. This chapter reviews such techniques and their relevance to the risk integration process. Special attention is given to decision analysis and decision conferencing.

References (Chapter 5)

[1] Adrian Gheorghe - "Comparative risk assessment of the health and environmental impacts of various energy systems", in Int. J. of Environment and Pollution, 4(3-4) (1994) 329-349.

[2] Sam Haddad, Adrian Gheorghe - " Issues in comparative risk assessemnt of different energy sources", in Int. J. of Global Energy Issues, Vol. 4(3) (1992).

[3] United Nations Environment Programme, The Environmental Impacts of Production and Use of Energy. Part IV. The comparative assessment of the environmental impacts of energy sources. Phase I. Comparative data on the emissions, residuals and health hazards of energy sources, Rep. ERS-14-85, UNEP, Nairobi (1985).

[4] Adrian Gheorghe, Decision Processes in Dynamic Probabilistic Systems, Kluwer Academic Publishers, Dordrecht, (1990).

[5] Adrian Gheorghe, Connectionist expert systems for analysis of health problems associated with industrial activity and electricity generation systems, Int. J. Environment and Pollution, 4 xxx 107-124.

[6] International Chernobyl Project - Input from the Commission of the European Communities to the Evaluation of the Relocation Policy Adopted by the Former Soviet Union", Report EUR 14543 EN, Commission of the European Communities, 1992.

APPENDIX A

APPENDIX A: COMPILATION OF U.S. EMISSION FACTORS

"New Source Performance Standards for Air Pollutants" have been developed by the U.S. Environmental Protection Agency (TABLE A-1).

The industries for which *Pretreatment and Effluent Guidelines and Standards for Water Pollutants* have been developed by the U.S. Environmental Protection Agency are listed in TABLE A-2.

Pollutants and activities for which *Hazardous Air Pollutant Emission Standards* have been developed by the U.S. Environmental Protection Agency are listed in TABLE A-3.

Emission coefficients for *problematic air pollutants* from various energy technologies are given in TABLE A-4.

A comparison of national emission standards for fossil electricity generating plants is presented in TABLE A-5, which gives *emission coefficients* for five *conventional air pollutants* (i.e., SO_X, NO_X, CO, HC, and TSP) for a range of energy systems. These are compiled from a report prepared for the U.S. Department of Energy (The Aerospace Corporation and Mueller Associates, Inc., 1983). Detailed documentation needed to define the bases for these numbers are contained in that report. Although these data provide some perspective on the coefficients for similar activities elsewhere, the true coefficients will differ, perhaps in major ways, for some or all of the following reasons:

a) processes vary in their engineering characteristics (e.g., size, efficiency and temperature)
b) fuel supplies have different characteristics (e.g., heat, sulfur and ash content)
c) pollution control equipment have different impacts (e.g., efficiency or on types of pollutants scrubbed).

Thus, extrapolation or direct application of these coefficients to other countries may introduce large errors unless these factors are examined.

TABLE A-1. Industries for which "New Source Performance Standards for Air Pollutants" have been developed by the U.S. Environmental Protection Agency

Industry	Pollutants Regulated
Fossil-fueled steam generators	PM, SO_2, NOx
Incinerators larger than 50 TPD	PM
Portland cement plants	PM, Opacity
Coal preparation facilities	PM, Opacity
Nitric acid plants	NO_x, Opacity
Primary aluminum smelters	F, Opacity
Sulfuric acid plants	SO_x, Acid Mist, Opacity
Asphalt concrete plants	PM, Opacity
Sewage sludge incineration	PM, Opacity
Iron and steel plants	PM, Opacity
Electric arc furnaces	PM, Opacity
Ferro alloy production facilities	PM, CO
Secondary brass and bronze ingot	PM, Opacity
Kraft pulp mills	PM, Total Reduced Sulfur
Petroleum refineries	PM, Opacity, CO, SO_2
Storage vessels for petroleum	VOC
Secondary lead smelters and refining	PM, Opacity
Primary copper, lead and zinc	PM, SO_2, Opacity
Phosphate fertilizer industry	F
Grain elevators	PM, Opacity
Ammonium sulfate manufacture	PM, Opacity
Lead acid battery manufacture	Pb, Opacity
Stationary gas turbines	NO_x, SO_2,
Glass manufacturing	PM
Phosphate rock plants	PM, Opacity
Synthetic organic chemicals	VOC
Pressure-sensitive tape and label coating	VOC
Auto and light truck surface coating operations	VOC
Asphalt processing and asphalt roofing manufacture	PM, Opacity
Rotogravure printing	VOC
Bulk gasoline terminals	VOC
Beverage can coating	VOC

Acronyms: PM = Particulate Matter; VOC = Volatile Organic Carbon

TABLE A-2. Industries for which "U.S. Environmental Protection Agency Pretreatment and Effluent Guidelines and Standards for Water Pollutants" have been developed

Industry	Pollutants Regulated
Beet sugar	• BOD, TSS, pH
Cane sugar	• BOD, TSS, pH
Fiberglass insulation mfg.	• Phenol, COD, BOD, TSS, pH
Sheet, plate and laminated glass	• TSS, pH, O&G, P, F, Pb, Ammonia
Rubber processing	• TSS, O&G, pH, BOD, COD
Asbestos mfg.	• COD, TSS, pH
Meat products	• COD, TSS, O&G, Fecal Coliform, Ammonia
Phosphate mfg.	• TSS, Phosphorus, As, pH, F
Fruit & vegetable processing	• BOD, TSS, pH
Plastics & synthetics	• BOD, COD, TSS, pH, Cr, Zn, Phenols, O&G
Nonferrous metals	• TSS, F, Ammonia, Al, Cu, COD, pH, O&G, As, Cu, Pb, Cd, Se, Zn
Timber products	• BOD, TSS, pH, Phenols, O&G, Cu, CR, As
Organic chemicals	• COD, BOD, TSS, pH, Phenols, Cyanide
Leather tanning & finishing	• BOD, TSS, O&G, Cr, pH, Sulfide
Petroleum refining	• BOD, TSS, COD, O&G, pH, Phenols, Ammonia, Sulfide, Cr
Pulp, paper & paperboard mfg.	• BOD, TSS, pH, Pentachlorophenol, Trichlorophenol, Zn
Builders' paper & roofing felt	• BOD, TSS, pH, Pentachlorophenol, Settleable Solids, Trichlorophenol
Iron & steel mfg.	• TSS, O&G, Ammonia, CN, Phenols, pH, Benzene, Naphthalene, Benzo-(a)-pyrene, TRC, Pb, Zn, Ni, Cr, Tetrachloroethylene
Textiles	• BOD, TSS, COD, O&G, Cr, pH, Phenol, Sulfide, Color, Fecal Coliform
Steam electric power plants	• TSS, O&G, Cl, Cu, Fe, Cr
Paint formulating	• No discharge of process waste
Ink formulating	• No discharge of process waste
Paving & roofing materials	• O&G, pH, TSS, BOD
Offshore oil & gas extraction	• Produced water, deck drainage, Drilling mud, Drill cuttting, Well treatment, Sanitary, Domestic, Produced sand
Mineral mining & processing	• pH, TSS, F, Fe
Coal mining & processing	• Fe, Mn, TSS, pH, Settleable Solids
Pharmaceutical mfg.	• CN, COD, BOD, TSS, pH
Metal finishing	• CN, Cd, Cr, Cu, Pb, Ni, Ag, Zn, TTO, O&G, TSS, pH
Coil coating	• Cr, CN, Zn, Fe, O&G, TSS, pH, P, Mn, TTO
Porcelain enameling	• Cr, Pb, Ni, Zn, Al, Fe, O&G, TSS, pH, Ammonia, Phenols, CN

TABLE A-2. (continued)

Industry	Pollutants Regulated
Copper forming	• Cr, Cu, Pb, Ni, Zn, O&G, TSS, F, pH, TTO, Cd, As
Aluminum forming	• Cr, CN, Zn, Al, O&G, TSS, pH, TTO,
Ore mining & dressing	• TSS, Fe, pH, Al, COD, As, Zn, Ra226, NH, U, Cd, Cu, Pb
Explosives manufacturing.	• COD, BOD, TSS, pH, O&G
Hospitals	
Gum & wood chemicals mfg.	• BOD, TSS, pH
Photographic processing	• Ag, CN, pH
Pesticide manufacturing.	• COD, BOD, TSS, Organic Pesticides, pH
Electroplating	• CN, Pb, Cd, Ni, Cr, Zn, Total Metals, TSS, pH Ag, TTO
Dairy processing	• BOD, TSS, pH
Grain mills	• BOD, TSS, pH
Canned & preserved seafood	• BOD, TSS, O&G, pH processing
Cement manufacturing.	• TSS, Temperature, pH
Feedlots	• Fecal Coliform, BOD
Soap & detergent manufacturing.	• BOD, COD, TSS, O&G, pH, Surfactants
Fertilizer manufacturing.	• P, F, TSS, Ammonia, N
Phosphate manufacturing	• P, F, pH, TSS
Ferroalloy manufacturing.	• TSS, Cr, Mn, pH, CN, Phenols, Ammonia
Asbestos products manufacturing	• TSS, pH, COD
Electrical & electronic components	• TTO, F, pH, As, TSS
Inorganic chemicals	• TSS, pH, Zn, Hg, Cu, Pb, Ni, Cl, TOC, CN, Cr, Fe, COD, Se, Ba, Sulfide, A

Acronyms: TSS = Total Suspended; COD = Chemical Oxygen Demand; BOD = Biological Oxygen Demand; O&G = Oil and Grease; TTO = Total Toxic Organics; TOC = Total Organic Carbon.

TABLE A-3. Pollutants and Activities for which Hazardous Air Pollutant Emission Standards have been Developed by the U.S. Environmental Protection Agency

Asbestos
- *Asbestos mills*
- *Roadway surfacing*
- *Manufacture of: cloth, cord, wicks, tubing, tape, twine, rope, thread, yarn, roving, lap, or other textile materials, cement products, fireproofing and insulating materials, friction products, paper, millboard and felt, floor tile, paints, coatings, caulks, adhesives, plastics, rubber materials, chlorine, shotgun shells, and asphalt concrete.*
- *Demolition and renovation*

Beryllium
- *Extraction plants, ceramic plants, foundries, incinerators, propellant plants, rocket motor test sites and machine shops.*

Mercury
- *Stationary sources which process mercury ore to recover mercury, use mercury chlor-alkali cells to produce chlorine gas and alkali metal hydroxide, and incinerate or dry waste water treatment plant sludge.*

Vinyl Chloride
- *Plants which produce ethylene dichloride by reaction of oxygen and hydrogen chloride, vinyl chloride by an process, and or one or more polymers containing any fraction of polymerized vinyl chloride.*

Benzene
- *Fugitive emission source, coke by-product plants*

Radionuclides
- *DOE facilities, NRC-licensed facilities, elemental phosphorus plants.*

Inorganic Arsenic
- *Low and high arsenic copper smelters.*

TABLE A-4. Emission coefficients for problematic air pollutants from various energy technologies

Air pollutants, tons/10^{12} BTU						
Technical Activity: - Nuclear	**SOx**	**NOx**	**CO**	**HC**	**TSP**	**Comments**
Open Pit Uranium Mining	0.43	0.25	0.00	0.02	0.27	Open pit mining of ore for fuel
Underground Uranium Mining	0.02	0.32	0.19	0.03	0.01	Underground mining of ore for fuel
Uranium Milling	0.01	0.41			5.40	Milling ore to yellow cake (U_3O_8)
Hexafluoride Conversion	1.30	0.46	0.01	0.04		Yellow cake to UF_6
Gaseous Diffusion	197.00	51.80	1.30	0.50	51.80	Enrichment to 4% U^{235}
Gas Centrifuge Enrichment	0.46	0.37	0.01		0.02	Enrichment to 2-4% U-235
Fuel Fabrication	1.10	0.28	0.01			UF_6 to UO_2 fuel elements
Commercial Waste Repository	0.27	0.42	0.38	0.03	0.02	Construction & Operations Emissions

TABLE A-4. (continued)

Air pollutants, tons/10^{12} BTU						
Technical Activity: - Coal	**SOx**	**NOx**	**CO**	**HC**	**TSP**	**Comments**
Eastern Underground Mining Emissions	0.03	0.31	0.08	0.02	0.02	With preparation plant; diesel
Eastern Surface Mining	2.55	3.50	7.30	2.27	1.81	With preparation plant
Western Surface Mining	0.32	4.80	0.97	0.30	0.96	With prep. plant; TSP incl. fugitive dust
Beneficiation	0.01	0.60	0.20	0.20	0.90	Cleaning process
Dedicated Rail, eastern	3.70	3.20	3.40	2.50	102.90	4 diesels, 90 trips/yr
Dedicated rail, western	5.00	4.40	4.60	3.60	140.00	4 diesels, 90 trips/yr
Conventional Rail, eastern	2.60	2.90	0.50	2.00	102.00	1 diesel, 20 trips/yr (other cargo)
Conventional Rail, western	3.50	4.00	3.70	2.70	138.40	1 diesel, 20 trips/yr (other cargo)
Barge Transport, eastern	0.52	7.71	1.68	0.62	0.55	1 diesel tug, 22040 miles/yr.
Barge Transport, western	1.47	22.03	4.79	1.76	1.57	1 diesel tug, 26889 miles/yr
Truck Transport, eastern	0.29	1.87	2.95	0.47	35.16	1 trailer 1.2 x 106 net ton miles
Coal-Oil Power Plant	1297.	648.00	40.00	18.00	144.00	40/60 Mix (by wt) Coal/Oil
Fluidized Bed Bituminous	1440.	366.00	56.00	15.00	138.00	Steam plant with emission controls
Fluidized Bed, Subbitum	1700.	582.00	90.00	30.00	146.00	Steam plant with emission controls
Coal-Oil Power Plant	1297.	648.00	40.00	18.00	144.00	40/60 Mix (by wt) COAL / OIL
Coal-fired Plant, Eastern controls	850.00	850.00	60.00	18.00	42.00	Mine-mouth steam plant; emission controls
Coal-Fired Plant, Western	600. 00	850. 00	90.0 0	30.0 0	40.0 0	Conv. steam plant; emission controls

TABLE A-4. (continued)

Air pollutants, tons/10^{12} BTU						
Technical Activity: - Petroleum	**SOx**	**NOx**	**CO**	**HC**	**TSP**	**Comments**
Primary Oil Extraction	13.60	18.60	0.50	10.60	3.50	Emissions from drilling / production
Enhanced Oil Recovery	207.00	71.00	4.00	2.00	24.00	Recovery via steam injection
Offshore Oil Extraction	11.79	31.92	6.91	2.55	2.28	18 platforms, 4000 bbl/day
Crude Oil Storage				2.27		Lined salt-domes caverns
Oil-Fired Power Plant	3720.	432.00	49.30	9.80	410.00	Steam plant with emission controls

TABLE A-4. (continued)

Air pollutants, tons/10^{12} BTU						
Technical Activity: - Gas	**SOx**	**NOx**	**CO**	**HC**	**TSP**	**Comments**
Onshore Gas Extraction	1425.	84.70	1.90	0.60	1.90	120 gas wells
Offshore Gas Extractions	300.00	0.15	0.06	0.01		18 well platforms 88.7 x 106 cu. ft/day
Natural Gas Purification	0.01	40.90	0.00	0.36	0.16	Treatment prior to transmission
Natural Gas Pipeline	0.01	4.00	1.52	0.28		600 mile underground pipe
Liquified Nat. Gas tanker	7.42	5.84	0.41	0.52	2.44	63,460 dead wt ton tanker
Underground Gas Storage	0.19	136.98	3.92	9.45		5000 acres 6 x 1010 scf/yr capacity
Gas Fired Power Plant	0.79	930.00	22.40	0.03	42.90	Conventional steam plant
Residential Wood Stoves	32.30	134.65	29.098	28.15	565.00	Transport and fuel gas emissions
Industrial Wood-Fired Boiler	70.00	162.00	1300	325.00	79.60	Steam boiler with emission controls

Note: 1. Compiled from the Aerospace Corp. and Mueller Associates, Inc. (1983).

Emission Standards for Energy Facilities in OECD Countries

TABLE A-5 gives emission standards for electric generating plants for OECD countries (OECD 1984). The base reporting protocols for these coefficients vary among the different countries. As discussed by OECD, simply reporting the standards on one uniform basis (i.e., ng/J input) may introduce error because of underlying assumptions that must be made (e.g., temperature and moisture content of the flue gas). There may be other variations such as actual vs. normalized stack conditions, or weighted vs. rolling averages. Consequently, comparisons among the different coefficients should be made with caution.

TABLE A-5. Comparison of National Emission Standards for Fossil Electricity Generating Plants ‡

Solid Fuel	**Pollutant (tons/10^{12} BTU)**		
Country	**TSP**	**SOx**	**NOx**
Australia		122	
Belgium		171	
Canada	50	299	299
Denmark		73	
Germany	24	172	386
Greece	65		
Japan	49	267	201
Netherlands	23	267	313
New Zealand	60		
Sweden	17	116	325
United Kingdom	56		
United States	15	603	302

‡Adapted from OECD 1984.

TABLE A-5. (continued)

Liquid Fuel	**Pollutant (tons/10^{12} BTU)**		
Country	**TSP**	**SOx**	**NOx**
Australia			
Belgium		2146	
Canada	50	299	150
Denmark		42	
Germany	21	195	194
Greece	65		
Japan	21	235	115
Netherlands			
New Zealand			
Sweden		116	
United Kingdom			
United States	394	244	

TABLE A-5. (continued)

Gaseous Fuel	Pollutant (tons/10^{12} BTU)		
Country	**TSP**	**SOx**	**NOx**
Australia	122		
Belgium			
Canada	50	299	100
Denmark			
Germany	2	12	123
Greece	65		
Japan	17	191	43
Netherlands			
New Zealand			
Sweden			
United Kingdom			
United States		340	86

APPENDIX B: SAMPLE CALCULATIONS

a) Emission From an Elevated Continuous Point Source

Assumptions:

- A 60 m stack with a 2.0 m exit diameter is emitting 45 m^3/s gas with 120 ppm by volume of sulfur dioxide.
- Exit temperature of gases emitted is 110°C.
- Wind velocity is 3 m/s, ambient temperature is 15°C and insolation is moderate.

Calculate the maximum ground level concentration and the distance corresponding to the location where this maximum is occurring.

Solution:

Step 1: Calculate the effective height, H_S, of the stack according to the formula:

$$H = H_s + \Delta H$$

Stack height H_S = 60 m
Plume rise ΔH ....
Critical downwind distance x*

$$x^* = 2.16\,F2/5\,H3/5 = 104.3 \quad \text{m}$$

$$F = 9.81\frac{45}{TC}\frac{110-15}{110+273} = 34.85$$

$$\Delta H = 38.6\frac{0.4 + 6.14\ 10^{-3}v + 2.02\ 10^{-4}x^2}{\left(1+7.67\ 10^{-3}x\right)^2}$$

Step 2: Distance x_{max} of maximum ground level concentration
x_{max} such that: $\sigma_z = \frac{H}{\sqrt{2}}$ from Pasquill conditions B, by trial and error (or graphical interpolation).
σ_Z = 115 m, σ_y = 150 m, H= 163 m, x_{max} = 840 m

Step 3: Maximum ground level concentration c_{max}

$$C_{max} = \frac{2Q}{\pi UeH^2}\cdot\frac{\sigma_z}{\sigma_y}$$

$$C_{max} = \frac{2\quad 45\quad 120\ 10^{-6}}{TC\ 3\ e\ 163^2}\frac{115}{150}1.2\ 10^{-*}$$

C_{max} = 0.012 ppm (by vol.)

corresponding to: $C_{max} = 34$ kg/m^3 (density $SO_2 = 2.9$ kg/m^3)

b) Emission from Elevated Instantaneous Point Sources (Puff)

The model used in this case is the following:

$$C = \frac{\Delta M}{(2\pi)^{3/2} \cdot \sigma_x \sigma_y \sigma_z} \left\{ \exp\left[-\frac{(x-ut)^2}{2\sigma_x^2} - \frac{y^2}{2\sigma_y^2} \right] \right\} \cdot \left[\exp\frac{-(z-H)^2}{2\sigma_z^2} + \exp\frac{-(z+H)^2}{2\sigma_z^2} \right]$$

where:

ΔM	=	quantity release as a puff [kg]
t	=	time elapsed after release [s]
C	=	ground level concentration on the centerline of the plume [g/m^3], at location of coordinates x, y, z
H	=	final rise of plume above ground H_s [m]
u	=	wind velocity [m/s]
σ_y, σ_z	=	horizontal and vertical dispersion parameters [m] (functions of x).
x, y, z	=	coordinates of the bottom of the release stack (x - downwind, y - crosswind, z - vertical) [m].

where σ_x is the diffusion parameter in the x direction, and x_o and y_o represent the coordinates of the center of the puff in a horizontal plane.

Numerical Example

Let us assume 500 m^3 chlorine is suddenly released from a chemical plant at 35 m above ground level. Wind velocity is 4 m/s and insolation is slight. Chlorine density is 3.2 kg/m^3. what will be the maximum concentration 300 m downwind:

a) at the center of the cloud;

b) at ground level.

Solution:

Stability category is Pasquill "class C" and corresponding dispersion coefficients for x = 300 m are: $\sigma_x = \sigma_y = 37$ m, and $\sigma_z = 23$ m.

The maximum concentration C_{max} occurs at x = 300 m, when the center of the cloud has been moved by 300 m, i.e. after x/u = 300 (m) / 4 (m/s) = 75 s.

- cloud center, x = 300 m

- ground level, x = 300 m

$$C_{max} = \frac{500 \times 3.2}{(2\pi)^{3/2} \times 37 \times 37 \times 23} \cdot \exp\frac{0}{2(37)^2} \cdot \left[1 + \exp\frac{-(70)^2}{2(23)^2} \right]$$

$$C_{max} = 3.2\ g/m^3$$

APPENDIX C: ADDITIONAL GRAPHS FOR THE ESTIMATION OF THE DIFFUSION PARAMETERS (σ_Y and σ_Z)

In Gaussian plume modeling, the rate of plume dilution is described by empirical dispersion coefficients, which specify the horizontal extent (σ_y) and vertical extent (σ_z) of the plume versus downwind distance x from the source under different weather conditions, as characterized by six atmospheric stability categories A through F. The definition of Pasquill stability categories can be found in Chapter 1, (TABLE 1.8 and TABLE 1.10) and is not reproduced at this place.

For computer calculation, Briggs' sigme curves are convenient to use. The value of σ_y, for example, is given by:

$$\sigma_y = a\,x\,(1 + b\,x)^{-1/2} \tag{1.1}$$

The set of equations corresponding to the different Pasquill stability categories, as well as the values of the parameters *a* and *b*, to be used for calculating the dispersion coefficients σ_y and σ_z, were also presented in Chapter 1 (TABLE 1.9).

In this APPENDIX C, consisting of *Figure C-1* to *Figure C-12*, the parameters σ_y and σ_z for various models discussed in Chapter 2 are presented graphically in a standardized form for distances up to 10 km. The conditions under which the curves are valid are also indicated (e.g., height of release, type of terrain, sampling time).

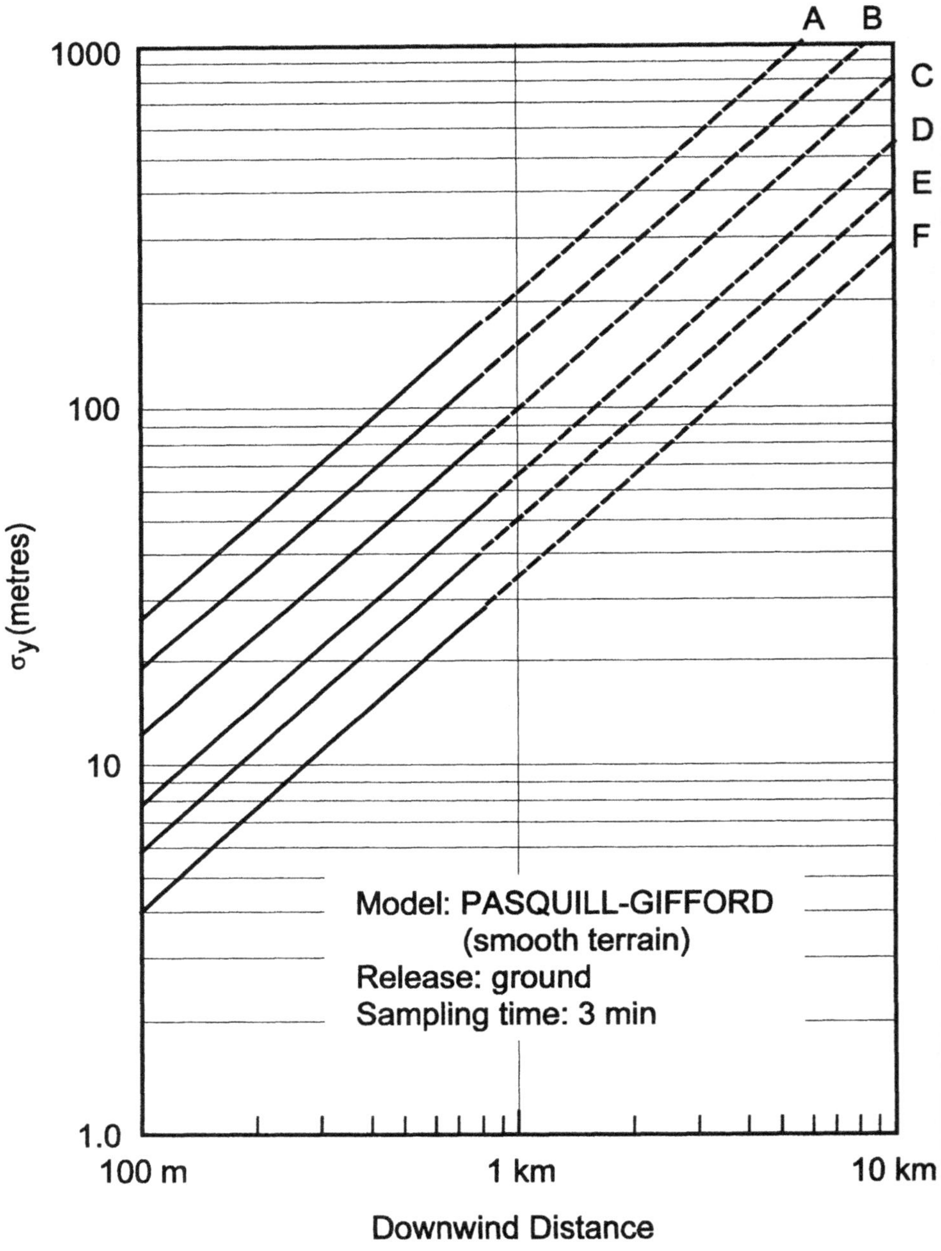
A
B
C
D
E
F
1000
100
10
1.0
σy (metres)
Model: PASQUILL-GIFFORD
(smooth terrain)
Release: ground
Sampling time: 3 min
100 m
1 km
10 km
Downwind Distance

Figure C-1.

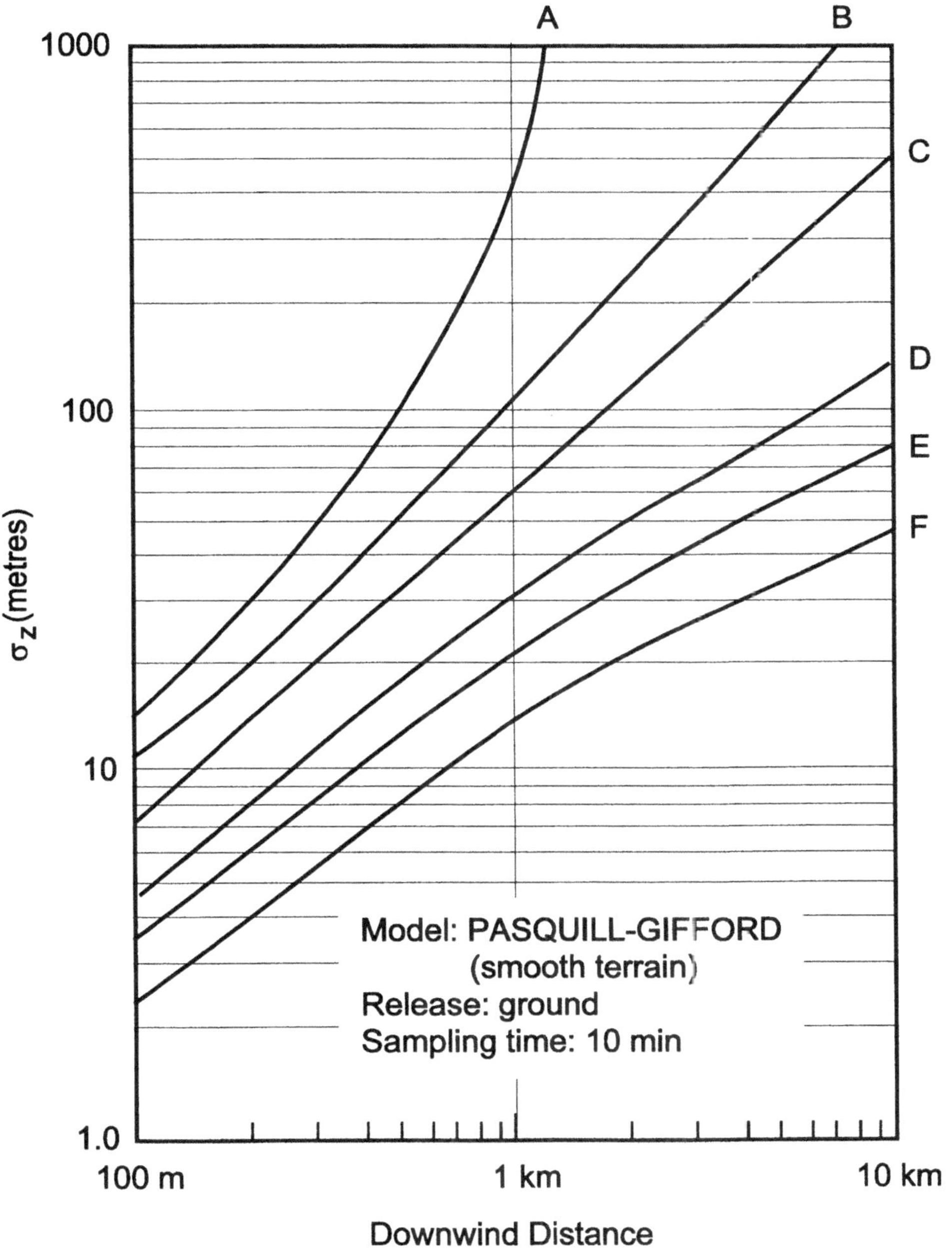
A
B
C
D
E
F
1000
100
10
1.0
σz (metres)
100 m
1 km
10 km
Downwind Distance
Model: PASQUILL-GIFFORD
(smooth terrain)
Release: ground
Sampling time: 10 min

Figure C-2.

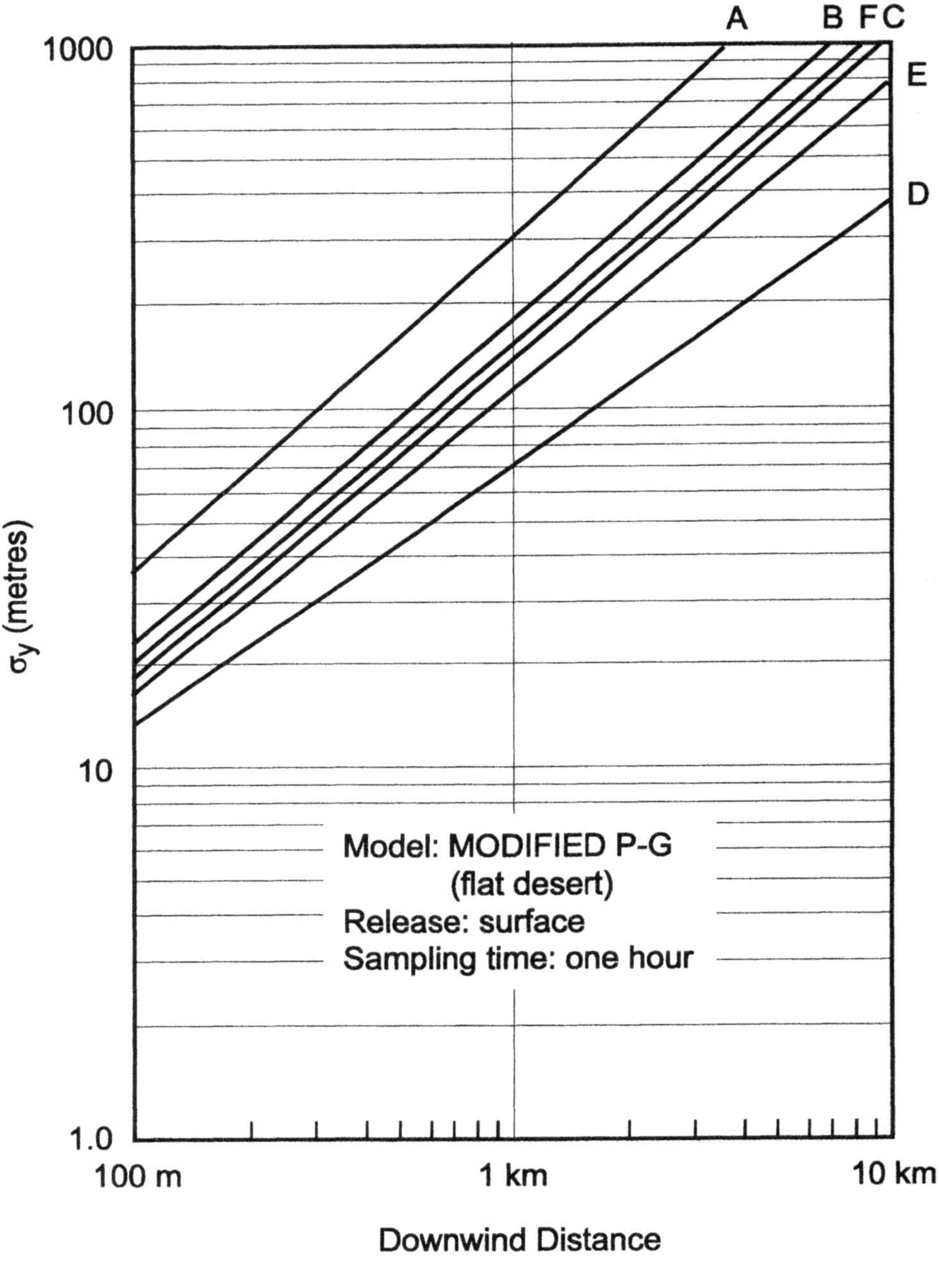
A
B FC
E
D
1000
100
10
1.0
σy (metres)
Model: MODIFIED P-G
(flat desert)
Release: surface
Sampling time: one hour
100 m
1 km
10 km
Downwind Distance

Figure C-3.

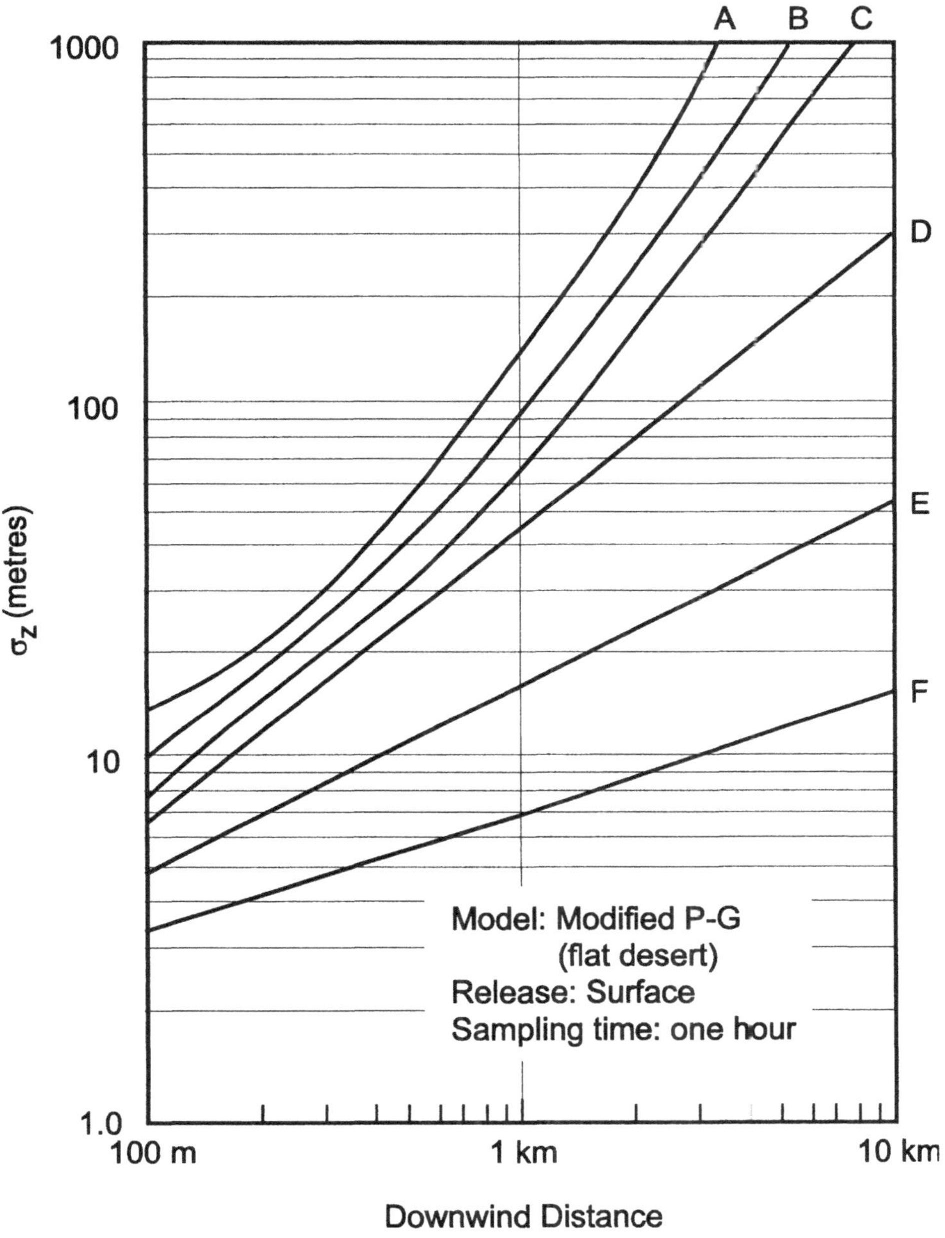
A
B
C
D
E
F
1000
100
10
1.0
σ_z (metres)
100 m
1 km
10 km
Downwind Distance
Model: Modified P-G
(flat desert)
Release: Surface
Sampling time: one hour

Figure C-4.

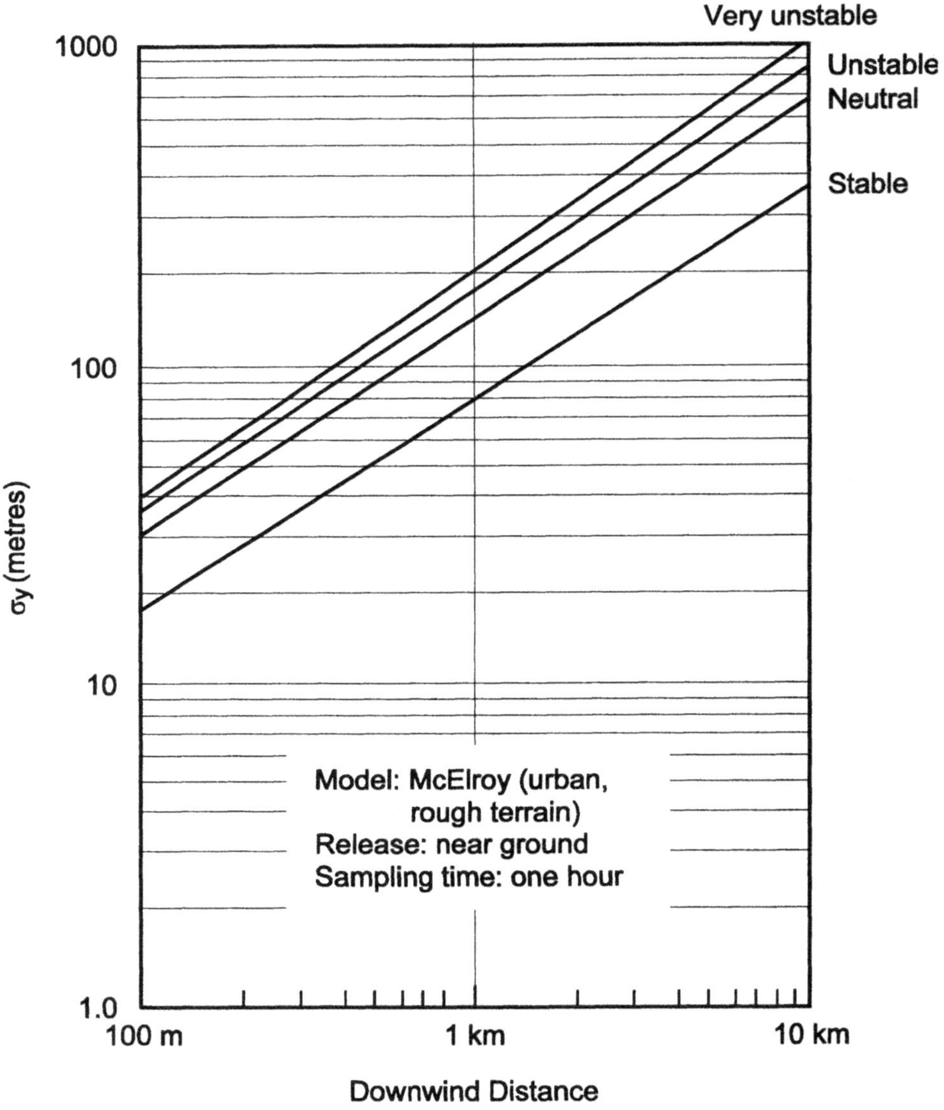
Very unstable
Unstable
Neutral
Stable
1000
100
10
1.0
σy (metres)
Model: McElroy (urban,
rough terrain)
Release: near ground
Sampling time: one hour
100 m
1 km
10 km
Downwind Distance

Figure C-5.

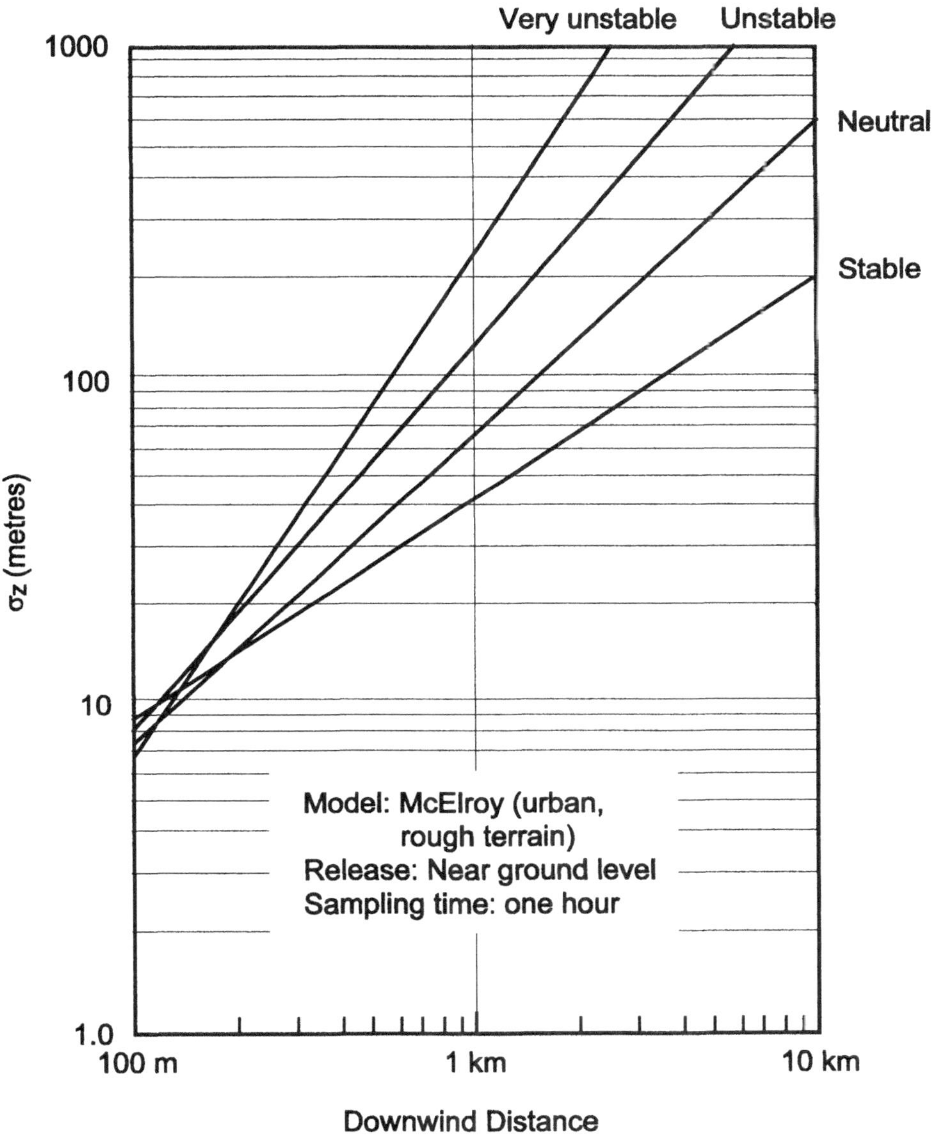
Very unstable
Unstable
Neutral
Stable
1000
100
10
1.0
σz (metres)
Model: McElroy (urban,
rough terrain)
Release: Near ground level
Sampling time: one hour
100 m
1 km
10 km
Downwind Distance

Figure C-6.

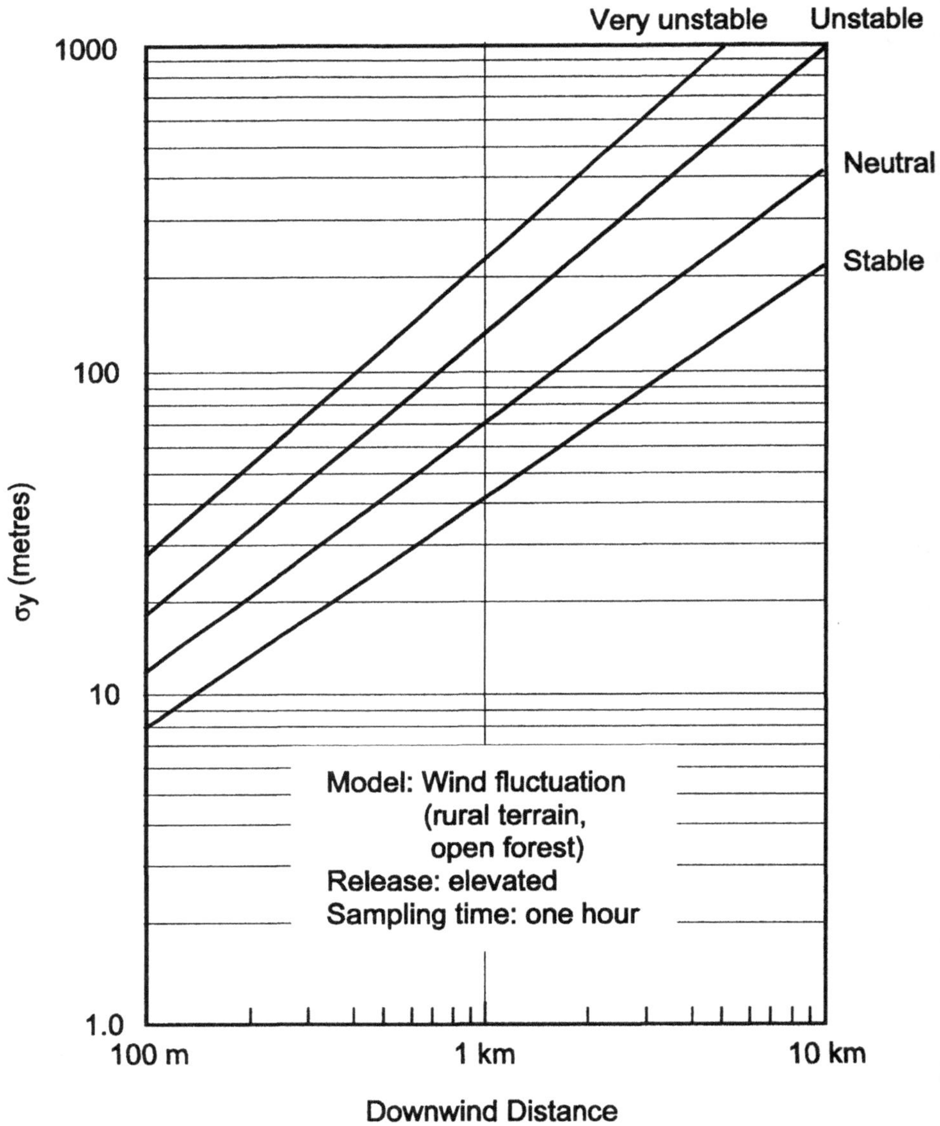
Very unstable
Unstable
Neutral
Stable
1000
100
10
1.0
σy (metres)
100 m
1 km
10 km
Downwind Distance
Model: Wind fluctuation
(rural terrain,
open forest)
Release: elevated
Sampling time: one hour

Figure C-7.

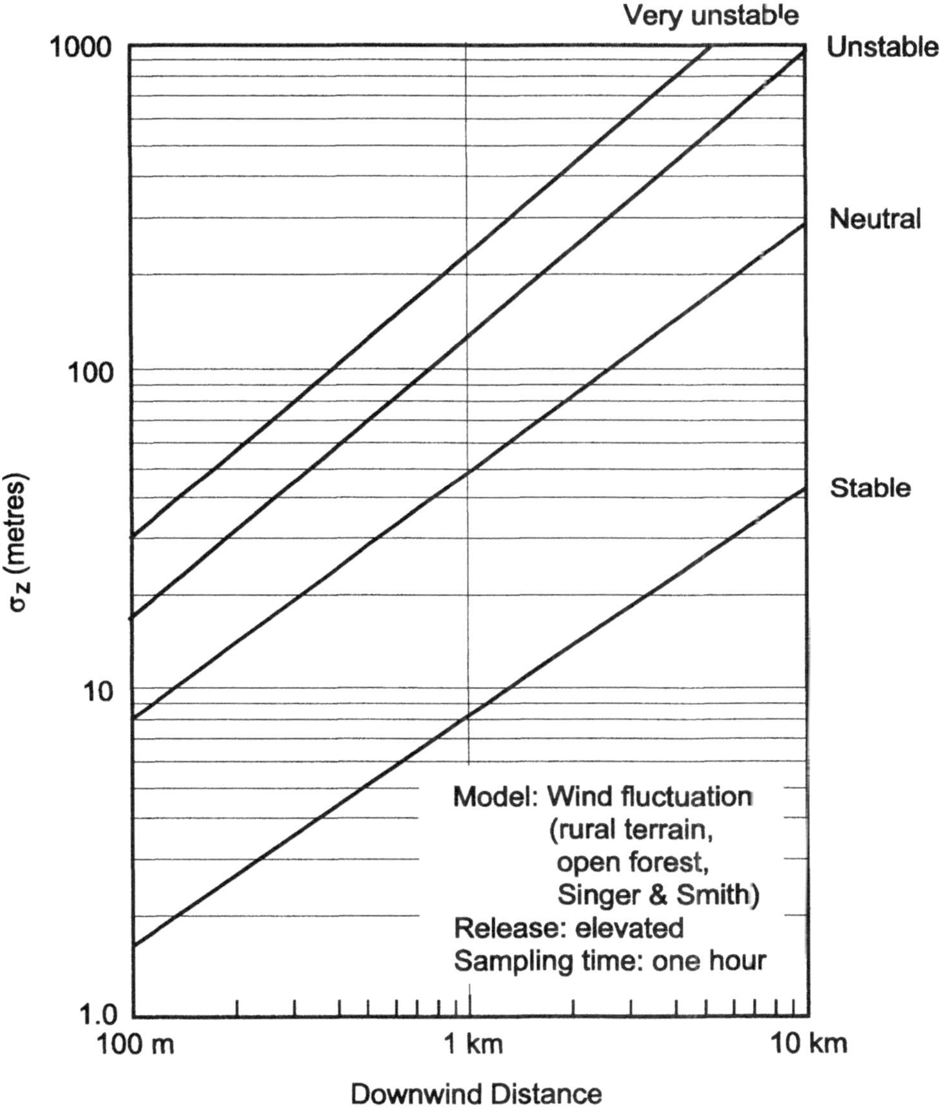
Very unstable
Unstable
Neutral
Stable
1000
100
10
1.0
σz (metres)
Model: Wind fluctuation
(rural terrain,
open forest,
Singer & Smith)
Release: elevated
Sampling time: one hour
100 m
1 km
10 km
Downwind Distance

Figure C-8.

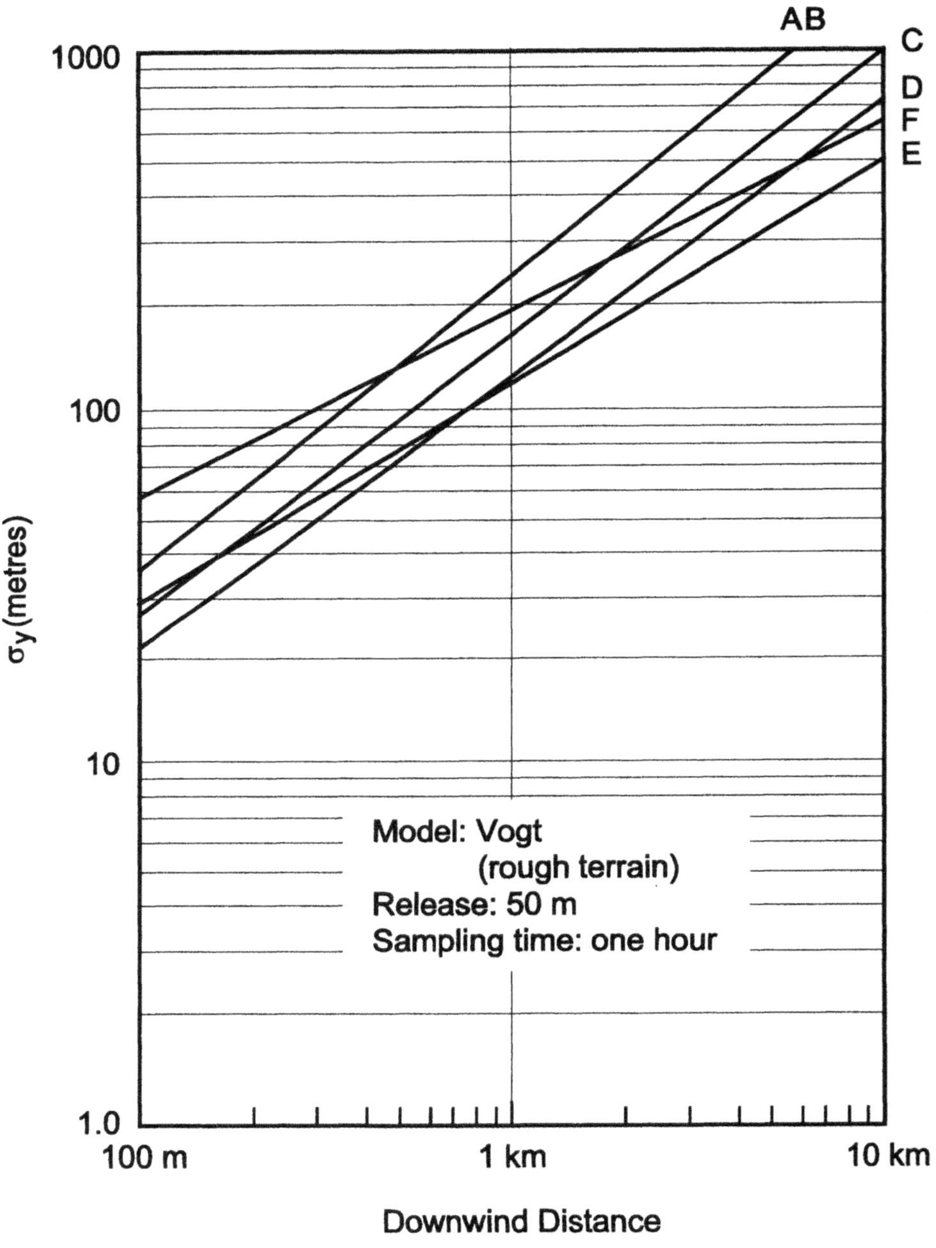
AB
C
D
F
E
1000
100
10
1.0
σy (metres)
Model: Vogt
(rough terrain)
Release: 50 m
Sampling time: one hour
100 m
1 km
10 km
Downwind Distance

Figure C-9.

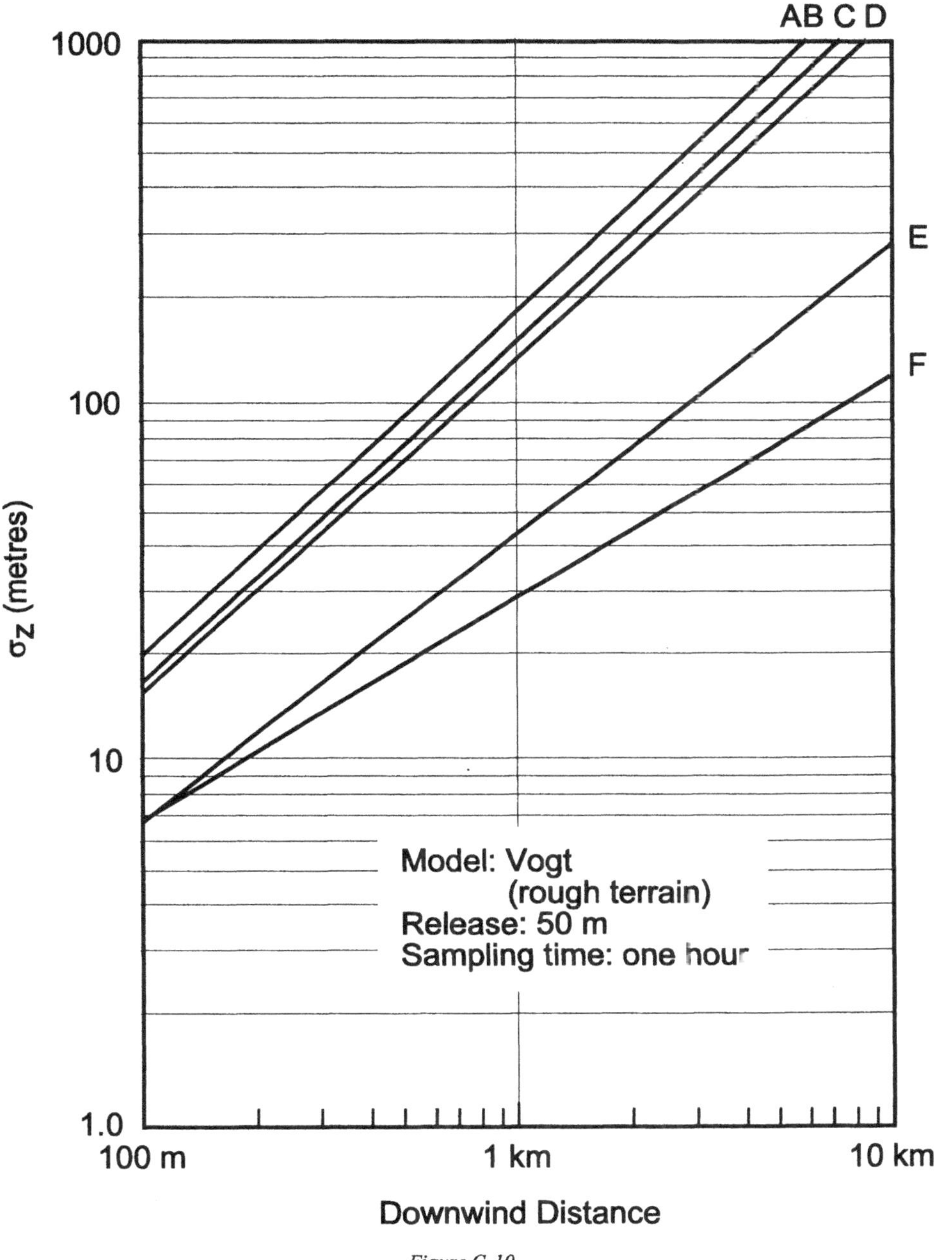
AB C D
E
F
1000
100
10
1.0
σz (metres)
100 m
1 km
10 km
Downwind Distance
Model: Vogt
(rough terrain)
Release: 50 m
Sampling time: one hour

Figure C-10.

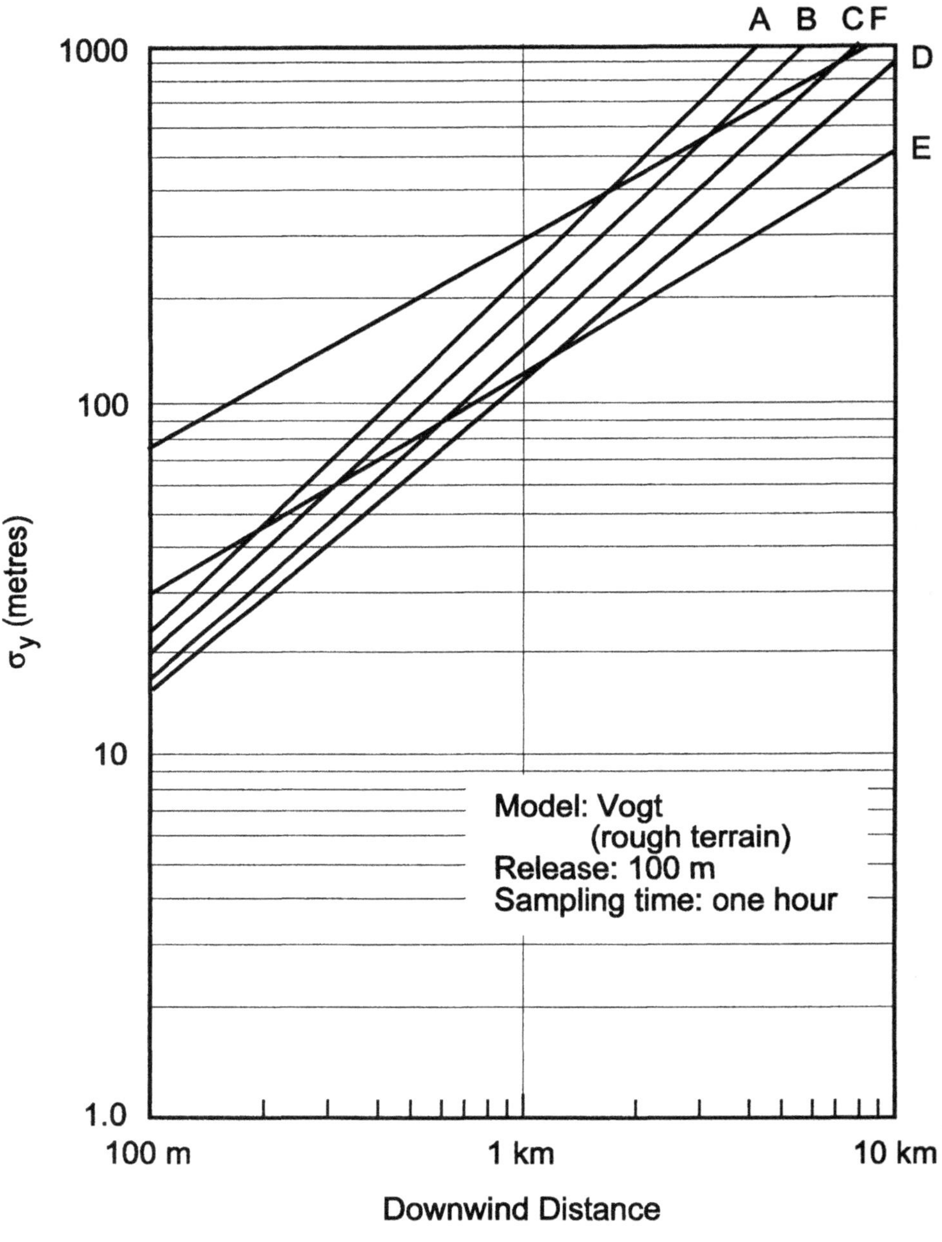
A B CF
D
E
1000
100
10
1.0
σy (metres)
Model: Vogt
(rough terrain)
Release: 100 m
Sampling time: one hour
100 m
1 km
10 km
Downwind Distance

Figure C-11.

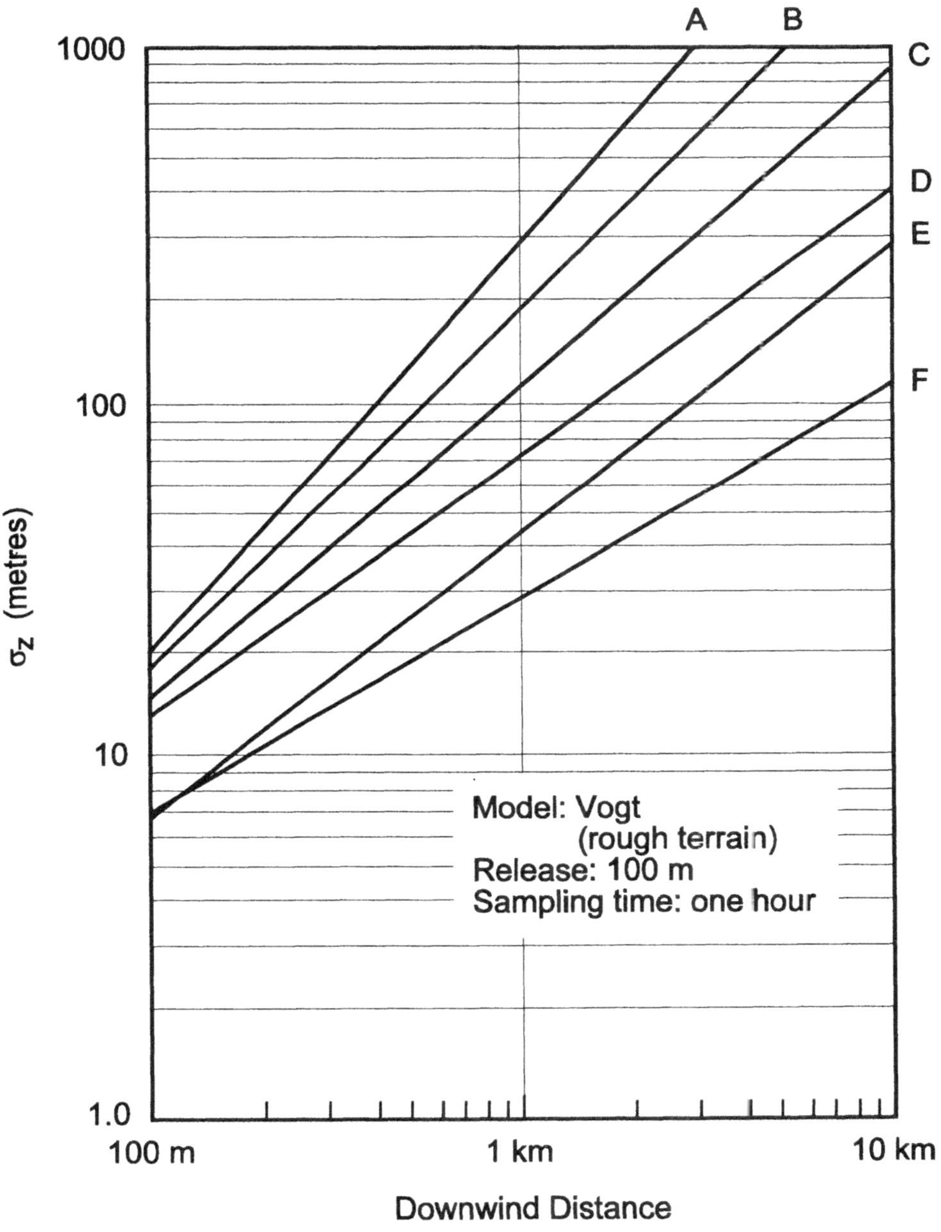
A
B
C
D
E
F
1000
100
10
1.0
σz (metres)
100 m
1 km
10 km
Downwind Distance
Model: Vogt
(rough terrain)
Release: 100 m
Sampling time: one hour

Figure C-12.

SUBJECT INDEX

A

B

C

D

F

G

M

N

Q

R

Y

Z

INTEGRATED REGIONAL RISK ASSESSMENT

Vol. II: Consequence Assessment Of Accidental Releases

CONTENTS (Vol. II)

9 780792 337171

GPSR Compliance

The European Union's (EU) General Product Safety Regulation (GPSR) is a set of rules that requires consumer products to be safe and our obligations to ensure this.

If you have any concerns about our products, you can contact us on ProductSafety@springernature.com

In case Publisher is established outside the EU, the EU authorized representative is:

Springer Nature Customer Service Center GmbH
Europaplatz 3
69115 Heidelberg, Germany

Batch number: 10761697

Printed by Printforce, the Netherlands